DATE DUE

DEMCO 38-296

Gel Electrophoresis of Proteins

The Practical Approach Series

SERIES EDITOR

B. D. HAMES
School of Biochemistry and Molecular Biology
University of Leeds, Leeds LS2 9JT, UK

See also the Practical Approach web site at **http://www.oup.co.uk/PAS**
★ **indicates new and forthcoming titles**

Gel Electrophoresis of Proteins

A Practical Approach

THIRD EDITION

Edited by

B. D. HAMES

School of Biochemistry and Molecular Biology
University of Leeds, Leeds LS2 9JT, UK

Oxford   New York   Tokyo
OXFORD UNIVERSITY PRESS
1998

Gel electrophoresis of
proteins

Clarendon Street, Oxford OX2 6DP

New York

gota   Bombay   Buenos Aires   Calcutta
lhi   Florence   Hong Kong   Istanbul
d   Melbourne   Mexico City   Mumbai
Paris   Sao Paulo   Singapore   Taipei   Tokyo   Toronto   Warsaw

and associated companies in
Berlin   Ibadan

Oxford is a trade mark of Oxford University Press

Published in the United States
by Oxford University Press Inc., New York

© Oxford University Press, 1998

A catalogue record for this book is available from the British Library

Library of Congress Cataloging in Publication Data

Gel electrophoresis of proteins: a practical approach/edited by
B.D. Hames.—3rd ed.
(Practical approach series; 197)
Includes bibliographical references and index.
1. Proteins—Analysis.   2. Gel electrophoresis.   I. Hames, B.D.
II. Series.
QP551.G334   1998
572'.6—dc21   98–18884   CIP

ISBN 0 19 963641 9 (Hbk)
0 19 963640 0 (Pbk)

Typeset by Footnote Graphics, Warminster, Wilts
Printed in Great Britain by Information Press Ltd., Eynsham, Oxon

Preface

When the first edition of this book was published in 1981, I had no idea that it would lead to a major series of laboratory manuals, the *Practical Approach* series, currently over 190 volumes in size and still growing. The popularity of the series has led to second editions of many books. The second edition of *Gel Electrophoresis of Proteins: A Practical Approach* was duly published in 1990. Since that time, the field has continued to develop apace, so much so that some current major techniques, such as capillary electrophoresis, were not even mentioned in the second edition. Clearly, it is time for a third edition!

The goal in devising and editing the third edition of *Gel Electrophoresis of Proteins: A Practical Approach* has been to reflect the laboratory usage of the wide variety of methods that fall under this general subject heading. I make no apology for revisiting the basic techniques of one-dimensional polyacrylamide gel electrophoresis, including SDS–PAGE, since these methods have continued to be widely used on a regular basis in enormous numbers of laboratories around the world, but even here there have been some very useful developments in technology. These are fully described in the first chapter which has been written by new authors to bring fresh insights to the subject. Other techniques have also undergone major developments in the intervening years, as will be clear to readers of the chapters on isoelectric focusing, two-dimensional gel electrophoresis, preparative gel electrophoresis, and polypeptide detection systems. However, my aim was also to cover important emerging techniques. Thus there are completely new chapters on capillary gel electrophoresis, sequence analysis of gel-resolved proteins, fluorophore-labelled saccharide electrophoresis, and analysis of protein:protein interactions by gel electrophoresis. In total, only four of the chapter subjects covered in the previous edition are retained in the ten chapters of this new edition. Furthermore, eight of the ten chapters in this book are written by new authors. Thus the third edition is in every sense a new text, and very definitely a major update in the field.

I am greatly indebted to the authors of each chapter who accepted with good grace the very large number of editing changes I requested in that elusive quest for perfection. Given that the authors are active researchers with intimate knowledge of the methods (and pitfalls!) they describe, I am certain that the book will help colleagues to apply the new methods successfully in their own work. However, I have no illusions that this is the end of the story. The versatility and power of gel electrophoretic techniques in biological research ensure that still further developments and applications will continue to arise in future.

Leeds
January 1998

B. David Hames

Contents

2. Protein detection methods   53

Carl R. Merril and Karen M. Washart

3. Preparative gel electrophoresis   93

Kelvin H. Lee and Michael G. Harrington

4. Polymer solution-mediated size separation of proteins by capillary SDS electrophoresis   105

András Guttman, Paul Shieh, and Barry L. Karger

5. Conventional isoelectric focusing in gel slabs, in capillaries, and immobilized pH gradients   127

Pier Giorgio Righetti, Alessandra Bossi, and Cecilia Gelfi

Contents

6. Two-dimensional gel electrophoresis

Samir M. Hanash

Contents

7. Peptide mapping   213

Anthony T. Andrews

8. Sequence analysis of gel-resolved proteins   237

Richard J. Simpson and Gavin E. Reid

9. Fluorophore-labelled saccharide electrophoresis for the analysis of glycoproteins

Peter Jackson

10. Analysis of protein:protein interactions by gel electrophoresis

Vincent M. Coghlan

Contents

Contributors

ANTHONY T. ANDREWS
University of Wales Institute, Cardiff (UWIC), School of Food and Consumer Science, Faculty of Business, Leisure and Food, Colchester Avenue, Cardiff CF3 7XR, UK.

ALESSANDRA BOSSI
University of Verona, Department of Agricultural and Industrial Biotechnologies, Strada Le Grazie, Cà Vignal, 37134 Verona, Italy.

VINCENT M. COGHLAN
Neurological Sciences Institute, Oregon Health Sciences University, 1120 NW 20th Avenue, Portland, OR 97209-1595, USA.

CECILIA GELFI
ITBA, CNR, Via Fratelli Cervi 93, Segrate 20090, Milano, Italy.

ANDRÁS GUTTMAN
Genetic BioSystems, Inc., San Diego, CA 92121, USA.

SAMIR M. HANASH
University of Michigan Medical School, C.S. Mott Children's Hospital, Department of Pediatric Hematology, R4451 Kresge I, Box 0510, Ann Arbor, MI 48109-0510, USA.

MICHAEL G. HARRINGTON
Proteomics Laboratory, Huntingdon Medical Research Institute, 99 North El Molina Avenue, Pasadena, CA 91101, USA.

GEORGE JACKOWSKI
Department of Clinical Biochemistry, University of Toronto, and Skye PharmaTech, Inc., 6354 Viscount Road, Mississauga, Ontario L4V 1H3, Canada.

PETER JACKSON
Biomethod Consultants, 8 Oslar's Way, Fulbourn, Cambridge CB1 5DS, UK.

BARRY L. KARGER
Barnett Institute, Northeastern University, Boston, MA 02115, USA.

KELVIN H. LEE
Chemical Engineering, Cornell University, Ithaca, NY 14853-5201, USA.

CARL R. MERRIL
Laboratory of Biochemical Genetics, National Institutes of Health, Bldg 10 Rm 2D54, 9000 Rockville Pike, Bethesda, MD 20892, USA.

Contributors

GAVIN E. REID

Joint Protein Structure Laboratory, Ludwig Institute for Cancer Research and the Walter and Eliza Hall Institute of Medical Research, PO Box 2008, Royal Melbourne Hospital, Parkville, Victoria 3050, Australia.

PIER GIORGIO RIGHETTI

University of Verona, Department of Agricultural and Industrial Biotechnologies, Strada Le Grazie, Cà Vignal, 37134 Verona, Italy.

QINWEI SHI

Protein Engineering, Spectral Diagnostics, Inc., 135-2 The West Mall, Toronto, Ontario M9C 1C2, Canada.

PAUL SHIEH

Supelco, Inc., Supelo Park, Bellefonte, PA 16823, USA.

RICHARD J. SIMPSON

Joint Protein Structure Laboratory, Ludwig Institute for Cancer Research and the Walter and Eliza Hall Institute of Medical Research, PO Box 2008, Royal Melbourne Hospital, Parkville, Victoria 3050, Australia.

KAREN M. WASHART

Laboratory of Biochemical Genetics, National Institute of Mental Health, NIH, Bethesda, MD 20892, USA.

Abbreviations

AMAC	2-aminoacridone
ANS	anilinonaphthalene sulfonate
ANTS	8-aminonaphthalene-1,3,6-trisulfonic acid
AP	alkaline phosphatase
BCIP	5-bromo-4-chloro-3-indolyl phosphate
BIS	N,N'-methylene-bisacrylamide
BN-PAGE	blue native polyacrylamide gel electrophoresis
BSA	bovine serum albumin
%C	percentage of the cross-linker relative to the total monomer (w/w)
CA	carrier ampholytes
CCD	charge-coupled device
CE	capillary electrophoresis
CID	collision-induced dissociation
cIEF	capillary isoelectric focusing
CPTS	copper phthalocyanine 3,4′,4″,4‴-tetrasulfonic acid
CTAB	cetyltrimethylammonium bromide
CZE	capillary zone electrophoresis
DAB	3,3′-diaminobenzidine
2DE	two-dimensional gel electrophoresis
DMSO	dimethyl sulfoxide
DNFP	2,4-dinitrofluorobenzene
DNP	2,4-dinitrophenyl
DSP	dithiobissuccinimidyl propionate
DTE	dithioerythritol
DTT	dithiothreitol
ESI	electrospray ionization
FACE	fluorophore assisted carbohydrate electrophoresis
FITC	fluorescein isothiocyanate
HPLC	high-performance liquid chromatography
HRP	horse-radish peroxidase
i.d.	internal diameter
IEF	isoelectric focusing
IPG	immobilized pH gradients
LIF	laser-induced fluorescence
MALDI-TOF	matrix assisted laser desorption ionization-time of flight
MB	methylene blue
MDPF	2-methoxy-2,4-diphenyl-3(2H)-furanone
MECC	micellar electrokinetic chromatography
M_r	molecular weight

MS	mass spectrometry *or* mass spectroscopy
MTT	methyl thiazolyl
m/z	mass/charge ratio
NAD	nicotinamide–adenine dinucleotide
NADP	nicotinamide–adenine dinucleotide phosphate
NBT	nitroblue tetrazolium
NP-40	Nonidet P-40
OSP	oligosaccharide profile
PAGE	polyacrylamide gel electrophoresis
PAGEFS	PAGE of fluorophore-labelled saccharides
PAP	peroxidase–anti-peroxidase
PBS	phosphate-buffered saline
PEG	polyethylene glycol
PEO	polyethylene oxide
pI	isoelectric point
PLP	pyridoxal $5'$-phosphate
PMS	phenazine tetrazolium
PNGase F	peptide-N^4-(acetyl-β-glucosaminyl)asparagine amidase
PSD	post-source decay
PVA	polyvinyl alcohol
PVDF	polyvinylidene difluoride
R_f	relative mobility
RP-HPLC	reversed-phase high-performance liquid chromatography
SDS	sodium dodecyl sulfate
SDS–PAGE	SDS polyacrylamide gel electrophoresis
SSA	sulfosalicylic acid
%C	g cross-linker per 100 g monomers
%T	g monomers per 100 ml
TCA	trichloroacetic acid
TEMED	N,N,N',N'-tetramethylethylenediamine
TFA	trifluoroacetic acid
UV	ultraviolet
WSD	wheat starch digest
WWW	World Wide Web

1

One-dimensional polyacrylamide gel electrophoresis

QINWEI SHI and GEORGE JACKOWSKI

1. Introduction

With the approach of the post-genome era and the popularity of recombinant DNA technology, there has been a resurgence in the use of polyacrylamide gel electrophoresis to identify and characterize various gene products. Electrophoresis is a relatively simple, rapid, and highly sensitive tool to study the properties of proteins. The separation of proteins by electrophoresis is based on the fact that charged molecules will migrate through a matrix upon application of an electric field usually provided by immersed electrodes. Generally the sample is run in a support matrix such as agarose or polyacrylamide gel. Agarose is mainly used to separate larger macromolecules such as nucleic acids whereas a polyacrylamide gel is widely employed to separate proteins. Polyacrylamide gel electrophoresis can be used to analyse the size, amount, purity, and isoelectric point of polypeptides and proteins. Therefore this technique has become the principle tool in analytical chemistry, biochemistry, and molecular biology.

Within the last 30 years, various polyacrylamide gel electrophoretic methods have been developed to resolve different problems. None of these methods is universally applicable due to the diversities of protein molecules and different experimental goals. One-dimensional polyacrylamide gel electrophoresis is a relatively simple and affordable technique. Among different variants of this technique, sodium dodecyl sulfate (SDS)–polyacrylamide discontinuous gel electrophoresis, originally described by Laemmli (1), is the most commonly used system in which proteins are fractionated strictly by their size. This procedure denatures proteins and hence cannot be used to analyse native proteins and proteins whose biological activity needs to be retained for subsequent functional testing. On such occasions, it is necessary to use a non-denaturing system.

In this chapter we present background information and key protocols for a range of different one-dimensional polyacrylamide gel electrophoresis systems, together with practical hints and tips for success and troubleshooting.

2. The polyacrylamide gel matrix

2.1 Chemical structure and mechanism of polymerization

The compounds used to form polyacrylamide are monomeric acrylamide and *N,N'*-methylene-bisacrylamide (bisacrylamide). The chemical structure of the monomers and the polymer are shown in *Figure 1*.

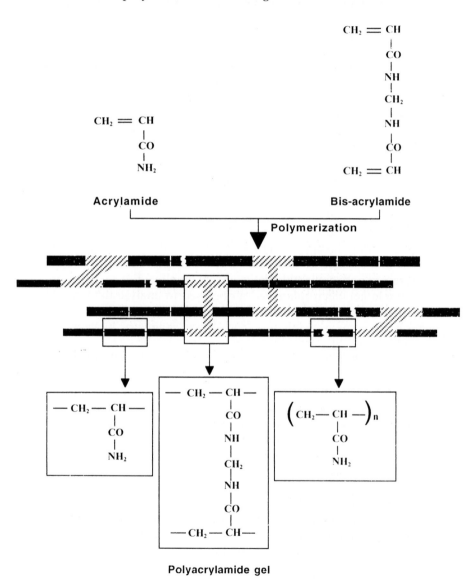

Figure 1. The chemical structure of acrylamide, bisacrylamide, and polyacrylamide gel.

2

The co-polymerization reaction of acrylamide and bisacrylamide is triggered by either chemical or photochemical free radical-generating systems and hence these systems are called initiators. Chemical polymerization is initiated by TEMED (tetramethylethylenediamine) and ammonium persulfate, while photochemical polymerization can be initiated by riboflavin-5′-phosphate or methylene blue.

When ammonium persulfate is dissolved in water it forms persulfate free radicals which, in turn, activate the acrylamide monomer. TEMED is added to serve as a catalyst to accelerate the polymerization reaction due to its ability to carry electrons. The activated acrylamide monomer can then react with inactivated monomer to produce a long polymer chain. The elongating polymer chains are randomly cross-linked by bisacrylamide to form a net of acrylamide chains.

Riboflavin can also be used to generate free radicals, sometimes in combination with ammonium persulfate. In the presence of oxygen and ultraviolet light, riboflavin undergoes photodecomposition and polymerization is initiated by the resulting free radicals. An ordinary fluorescent lamp placed near the gel mixture is adequate as the source of UV light.

A unique photopolymerization system, comprising a cationic dye (methylene blue, MB) and a redox couple (sodium toluene sulfinate, a reducer, and diphenyliodonium chloride, a mild oxidizer) was described recently by Lyubimova (2). It appears that this system can catalyse polymerization at an optimum rate under the most adverse conditions. This system is completely insensitive to any kind of positive and negative effectors, such as pH, detergents, and organic solvents (3, 4). When the persulfate/TEMED system was replaced by this new system in standard SDS–PAGE, a decreased resolution was observed (5). However, for acid–urea–Triton X-100 (AUT) polyacrylamide gel, the methylene blue system is superior to the persulfate/TEMED system in terms of resolution, simplicity, and reduced background staining (6).

2.2 Factors affecting polymerization

Various factors affect polymerization. Some factors alter the rate of polymerization while others change the properties of the gel such as its pore size, elasticity, and homogeneity. Understanding how these factors affect the polymerization is essential in order to obtain reproducible polyacrylamide gels with desired properties.

2.2.1 The rate of polymerization

The polymerization rate of an acrylamide gel can be monitored by spectrophotometry; the absorbance at 260 nm falls as the reaction proceeds. The following factors affect the rate of polymerization:

- type and concentration of initiators
- purity of reagent

3

- pH
- temperature
- light intensity (for photopolymerization)
- oxygen
- concentration of monomers

(a) Initiators. The rate of polymerization is affected by the type of initiators (7). Under optimum conditions the speed of polymerization decreases in the following order:

methylene blue system > persulfate system > riboflavin system.

For chemical polymerization, a higher concentration of initiator results in faster polymerization and vice versa. For photopolymerization, defined concentrations of the catalysts are required; polymerization of gels is less satisfactory if higher or lower concentrations are used.

(b) Purity of reagents. Contaminants in the acrylamide, bisacrylamide, initiators, buffers, and necessary additives (such as SDS) may inhibit or accelerate polymerization.

(i) Metals such as copper can inhibit gel polymerization.

(ii) TEMED containing oxidation products is yellow in colour and gradually loses catalytic activity with time.

(iii) Ammonium persulfate is very hygroscopic and begins to break down immediately when dissolved in water. Therefore, ammonium persulfate solutions should be prepared fresh daily.

(iv) Contaminants such as non-buffer ions, metals and breakdown products in buffers, and gel additives can also affect the rate of polymerization.

Thus, electrophoresis-purity reagents (or better) are highly recommended for preparation of all solutions.

(c) pH. The effect of pH on polymerization efficiency depends on what initiators are used. Caglio and Righetti (8) reported that the persulfate system gives optimum efficiency only in the pH 7–10 range; at progressively acidic pH values, the rate drops markedly. No polymerization occurs at pH 4. In the riboflavin system, the opposite behaviour is observed. The optimum range is between pH 4–7 with a peak at about pH 6.2. The reaction slowly declines at acidic pH values and is strongly quenched at alkaline pH values. No polymerization occurs at pH 10. In contrast, the methylene blue system performs extremely well in the pH 4–8 range with somewhat lower efficiencies at pH 9–10.

(d) Temperature. Higher temperature drives the polymerization faster whereas lower temperature slows down the polymerization process (9, 10). The optimal temperature is between 23–25 °C. To achieve reproducible gels, the gels should be polymerized each time at the same temperature. Stock solutions are usually kept at 4 °C so it is important to equilibrate the final solution to room temperature before initiating polymerization. Leaving the solution at room temperature for 10–20 min (depending on the volume) is sufficient to achieve this.

(e) Light intensity. For photopolymerization, light intensity also plays an important role in terms of the rate of polymerization. The polymerization rate is roughly proportional to the intensity of absorbed light. Optimum efficiency can only be achieved at the correct light intensity, both for the riboflavin and methylene blue systems (11). When light intensities are too high, the rate of catalyst consumption is so rapid that the catalyst is depleted quickly from the system before the monomers are properly polymerized. On the other hand, if the light intensities are too low, prolonged polymerization will occur. The light intensity is affected by the power of the light source, the distance of the light source from the gel mixture, and the gel thickness. This sensitivity of photopolymerization to the light intensity is in contrast to the persulfate/TEMED system where a large excess of persulfate still guarantees very high polymerization efficiency.

(f) Oxygen. The presence of oxygen inhibits the polymerization of acrylamide since oxygen traps free radicals. Therefore oxygen dissolved in a gel mixture should be removed by degassing under vacuum before polymerization. Cold solutions contain more oxygen and hence require more time to be degassed completely. Therefore, the gel solution should be brought to room temperature before evacuation begins. For ammonium persulfate/TEMED initiated polymerization reactions, longer degassing time leads to faster polymerization. Generally, degassing should be carried out for at least 10 min at room temperature and at a vacuum of 125 torr or better. Although polymerization can occur without prior degassing of gel solutions, the reproducibility of such gels are poor. The photodecomposition of riboflavin requires a small amount of oxygen so that complete degassing actually inhibits polymerization. Therefore, for polymerization initiated by the riboflavin/TEMED system, degassing should not exceed 5 min.

(g) Monomer concentration. The concentration of monomer also affects the rate of polymerization. A higher concentration results in faster polymerization.

2.2.2 Porosity of the gel

The size of the pores in the gel is governed primarily by the amount of total acrylamide used per unit volume and the degree of cross–linkage. The latter is determined by the relative percentage of bisacrylamide used. As the proportion of cross-linker is increased, the pore size decreases. It has been suggested that the average pore size reaches a minimum when the amount of bisacryl-amide represents about 5% of the total acrylamide (12). As the proportion of cross-linker is increased above the value required for minimum pore size, the acrylamide polymer chains become cross-linked to form increasingly large bundles with large spaces between them so that effective pore size increases again. For this reason, more than 20% of cross-linker is used to prepare matrices with high porosity.

The rate of polymerization also affects the pore size. Increasing the rate of polymerization results in a less porous gel whereas reducing the rate of poly-merization often produces a gel with greater porosity. Therefore any factors that alter the rate of polymerization also influence the pore size. For example, a smaller pore size may be obtained at a higher polymerization temperature, whereas a large pore structure can be produced at lower temperature since the rate of polymerization is strongly influenced by temperature.

The efficiency of monomer to polymer conversion also influences the porosity of gels. Poorly polymerized gels that result from using poor quality acrylamide, extreme pH, and some gel additives have greater porosity due to incomplete conversion of monomer to polymer. Urea, which is often added to polyacrylamide gel to fractionate small proteins and polypeptides (see Section 7.4), causes the formation of smaller pore size gels by accelerating the persulfate-driven reaction and boosting the monomer conversion to near completion.

The presence of polymers, such as polyethylene glycol, in the gel solution causes lateral chain aggregation and thus results in macroporous gel forma-tion (4, 13).

2.2.3 Homogeneity of the gel

Several factors can affect the homogeneity of polyacrylamide gels:

(a) Lack of thorough mixing of polymerization initiator with the monomer solutions causes swirls in the resulting gel. However, too much mixing introduces excess oxygen to the solution which will inhibit polymerization. Therefore, gentle but thorough mixing is required.

(b) Gel thickness also affects gel homogeneities since above a critical gel thickness (3 mm) gravity-induced gel inhomogeneities occur (14). Below this critical thickness, the gel layer appears to be devoid of any convective flows. Interestingly, when the gel thickness is above 3 mm the presence of density gradients suppresses almost entirely such gel inhomogeneities (14).

(c) Because persulfate-driven polymerization is very sensitive to oxygen, if the top of the gel mixture is not properly sealed during polymerization, the absorption of oxygen always results in a non-homogeneous gel.

(d) The rate of polymerization also affects the uniformity of the gels. Poly-merization that is too fast (< 10 min) or too slow (> 60 min) leads to a non-uniform polymerization. The rate of polymerization is most easily controlled by adjusting the concentration of initiators. As a general rule, the lowest catalyst concentrations that allow polymerization in the optimal period of time should be employed.

2.3 Resolving range of polyacrylamide gels

The composition of acrylamide mixtures is defined by the letters T and C according to Hjerten (15). T denotes the total percentage concentration of both monomers (acrylamide plus bisacrylamide) in grams per 100 ml. C denotes the percentage (by weight) of the cross-linker relative to the total monomer.

The choice of acrylamide gel concentration is critical for optimal separation of proteins by zone electrophoresis. Gels with concentrations of acrylamide less than about 3% are almost fluid and very difficult to handle although this can be remedied by the inclusion of 0.5% agarose. At the other extreme, polyacrylamide gels will form up to about 35% acrylamide, but their use is limited by the mechanical properties of the gel that results, particularly its extreme brittleness. In practice the effective range of a polyacrylamide gel is between 5–20% for a uniform gel concentration and 3–30% for a gradient concentration gel. A guide for choosing acrylamide concentration for a uniform gel is given in *Table 1*. This applies to both the continuous and discontinuous buffer systems.

Unfortunately, when using uniform concentration gels, a gel prepared with a pore size large enough to resolve large proteins is unlikely to resolve smaller polypeptides well. This problem has been overcome by using pore gradient gels, in which the size of the pores changes from the top to the bottom of the gel. Thus a gradient gel separates polypeptides over a larger molecular weight range than a uniform concentration gel and with higher resolution. The choice of gradient gel range is of course dependent on the size of proteins being fractionated; *Table 2* provides a guide.

It must be emphasized that the effective resolving range of a polyacrylamide gel is not determined only by the pore size but also by other factors such as buffers, gel additives, pH, and so on. For example, buffers containing Tricine instead of glycine are known to allow the resolution of small

Table 1. Molecular weight separation guide-lines for uniform concentration gels

Acrylamide concentration		M_r range of sample polypeptides
%T	%C	
5	2.6	25 000 to 300 000
10	2.6	15 000 to 100 000
10	3.0[a]	1000 to 100 000
15	2.6	12 000 to 50 000

[a] When using Tricine instead of glycine in running buffer (16).

Table 2. Molecular weight separation guide-lines for gradient gels[a]

Acrylamide concentration		M_r range of sample polypeptides
%T	%C	
3.3–12	2.0[b]	14 500 to 2 800 000
3–30	8.4	13 000 to 1 000 000
5–20	2.6	14 000 to 210 000
8–15	1.0	14 000 to 330 000

[a] The list is not exhaustive; other gradient gel ranges may also be used.
[b] See ref. 34.

polypeptides at lower acrylamide concentrations (16). Changing the pH of the resolving gel can alter the molecular sieving properties of SDS–PAGE (17). Some of the gel additives that affect resolution are listed below:

(a) Urea can be added to gel mixtures to be initiated using the persulfate/TEMED system to produce polyacrylamide gels with smaller pore size so that small peptides can be resolved (18, 19).

(b) Glycerol changes gel sieving properties by its inhibitory effect on acrylamide polymerization and contribution of viscosity to friction in the gel (20). This has been used to facilitate the separation of certain proteins such as myosin heavy chain (21) and to improve protein band edge tailing (22) (see also Section 5.3.2). Glycerol also helps gradient formation (see Section 5.3.3) when present in gradient gels (23).

(c) Polymers such as polyethylene glycol, dextrans, methylcelluloses, Ficoll, and polyvinylpyrrolidone induce the formation of large pore gels. This facilitates the separation of giant proteins (13). It was also reported that water soluble polymers can be used to sharpen the protein bands and enhance the separations in certain molecular ranges when present in uniform or gradient SDS gels (24–26). Although the mechanisms remain unclear, some researchers have suggested that both viscosity and volume exclusion are contributory factors.

(d) Detergents are often added to polyacrylamide gel systems. SDS is the best known example, but other detergents that have been used include Triton X-100, Nonidet P-40, Sarkosyl, and various cationic detergents. Detergents not only alter the electrophoretic behaviour of proteins but also inhibit gel polymerization (3).

3. Gel apparatus

3.1 Electrophoresis apparatus

Polyacrylamide gels may be run on either disc or slab gel systems. Presently, slab gel systems are the most widely used. Several designs for a slab gel apparatus have been published, but one of the most popular designs has been that of Studier (27). Slab gel apparatuses are also available commercially from a number of suppliers in both standard and mini vertical formats. All conditions described in this chapter are designed for use with the Bio-Rad Mini-Protean II Dual Slab Cell gel apparatus (*Figure 2*), but the gel volumes can readily be scaled-up to prepare standard gels for use with other apparatus. The dimensions of the mini slab gels for this Bio-Rad apparatus are 7.3 × 8 cm^2. The thickness of the gel is achieved by placing two spacers (1.0 or 1.5 mm in thickness) between the glass plates. The two spacers can be easily positioned by placing the Teflon sheet (supplied with the Bio-Rad Mini gel apparatus) between the spacers. Combs with the same thickness containing either 10 slots (5 × 13 mm) or 15 slots (3 × 13 mm) are used to form gel wells. Compared to standard gel systems, the mini system minimizes reagent consumption, and reduces electrophoretic run time; a typical run time for SDS–PAGE is only 45 minutes. Assembly of the glass plates to form the gel mould is achieved using two one-piece clamps. The clamps hold the glass plates apart by the required distance and the sandwich is then locked onto the casting stand. The gel is then poured. After polymerization, the gel sandwich is transferred from the casting stand to the upper buffer chamber. The entire inner glass plate of the gel sandwich is in contact with the upper buffer, creating even heat distribution for 'smile-free' separations. Further cooling of the gels can be achieved by using more lower buffer so that almost entire gel sandwich is immersed in the buffer.

Figure 2. Bio-Rad Mini-Protean II Dual Slab Cell gel apparatus.

3.2 Power pack

A power pack capable of supplying about 500 V and 100 mA is sufficient for gel electrophoresis. Although it is not too important whether to use constant voltage or constant current, for versatility it is worthwhile obtaining a power pack with both functions. A number of power packs are available commercially. Some are suitable for running minigels, such as the PowerPac 300 (300 V/400 mA) from Bio-Rad and Hoefer SX 250 (250 V/200 mA) from Pharmacia.

3.3 Apparatus for gradient gels

Polyacrylamide gradient gels are prepared using a linear gradient maker and a peristaltic pump as illustrated in *Figure 3*. The gradient gels are cast usually from top to bottom in which the more dense solution is placed in the downstream compartment (B in *Figure 3*). Gradient gels can also be cast from bottom to top in which the low concentration gradient solution is pumped first while filling up the chamber from the bottom, but this can result in undesirable mixing during gel pouring unless great care is taken. In each case, constant mixing of solution in compartment B is required to form a linear gradient. A piston-type gradient maker without a pump can also be used to form gradient gels especially for small volume gels.

3.4 Photopolymerization equipment

A 15 W daylight fluorescent lamp placed 10 cm from the gel can be used for riboflavin catalysed polymerization. Two light boxes, each with two 12 W neon tubes, also placed 10 cm from the gel on both sides, or a 500 W halogen lamp at a 40 cm distance are sufficient to activate the methylene blue photopolymerization system.

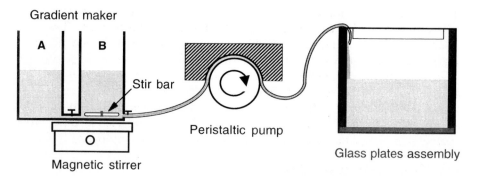

Figure 3. Apparatus for making a gradient gel. In the arrangement shown ('bottom to top' casting), the more dense solution is placed in compartment B of the gradient maker.

4. Choice of a gel system

The choice of a detailed methodology of zone electrophoresis in polyacryl-amide gels depends on what samples are being analysed and what information is desired.

Originally, analytical zone electrophoresis in polyacrylamide gels used cylindrical rod gels in glass tubes but now flat slab gels, 0.5–1.5 mm thick, have become the choice of format for PAGE. The main advantage is that many samples, including molecular weight markers, can be electrophoresed under identical conditions in the same gel, allowing direct comparison of the band patterns of different samples.

Zone electrophoretic systems in which the same buffer ions are present throughout the gel and running buffer at constant pH, are referred to as continuous buffer systems. In these systems the protein sample is loaded directly onto the resolving gel. In contrast, discontinuous buffer systems employ different buffer ions and pH in the gel compared to those in the running buffer. In these systems the protein sample is loaded onto a stacking gel polymerized on top of a resolving gel.

Furthermore, two fundamentally different types of gel system exist, non-dissociating and dissociating. A non-dissociating buffer system is one that is designed to separate native proteins under conditions that preserve protein function and activity. In contrast, a dissociating system is designed to denature proteins into their constituent polypeptides and hence examines the poly-peptide composition of samples.

4.1 Non-dissociating systems

Since a non-dissociating gel system (Section 6) separates native proteins, separation is based not only on protein size but also on protein charge and shape. For this reason, accurate estimation of molecular weight becomes impossible for many proteins by non-dissociating gel electrophoresis. Thus, identification of a specific protein by its molecular weight in a protein mixture without a specific detection method is very difficult (although improved molecular weight determination can be achieved by using gradient gels) (28). Therefore, using a non-denaturing gel system is recommended only if one needs to analyse native proteins rather than denatured ones, particularly if the biological activity (for example, enzyme activity, binding activity, and so on) of a protein needs to be retained for subsequent steps. In non-dissociating gel systems, the choice of buffer pH depends on the isoelectric points of proteins under study. If the pI is unknown, a charge-shift method such as blue native polyacrylamide gel electrophoresis (Section 7.1) may be employed.

4.2 Dissociating systems

The SDS–polyacrylamide (SDS–PAGE) discontinuous gel system (Section 5.3.1) is the most popular system for routine analysis of proteins. Although a standard slab gel gives excellent resolution and is relatively insensitive to sample overloading, minigels require much less time to run, stain, and destain without loss of resolution making it the usual choice of gel size. In addition, the minigel system needs less gel reagents and buffers, and smaller amounts of samples. It is also easier to store and thus is generally more convenient.

In the SDS–PAGE system the protein mixture is denatured by heating at 100°C in the presence of excess SDS and a thiol reagent. Under these conditions, all proteins are dissociated into their individual polypeptide subunits. Most polypeptides bind SDS in a constant weight ratio and form SDS: polypeptide complexes with essentially identical charge densities. Thus, proteins are separated in polyacrylamide gels of the correct porosity strictly according to their size.

In an SDS–PAGE discontinuous buffer system, the samples are not loaded directly onto the resolving gel but onto a large pore gel, called a stacking gel, polymerized on top of the small pore resolving gel. The stacking gels have different buffer ions and a different pH compared to those in the running buffer and have the effect of concentrating relatively large volumes of protein samples into narrow bands before they enter the resolving gel (the 'stacking' effect). This phenomenon is described more fully in Section 5.3.1. The fact that relatively large volumes of dilute protein samples can be applied to the discontinuous gel and good resolution of sample components can still be obtained has made this system the choice for high-resolution fractionation of protein mixtures.

SDS–PAGE is used mainly for the following purposes:

- estimation of protein size
- estimation of protein purity
- protein quantitation
- monitoring protein integrity
- comparison of the polypeptide composition of different samples
- analysis of the number and size of polypeptide subunits
- when certain post-electrophoretic applications, such as Western blotting, are to be used.

Some proteins tend to precipitate in stacking gels due to the lower pH of such gels or the protein stacking effect which concentrates protein samples. In this case the SDS–PAGE continuous buffer (Section 5.3.2) system should be used.

SDS–PAGE can be carried out with uniform or gradient gels. If a higher resolution or separation of proteins with a wide range of molecular weight is intended, especially for high molecular weight proteins (33, 34), a gradient SDS–polyacrylamide gel system (Section 5.3.3) is recommended. However, even with a gradient gel, proteins smaller than about 12 kDa are poorly separated and an alternative system must be used. Most of these involve the combined use of SDS and urea or replacing the tracking ion, glycine, with Tricine in the electrode buffer (Section 7.4). Since SDS–PAGE separates proteins only by their size, it is less effective at resolving proteins with similar molecular weights or variants of the same protein caused by post-translational modifications.

In certain cases, SDS causes protein aggregation or precipitation. In other cases, SDS–PAGE does not give optimal resolution (e.g. of proteins and their post-translationally modified forms) or accurate molecular weight determination. This has led to the use of cationic detergents for polyacrylamide gel electrophoresis (Section 7.5). Cationic detergents such as cetyltrimethylammonium bromide (CTAB) and benzyldimethyl-*n*-hexadecylammonium chloride (16-BAC), when bound to proteins, impart positive charges on them. Therefore, proteins migrate in the opposite direction compared to SDS–PAGE. In general, this type of gel separates proteins based on their sizes and can be used when SDS–PAGE does not perform well. Finally, acid–urea PAGE is a system developed to resolve basic proteins such as histones and is described in Section 7.2.

4.3 Transverse gradient gels

Sometimes it is necessary to optimize acrylamide or bisacrylamide concentration for the optimal separation of proteins or other purposes, such as high efficiency transfer of different sized proteins to membranes. Instead of testing multiple gels of fixed concentration, transverse pore gradient gels, which allow the sample to be resolved over a wide range of polyacrylamide concentrations on a single gel, can be employed (Section 7.3). In addition, transverse gradient gels can be used to monitor the effect of added components, such as urea or glycerol, on protein migration. In these cases, the variable component in the transverse gradient gel is not the gel concentration but the additive being tested.

5. SDS–polyacrylamide gel electrophoresis

5.1 Introduction

In SDS–PAGE, the protein mixture is denatured by heating at 100°C in the presence of excess SDS and a thiol reagent is employed to break disulfide bonds. Under these conditions, all reduced polypeptides bind the same amount of SDS on a weight basis (1.4 g SDS/g polypeptide) independent of the amino acid composition and sequence of the protein. The SDS:protein complex forms a rod with its length roughly proportional to the molecular weight of the protein. All proteins are now negatively charged with similar charge density and thus can be separated on the basis of their size only.

5.2 Reagent preparation

5.2.1 Acrylamide monomer

Unpolymerized acrylamide and bisacrylamide are strong neurotoxins and suspected carcinogens and should be handled accordingly. Electrophoresis purity
acrylamide and bisacrylamide are very stable at room temperature and can be
stored for at least one year. Monomer stock solutions should be prepared in
distilled water (see *Table 3*) and filtered through a 0.45 µm filter. They should
be stored at 4 °C in a dark glass bottle or clear bottle wrapped with aluminium
foil for no longer than about two months to avoid hydrolysis of acrylamide to
acrylic acid.

5.2.2 SDS stock solution

A 10% SDS stock is prepared by dissolving 10 g of electrophoresis grade SDS
in water to 100 ml. Gentle warming may be required to help dissolve the SDS.
The solution should be stable at room temperature for several months but
precipitates in the cold.

5.2.3 Polymerization initiator solutions

Although the ammonium persulfate/TEMED system is employed in almost
all recipes throughout this chapter, the usage of two photopolymerization
systems is also described below as alternatives.

TEMED is subject to oxidation (the oxidation products are characterized
by a yellow colour) and is also very hygroscopic. Water will accumulate over
time which will accelerate the oxidative process. Therefore, prolonged storage
causes gradual reduction in activity. Nevertheless, TEMED can be stored in a
tightly closed dark glass bottle at room temperature for at least six months. It
is used as supplied.

Ammonium persulfate is also very hygroscopic and begins to break down
immediately after being dissolved in water. Although it has been suggested by
some researchers that this solution is stable for a week if kept at 4 °C and in
darkness, it should be prepared fresh daily (10%, w/v in distilled water).
Ammonium persulfate/TEMED initiated reactions should be allowed to proceed for at least 2 h before running a gel to ensure a complete polymerization.

Riboflavin can be stored dry at room temperature for at least one year.

Table 3. Preparation of acrylamide stock solutions

Stock	Acrylamide (g/100 ml)	Bisacrylamide (g/100 ml)
30%T, 2.6%C	29.22	0.78
40%T, 2.6%C	38.96	1.04
50%T, 2.6%C	48.7	1.3

Table 4. Summary of polymerization systems

	Concentration		Storage
	Stock	**Final**	
Chemical polymerization			
Persulfate system:			
Ammonium persulfate	10% (w/v)	1.5–7.5 µl[a]/ml	Made fresh
TEMED	As supplied	0.5–1 µl[a]/ml	RT,[b] six months
Photopolymerization			
Riboflavin system:			
Riboflavin-5′-phosphate	4 mg/100 ml	5 µg/ml	4°C, one month
TEMED	As supplied	0.5–1 µl[a]/ml	RT,[b] six months
Methylene blue system:			
Methylene blue	2–10 mM	30–100 µM	RT,[b] one year
Sodium toluene sulfinate	20–250 mM	0.5–1 mM	4°C, one week
Diphenyliodonium chloride	1–10 mM	20–50 µM	4°C, one week

[a] Volume of stock solution.
[b] Room temperature.

Riboflavin in solution (20 mM or 0.004%, w/v in distilled water) is stable for at least one month if kept in the dark at 4°C.

Methylene blue (2 mM), sodium toluene sulfinate (20 mM), and diphenyliodonium chloride (1 mM) stock solutions are all prepared in distilled water. If kept refrigerated in the dark, these solutions are usable for only one week (except for the dye which is stable for up to one year).

A summary of the usage of these three polymerization systems is given in *Table 4*.

5.2.4 Discontinuous buffer system

Details of buffer composition and gel mixture preparation for the discontinuous buffer system are based on the method of Laemmli (1).

(a) 4 × stacking gel buffer: 0.5 M Tris–HCl pH 6.8. To 12.1 g Tris base, add 170 ml distilled water and adjust to pH 6.8 with 6 M HCl. Cool the solution to room temperature and readjust to pH 6.8 with 6 M HCl. Add distilled water to 200 ml and store at 4°C.

(b) 4 × resolving gel buffer: 1.5 M Tris–HCl pH 8.8. To 36.3 g Tris base, add 170 ml distilled water and adjust to pH 8.8 with 6 M HCl. Cool the solution to room temperature and readjust to pH 8.8 with 6 M HCl. Add distilled water to 200 ml and store at 4°C.

(c) 5 × running buffer. To 15 g Tris base, 72 g glycine, and 5 g SDS add distilled water to 1 litre. The pH should be 8.3 without adjustment. Store at room temperature and dilute to 1 × before use.

5.2.5 Continuous buffer system

The continuous buffer system is essentially as described by Weber and Osborn (29).

(a) Separating buffer: 0.5 M sodium phosphate pH 7. Mix 250 ml of 0.5 M NaH_2PO_4 with 500 ml of 0.5 M Na_2HPO_4. Adjust to pH 7 with 0.5 M NaH_2PO_4. Store at 4°C.

(b) 5 × running buffer pH 7. Dissolve 1 g SDS in 200 ml separating buffer.

5.3 Gel preparation

5.3.1 SDS–PAGE discontinuous buffer system

SDS–PAGE with a discontinuous buffer system is the most popular electrophoretic technique used to analyse polypeptides. It gained its popularity mainly due to its excellent powers of resolution that is derived from the use of a stacking gel. The stacking gel has a different pH (pH 6.8) compared to both running buffer (pH 8.3) and resolving gel (pH 8.8) and contains no glycine ions. It also has a large pore size to reduce its sieving power and thus enhance protein stacking. Stacking results from the formation of a limited high voltage gradient in which the sample proteins are confined to a thin and highly concentrated zone of intermediate mobility between leading chloride ions and trailing glycine ions. Separation of the stacked proteins is then accomplished as the proteins enter the resolving gel because of the decreased mobility of proteins and increased mobility of the trailing glycine ions. The former is achieved by the increased gel concentration so that molecular sieving is enhanced, and the latter is achieved by an increase in the pH from 6.8 in the stacking gel to 8.8 in the resolving gel (since the mobility of glycine ions is pH-dependent). It must be emphasized that the stacking effect does not apply to all sizes of proteins. It is less effective for proteins smaller than 12 kDa when using glycine as trailing ions. On the other hand, giant proteins tend to aggregate when they are stacked.

The method to cast minigels for the SDS–PAGE discontinuous buffer system is described in *Protocol 1*. Pre-cast SDS–polyacrylamide gels for the discontinuous buffer system can also be obtained commercially from suppliers such as Bio-Rad and Pharmacia.

5.3.2 SDS–PAGE continuous buffer system

Although the SDS discontinuous system is usually the choice for high-resolution fractionation of proteins, the SDS continuous buffer system is still used in certain situations for the following reasons. First, the system is simple and its precise buffer composition and pH is known; the pH remains constant throughout the separation. Secondly, some very large proteins, such as titin, tend to aggregate during electrophoresis in gel systems that employ stacking

---

Protocol 1. Casting minigels for the SDS–PAGE discontinuous buffer system[a]

Equipment and reagents

- Mini slab gel apparatus (e.g. Bio-Rad Mini-Protean II Cell)
- Stacking gel buffer (see Section 5.2.4)
- Resolving gel buffer (see Section 5.2.4)
- 10% (w/v) SDS
- Acrylamide:bisacrylamide mixture: 30%T, 2.6%C (see *Table 3* and Section 5.2.1)
- TEMED (see Section 5.2.3)
- 10% (w/v) ammonium persulfate: prepare daily in distilled water (see Section 5.2.3)
- Water-saturated isobutanol
- 95% ethanol

Method

1. Clean the glass plates of the minigel apparatus by soaking in chromic acid for 1 h or overnight, followed by rinsing with water. Put the plates down onto clean tissue paper, with the sides which are to be in contact with the gel upmost. Swab the plates with tissue paper soaked in 95% ethanol. Allow the plates to air dry. Assemble the spacers and two glass plates in the clamp. Tighten the clamp after the glass plates and spacers are aligned on the casting stand. Snap the sandwich onto the casting stand to seal the bottom of the assembly. Mark the glass plate with a marker pen to indicate the desired upper limit of the resolving gel (5.5 cm from the bottom edge of the glass plate sandwich).

2. Prepare the resolving gel mixture (see *Table 5*) and transfer this to the glass plate sandwich using either a 10 ml syringe or a 10 ml serological pipette up to the marker line. Degassing of solutions is found to be unnecessary with the Mini-Protean format although it is routinely done for larger format gels.

3. Overlay the gel mixture with water-saturated isobutanol (isoamyl alcohol and isopropanol may also be used here) to exclude oxygen from the surface. After a clear line forms between the resolving gel and the isobutanol to indicate gel polymerization, drain off the isobutanol and rinse the gel surface with distilled water. Remove any remaining water with filter papers without damaging the gel surface.

4. Prepare the stacking gel mixture (see *Table 6*) and overlay the resolving gel with this. Insert the Teflon comb, leaving approx. 5 mm between the top of the resolving gel and the bottom of the comb. Make sure that there are no air bubbles trapped beneath the comb. Let the monomer solution polymerize for at least 2 h before using the gel.

[a] Buffer system based on the method of Laemmli (1).

---

gels or discontinuous buffers to concentrate protein bands (30). Such aggregation artefacts can often be avoided by using the continuous buffering system.

Since stacking does not occur when using a continuous buffer system, the sample must be applied in the smallest possible volume to give a thin starting

Table 5. Resolving gel mixture[a]

Reagent (ml)	Final concentration (%) polyacrylamide gel					
	7.5	10	12.5	15	17.5	20
Water	7.71	6.38	5.05	3.71	2.38	1.05
Acrylamide:bisacrylamide (30%T, 2.6%C)	4.0	5.33	6.67	8.0	9.33	10.67
4 × resolving gel buffer	4.0	4.0	4.0	4.0	4.0	4.0
10% SDS	0.16	0.16	0.16	0.16	0.16	0.16
TEMED[b]	0.008	0.008	0.008	0.008	0.008	0.008
10% ammonium persulfate[c]	0.12	0.12	0.12	0.12	0.12	0.12
Total volume[a]	16.0	16.0	16.0	16.0	16.0	16.0

[a] Enough for two 1.5 mm thick Bio-Rad minigels.
[b] Add TEMED just prior to pouring the gel.
[c] This solution must be made fresh.

Table 6. Stacking gel mixture[a]

Reagent (ml)	Final concentration (%) polyacrylamide gel			
	3	3.5	4	5.7
Water	3.17	3.09	3.0	2.72
Acrylamide:bisacrylamide (30%T, 2.6%C)	0.5	0.58	0.67	0.95
4 × stacking gel buffer	1.25	1.25	1.25	1.25
10% SDS	0.05	0.05	0.05	0.05
TEMED[b]	0.005	0.005	0.005	0.005
10% ammonium persulfate[c]	0.025	0.025	0.025	0.025
Total volume[a]	5.0	5.0	5.0	5.0

[a] Enough for two 1.5 mm thick Bio-Rad minigels.
[b] Add TEMED just prior to pouring the gel.
[c] This solution must be made fresh.

zone. Additional zone sharpening can be obtained by loading the protein sample in a buffer which has a lower ionic strength than that of the gel and electrode buffer. The lower conductivity caused by lower ionic strength results in a high voltage field which accelerates protein migration in free solution. As they move into the gel the protein is slowed down due to the sieving effect and the drop in voltage gradient. Kubo (22) reported a modified procedure to the original method of Weber and Osborn (29) and showed that by including 10–15% (v/v) glycerol in a large pore size sample well gel on top of the resolving gel, the band 'edge tailing' (bands distorted by tailing on both edges) phenomenon usually associated with the continuous buffer system was eliminated and a sharper band comparable to the Laemmli procedure was achieved.

Pouring minigels for the SDS–PAGE continuous buffer system is described in *Protocol 2*.

Protocol 2. Casting minigels for the SDS–PAGE continuous buffer system

Equipment and reagents

- Mini slab gel apparatus (e.g. Bio-Rad Mini-Protean II Cell)
- Separating buffer: 0.5 M sodium phosphate pH 7 (see Section 5.2.5)
- 10% (w/v) SDS
- TEMED (see Section 5.2.3)

- Acrylamide:bisacrylamide mixture: 30%T, 2.6%C (see *Table 3* and Section 5.2.1)
- 10% (w/v) ammonium persulfate: prepare daily in distilled water (see Section 5.2.3)
- 95% ethanol

Method

1. Prepare the gel casting mould as described in *Protocol 1*, step 1.

2. Prepare the gel mixture (*Table 7*) followed by gentle but thorough mixing (eight to ten cycles of swirling). Swirling too little can result in non-uniform polymerization, whereas swirling too much may introduce too much oxygen into the solution.

3. Cast the gel by introducing the solution into the gel mould with either a 10 ml syringe or a 10 ml serological pipette in a steady stream to minimize the introduction of oxygen. Insert the well-forming comb without trapping air. This can be done by first placing one end of the comb into the gel, then slowly inserting the comb fully into the gel.

4. Allow at least 2 h for the acrylamide to polymerize before running the gel. Prepared slab gels may be stored for up to two weeks at 4°C with the comb left in place if they are first wrapped in damp paper towels and then sealed in plastic wrap.

Table 7. Continuous gel mixture[a]

Reagent (ml)	Final concentration (%) polyacrylamide gel					
	7.5	**10**	**12.5**	**15**	**17.5**	**20**
Water	10.64	8.97	7.31	5.64	3.97	2.31
Acrylamide:bisacrylamide (30%T, 2.6%C)	5.0	6.67	8.33	10.0	11.67	13.33
0.5 M sodium phosphate pH 7	4.0	4.0	4.0	4.0	4.0	4.0
10% SDS	0.2	0.2	0.2	0.2	0.2	0.2
TEMED[b]	0.01	0.01	0.01	0.01	0.01	0.01
10% ammonium persulfate[c]	0.15	0.15	0.15	0.15	0.15	0.15
Total volume[a]	20.0	20.0	20.0	20.0	20.0	20.0

[a] Enough for two 1.5 mm thick Bio-Rad minigels.
[b] Add TEMED just prior to pouring the gel.
[c] This solution must be made fresh.

5.3.3 Gradient gels

One of the main advantages of gradient gel electrophoresis, also called pore limit PAGE, is that the migrating proteins are continually entering areas of the gel with decreasing pore size such that the advancing edge of the migrating protein zone is retarded more than the trailing edge, resulting in marked sharpening of the protein bands. In addition, the gradient in pore size increases the range of molecular masses which can be fractionated simultaneously on one gel. Furthermore, it has been suggested that glycoproteins which behave anomalously in uniform SDS–polyacrylamide gels (see Section 5.6.4) show normal mobility in SDS gradient gels (31). Similarly, polyacrylamide gradient gels have been used to determine the molecular weight of native proteins and their constituent subunits simultaneously based on the distance they migrated through the gel by limited use of SDS (28).

When proteins reach their pore limit in a gradient gel, their migration rate approaches zero and the protein banding pattern will not change appreciably with additional electrophoresis, although migration does not cease completely (32). At this point, even in a non-dissociating buffer system, the influence of charge on the final migration position of a protein is eliminated so that the final migration position is a function only of protein size and shape. Of course the time required for a given protein to reach its pore limit is still a function of its charge.

The preparation of gradient gels using the Bio-Rad minigel apparatus is described in *Protocol 3*. In addition to a gradient in acrylamide concentration, a density gradient of glycerol or sucrose is often included to minimize mixing by convective disturbances caused by the heat evolved during polymerization for improved gradient formation (14, 23). In this case, the most concentrated acrylamide mixture contains the glycerol (or sucrose) so that the glycerol (or sucrose) gradient forms during pouring of the gel. Pre-cast gradient gels for SDS–PAGE can also be obtained commercially. An example of SDS gradient PAGE is given in *Figure 6*.

Protocol 3. Casting SDS–PAGE gradient minigels[a]

Equipment and reagents

- Mini slab gel apparatus (e.g. Bio-Rad Mini-Protean II Cell)
- Gradient maker (Bio-Rad)
- Magnetic stirrer
- Peristaltic pump
- Stacking gel buffer (see Section 5.2.4)
- Resolving gel buffer (see Section 5.2.4)
- 10% (w/v) SDS
- Acrylamide:bisacrylamide mixture: 50%T, 2.6%C (see *Table 3* and Section 5.2.1)
- TEMED (see Section 5.2.3)
- 10% (w/v) ammonium persulfate: prepare daily in distilled water (see Section 5.2.3)
- 95% ethanol

Method

1. Prepare the gel casting mould as described in *Protocol 1*, step 1.

2. Choose the most appropriate acrylamide gradient range of concentrations for the protein sample to be analysed (see *Table 2*). Now prepare two gel mixtures corresponding to the minimum and maximum concentrations in the desired gradient. Thus for a 3–30% gradient gel, prepare a 3%T and a 30%T gel solutions. Prepare these as described in *Table 8* but **without** TEMED. Now add glycerol to the high concentration mixture (to 15% final concentration) to facilitate gradient formation.

3. Set-up the gradient maker as illustrated in *Figure 3*. Add TEMED to both gel mixtures (see *Table 8*). Close all connections in the gradient maker and pour the lower concentration mixture into chamber A first. Open the connection between A and B momentarily and allow the solution to fill the connection tube and then close the connection so that air bubbles can be removed. Add the higher concentration mixture into chamber B. With the stirrer stirring in chamber B, partly open the connection between A and B and add clean glass beads to chamber A such that there is no flow of liquid between the two reservoirs.

4. Open all connections and at the same time turn the peristaltic pump on. Fill the gel sandwich with gel mixture at a flow rate of about 3 ml/min.

5. When the gel mixture has all been delivered, connect the outlet tubing from the gradient former to distilled water, reduce the flow rate to 0.5 ml/min, and overlay the gel with water.[b,c] Allow the gel to stand to polymerize.

6. After a clear line is formed between the resolving gel and the water overlay, pour off the water.

7. Prepare the stacking gel mixture (*Table 6*). Overlay the resolving gel with the stacking gel mixture.

8. Insert the sample comb, leaving about 5 mm between the top of the resolving gel and the bottom of the comb. Allow the stacking gel to polymerize for 2 h before running the gel.

[a] Buffer system based on the method of Laemmli (1).
[b] If a pump is not available, the gradient gel may be poured under gravity.
[c] Immediately after the gradient has been poured, wash out the gradient former with water to prevent gel polymerization in the apparatus.

5.4 Sample preparation

The amount of protein required per sample is dependent on the number of proteins present, the polyacrylamide gel format, and the method used for detection. In general, for Coomassie blue staining (see Chapter 2), about 0.2–2 μg of each polypeptide should be loaded onto a minigel well and 1–10 μg to a standard slab gel well to give optimal results, such that for a

Table 8. Resolving gel mixture for gradient gels[a]

Reagent (ml)	Final concentration (%) polyacrylamide gel						
	3	5	8	15	17.5	20	28
Water	2.69	2.53	2.29	1.13	0.93	0.73	0.09
Acrylamide:bisacrylamide (50%T, 2.6%C)	0.24	0.4	0.64	1.2	1.4	1.6	2.24
Resolving gel buffer	1.0	1.0	1.0	1.0	1.0	1.0	1.0
10% SDS	0.04	0.04	0.04	0.04	0.04	0.04	0.04
Glycerol[b]	–	–	–	0.6	0.6	0.6	0.6
TEMED[c]	0.002	0.002	0.002	0.002	0.002	0.002	0.002
10% ammonium persulfate[d]	0.03	0.03	0.03	0.03	0.03	0.03	0.03
Total volume[a]	4.0	4.0	4.0	4.0	4.0	4.0	4.0

[a] 8 ml is enough for one 1.5 mm thick Bio-Rad minigel.
[b] Glycerol is only included in the higher concentrations of acrylamide solution.
[c] Add TEMED just prior to pouring the gel.
[d] This solution must be made fresh.

complex mixture about 20–40 μg (50–100 μg) is usually sufficient for a minigel (standard gel) system. Overloading causes band distortion and underloading results in faint bands. For continuous buffer systems, the sample volume should be as small as possible since the depth of the sample starting zone has a large effect on protein band sharpness. For discontinuous buffer systems, the stacking effect sharpens all samples and so sample volume is limited only by the size of the sample wells.

5.4.1 Preparation of sample buffers

For SDS–PAGE, the protein sample is mixed with concentrated sample buffer and then denatured by heating (see Section 5.5). Various recipes exist for sample buffers and different concentrations are used by different researchers. A common stock concentration is twofold. *Table 9* gives recipes for 2 × sample buffer for both the SDS–PAGE discontinuous and SDS–PAGE continuous buffer systems.

Common components of these sample buffers are listed below:

(a) SDS. Used to solublize and denature the proteins for accurate molecular weight determination. The typical final concentration of SDS, after mixing sample buffer with the sample is 2% (w/v).

(b) 2-mercaptoethanol (or dithiothreitol). Employed to break disulfide bonds of proteins to ensure polypeptide denaturation and maximal binding of SDS. The final concentration is usually 5% (v/v). Since 2-mercapto-ethanol is very volatile and dissolved oxygen is able to reoxidize thiols into disulfides, the free thiols will be consumed upon aging of the sample buffer. Fresh 2-mercaptoethanol should be added to aged buffer to maintain thiol concentration.

Table 9. Composition of SDS–PAGE sample buffers

2 × sample buffer for discontinuous gel systems

Stacking buffer (0.5 M Tris–HCl pH 6.8)	2.0 ml
Glycerol	1.6 ml
10% SDS	3.2 ml
2-mercaptoethanol	0.8 ml
0.1% (w/v) bromophenol blue in water	0.4 ml
Store at 4 °C for up to three months	

2 × sample buffer for continuous gel systems

0.5 M sodium phosphate buffer pH 7	0.64 ml
Water	1.36 ml
Glycerol	1.6 ml
10% SDS	3.2 ml
2-mercaptoethanol	0.8 ml
0.1% (w/v) bromophenol blue in water	0.4 ml
Store at 4 °C for up to three months	

(c) Glycerol (can also be replaced with urea or sucrose). Mainly used to increase the density of the sample solution. Inclusion of 10% (v/v) glycerol in the sample solution is sufficient to keep samples at the bottom of the well so that samples do not undergo convective mixing with the running buffer.

(d) Buffer (to maintain the pH value). The buffer used should be the same buffer as for the sample well gel. In the case of the discontinuous buffer system, this means it will be the stacking gel buffer whereas for continuous gel systems, it is the resolving gel buffer.

(e) Bromophenol blue. Serves as a tracking dye so that the progression of electrophoresis can be monitored visually. It also aids loading of the sample by making it readily visible. The final concentration in the sample is usually 0.001–0.002% (w/v). It is good practice to use a fixed amount of dye for each sample regardless of its total volume, so that the same intensity of the dye is achieved for each lane.

When using the sample buffers described in *Table 9*, the maximum concentration of protein in the final solution should not be higher than 10 µg/µl to ensure that enough SDS is present. If the protein content exceeds this limit, extra SDS should be added. SDS can be added to the sample buffer to at least 5% final concentration without deleterious effects on the electrophoretic separation.

5.4.2 Concentrating proteins for PAGE
Protein samples too dilute for immediate electrophoretic analysis can be concentrated in the following ways:

(a) Lyophilization. The time required for lyophilization is dependent on sample volume and salt content. It is not recommended for samples containing detergents and more than 0.5 M salt.

(b) Ultrafiltration. Ultrafiltration will concentrate large molecules while keeping salt concentration unchanged. The time required for this method depends on sample volume, protein concentration, and membrane size cut-off. Protein recovery is dependent on filter retention. This method is especially useful for concentrating samples containing detergents, urea, and high salt.

(c) Ammonium sulfate precipitation. The time required for ammonium sulfate precipitation is largely independent of sample volume, but some proteins may give lower recoveries. Both lyophilization and ammonium sulfate precipitation may produce a sample with high salt concentration and so the sample should then be dialysed against 0.1 M sodium phosphate buffer (pH 7.2) for the continuous buffer system, or 0.0625 M Tris–HCl (pH 6.8) for the discontinuous buffer system. Potassium ions in particular must be removed since they precipitate the SDS used in SDS–PAGE.

(d) Dialysis against polyethylene glycol. Dialysis against a high concentration of polyethylene glycol ($M_r > 20\,000$) will reduce both the sample volume and salt concentration. Protein loss is expected for any type of method involving dialysis primarily due to membrane retention.

(e) Precipitation with TCA. Precipitation of proteins with trichloroacetic acid (TCA) is a common method for concentrating proteins. Apart from speed in concentrating large number of samples, this method also removes salts. However, TCA precipitation should be used with caution since the recovery of some proteins is low and precipitated protein is often difficult to redissolve completely in the sample buffer.

(f) Precipitation with dyes (35, 36). Dyes used for protein assay such as Coomassie brilliant blue G-250 or pyrogallol red-molybdate can form complexes with proteins under acidic conditions and these complexes can be recovered following centrifugation. This method allows protein recovery following a protein assay. The recovery is poor if a high concentration of SDS is present, but denaturants such as urea and guanidinium have little effect upon the efficiency of dye precipitation. However the mobility of the proteins in SDS–PAGE and subsequent detection in the gel by staining may be affected with this method.

(g) Phenol–ether precipitation (37). Phenol–ether precipitation results in quantitative recovery of protein from solutions containing as little as 10 ng/ml protein. Detergents and salts do not seem to affect protein recovery. This method (*Protocol 4*) is particularly useful for concentrating protein samples for SDS–PAGE but may not be used for native gel electrophoresis due to protein denaturation.

Protocol 4.   Phenol–ether precipitation

Equipment and reagents
- SpeedVac centrifugal vacuum evaporator system (Fisher)
- Phenol (ACS grade)
- Ether (ACS grade)
- 1 × sample buffer (see *Table 9*)

Method
1. Add an equal volume of phenol to the protein sample. Vortex for 20 sec then centrifuge at 12 000 *g* for 5 min. Discard the upper phase.
2. Add 2 vol. of ether to the phenol phase. Vortex for 20 sec then centrifuge at 12 000 *g* for 5 min. Discard the upper phase.
3. Repeat step 2.
4. Dry the lower aqueous phase using a SpeedVac evaporator system.
5. Solubilize the dried sample in water or 1 × sample buffer for SDS–PAGE.

5.4.3 Modifications to the standard sample preparation method

The majority of proteins can be solubilized with the standard sample buffer (*Table 9*). However some proteins, such as nuclear non-histone proteins and membrane proteins, require the presence of 8 M urea in the SDS sample buffer (30) to achieve complete solubilization. When urea is present there is no need to add glycerol or sucrose to the sample buffer since the urea increases the density of the sample anyway. Since urea in water exists in equilibrium with ammonium cyanate, the concentration of which increases with increasing temperature, the sample should contain Tris as the buffer to minimize modification of proteins by cyanate (cyanate reacts with amines) if the sample is to be heated.

Some membrane bound proteins undergo an aggregation reaction during heating in the standard sample buffer, but remain soluble either if heating is avoided or if heating is carried out in the absence of 2-mercaptoethanol (38). Dissolution of the aggregates can be achieved only by raising the pH to 13.5. A possible explanation for this type of aggregation may be that internal hydrophobic regions are exposed during heating in the presence of a reducing agent and it is these that aggregate. In other cases, heating can be replaced with almost equal efficiency by using SDS and urea at room temperature (39). Other problems related to sample solubilization are reviewed in ref. 40.

The identification of interchain disulfide linkages is critical to determining the subunit structure of a protein complex. This can usually be done by denaturing parallel samples of the proteins in the presence and absence of a thiol agent. Since thermal denaturation of samples in the presence of SDS without reducing agents (e.g. 2-mercaptoethanol or dithiothreitol) does not break disulfide bonds, disulfide bond linked subunits will migrate together and lead to the appearance of higher molecular weight bands compared to samples denatured in the presence of a reducing agent. However, it has been reported that denaturation in the absence of a reducing agent can cause formation of intermolecular disulfide bonds which are not present in the native protein and the degree of dimerization is dependent on the concentration of the protein when denatured (41). This dimerization artefact is suggested only if a normal sized (monomeric) protein is observed following incubation with sample buffer without heating or without a reducing agent. If heating is necessary, dependence of the dimer formation on protein concentration indicates artefactual disulfide bond formation.

In some cases, the use of diluted sample buffer (0.5 × final concentration) results in better separation of the proteins of interest compared to the standard conditions. For example, by mixing the sample (in 1 × sample buffer) with an identical volume of running buffer, Rossini was able to achieve better separation of myosin heavy chains (21).

For most proteins, neither the temperature nor the duration of the denaturation step seems to require precise control. However, control of temperature and duration does appear to be important for some large proteins. Thus, heating at 95°C for more than two minutes causes extensive degradation of titin and platelet-derived growth factor receptor; Granzier found that the optimal solubilization condition for titin in sample buffer was 60°C for only 60 seconds (34).

5.4.4 Molecular weight standards

In order to determine the molecular weight of sample polypeptides (Section 5.6) using SDS–PAGE, it is essential to include proteins of known molecular weight. Protein standards are also useful for testing different gel systems for their resolving range and reproducibility between runs. A number of well-established proteins that can be used as molecular weight standards are listed in *Table 10* (for a more complete list see ref. 42). As molecular weight standards, these proteins should not exhibit anomalous migration in SDS–polyacrylamide gels and should give clear sharp bands. These proteins can be obtained either individually from commercial sources such as Sigma or as kits of different molecular weight ranges from a number of suppliers. For example, *Table 10* includes three kits with different molecular weight ranges from Bio-Rad.

A variety of other types of protein molecular weight standards are also commercially available:

(a) Pre-stained protein standards prepared by chemical modification of proteins with chromophores are suitable for monitoring protein migration during electrophoresis and electrophoretic transfer onto membranes.

Table 10. Molecular weights of protein standards

Polypeptide	Molecular mass	Bio-Rad kits		
		High	Low	Broad
Rabbit muscle myosin heavy chain	200 000	x		x
Horse plasma α_2-macroglobulin	170 000			
E. coli β-galactosidase	116 250	x		x
Rabbit muscle phosphorylase b	97 400	x	x	x
Transferrin	76 000			
Bovine serum albumin	66 200	x	x	x
Catalase	57 500			
Glutamate dehydrogenase	53 000			
Hen egg ovalbumin	45 000	x	x	x
Rabbit muscle aldolase	40 000			
Porcine lactate dehydrogenase	36 000			
Bovine carbonic anhydrase	31 000		x	x
Soybean trypsin inhibitor	21 500		x	x
Horse heart myoglobin	16 949			
Hen egg lysozyme	14 400		x	x
Horse heart cytochrome c	12 500			
Bovine pancreas aprotinin	6500			x
Insulin β chain	3400			
Insulin α chain	2300			
Bacitracin	1423			

However, one of the drawbacks of this type of protein standard is that their molecular weights are not predictable after modification. Indeed, each batch lot of the same proteins may have different apparent sizes by gel electrophoresis.

(b) Biotin-labelled or ^{14}C-labelled SDS–PAGE standards allowing accurate molecular determination directly on Western blots.

(c) Protein ladders (e.g. Gibco BRL) consisting of equally spaced (in molecular weight terms) proteins prepared by controlled polymerization of a chemically modified protein.

(d) SDS–PAGE standards prepared to give even band intensities with no extraneous bands when detected by silver staining (see Chapter 2).

5.5 Sample loading and electrophoresis

Sample preparation and loading is described in *Protocol 5*. For the discontinuous buffer system, prepare sufficient volume of sample to fill the wells (*Table 11*). For the continuous buffer system, the sample should be applied in as small a volume as possible to give a thin starting zone. In the latter system, additional zone sharpening can be obtained by loading the protein in a buffer which has a lower ionic strength than that of the gel and electrode buffer. The proteins

Protocol 5. Sample loading and electrophoresis

Equipment and reagents

- Mini slab gel apparatus (e.g. Bio-Rad Mini-Protean II Cell)
- Power pack (see Section 3.2)
- Protein samples
- 2 × sample buffer (see *Table 9* and Section 5.4.1)
- Running buffer (see Sections 5.2.4 and 5.2.5)

Method

1. Set-up the gel apparatus as instructed in the supplier's manual. Add reservoir buffer to the lower then upper reservoirs of the electrophoresis apparatus. For the Bio-Rad minigel apparatus the minimum volume requirement for running buffer is 300 ml (120 ml for upper buffer, 180 ml for lower buffer). Up to 600 ml of running buffer may be added to the lower reservoir so that the entire sandwich is immersed in the lower buffer for improved cooling.

2. Remove any air bubbles trapped along the bottom edge of the gel to allow even current distribution. This can be done by using either a 50 ml syringe connected with a long bent needle (reusable) or a Pasteur pipette with a bent tip.

3. Prior to electrophoresis, mix 1 vol. of 2 × sample buffer with 1 vol. of protein sample, and heat the samples in boiling water for 2 min to denature the proteins. After heating, allow the sample to cool to room temperature and remove any insoluble materials by centrifugation to avoid protein streaking during gel electrophoresis.

4. Load the samples carefully into the gel wells with either a micro-syringe or a pipette connected to a Prot/Elec loading tip (Bio-Rad).[a]

5. Connect the electrophoresis apparatus to the power pack with the anode (+) linked with the bottom reservoir and the cathode (−) connected to the upper reservoir. Run the gel at a constant current at 20–30 mA/gel or a constant voltage at 150 V until the dye band reaches the gel bottom. This usually takes 45–90 min depending on the concentration of the gel and size of the protein(s) of interest.

[a] The samples may also be stored in the freezer at −20 °C. When cooled, SDS crystallizes out of solution and thus stored samples must be reheated before use.

will initially be in a zone of lower ionic strength and thus higher voltage. Hence, they move faster in free solution, slowing down as they move into the gel as a result of the sieving effect produced by the gel and the drop in voltage gradient. A minimum of 0.2 µg of a single band protein should be loaded per well to ensure detection by Coomassie blue. All samples loaded onto the gel

Table 11. Maximum sample volume per well for Bio-Rad Protean II minigels

	Gel thickness (mm)		
Number of wells	0.5	1	1.5
15	15 μl	30 μl	45 μl
10	30 μl	60 μl	90 μl
5	65 μl	130 μl	195 μl
1		950 μl	1400 μl

should be blue in colour. A yellow colour indicates that the sample pH is too low, in which case the pH must be adjusted to the correct value before loading.

5.6 Molecular mass estimation

5.6.1 Introduction

Under appropriate conditions, all reduced polypeptides bind the same amount of SDS on a weight basis (1.4 g SDS/g polypeptide) (44) to form rod-like particles, with lengths proportional to the molecular weight of the polypeptides (45). The SDS binds mainly to the hydrophobic regions of polypeptides, whereas the hydrophilic regions bind much less SDS (47). Not surprisingly, therefore, unreduced polypeptides containing intact disulfide bonds bind much less SDS than reduced proteins. Also noteworthy is the fact that increasing the ionic strength decreases the amount of SDS bound (46).

For accurate molecular weight determination by SDS–PAGE, all poly-peptides (standard and unknowns) must bind the optimum amount of SDS and assume the same conformation or shape. Clearly, the standard and unknown proteins must also be electrophoresed under the same conditions and at the same time (preferably on the same gel). Under these ideal conditions, the molecular weights of the unknown polypeptides can be estimated from their relative mobilities (43).

5.6.2 Molecular weight determination using uniform concentration gels

A typical separation using a uniform concentration gel is shown in *Figure 4*. The molecular weights of the sample polypeptides can be estimated by the following procedure (adapted from ref. 29). Molecular weight determination by this technique has an accuracy of $\pm$ 10% and it applies to both the continuous buffer (Section 5.3.2) and the discontinuous buffer (Section 5.3.1) systems.

(a) Following staining and destaining of the gel (see Chapter 2), place the gel on a glass plate and on top of a light box. Measure the distance travelled

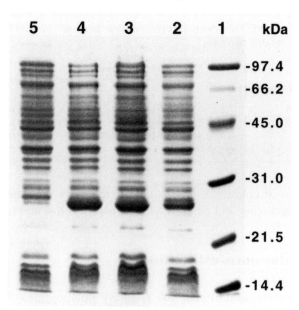

Figure 4. Example of proteins separated on a SDS discontinuous uniform polyacrylamide (12.5%T, 2.6%C) minigel stained with Coomassie brilliant blue R-250. Lane 1, molecular weight markers; lane 2 to lane 4, cell lysates from *E. coli* clones expressing human ventricular myosin light chain 1 (27 kDa); lane 5, cell lysate from control *E. coli*.

by the bromophenol blue dye and each polypeptide (standard and un-known) from the top of the resolving gel. Alternatively, the destained gel may be photographed and all measurements can be performed on the print.

(b) Calculate R_f (relative mobility) for each polypeptide from the formula: R_f = distance of protein migration/distance of dye migration.

(c) Plot R_f versus log M_r for the standards on semi-log paper (*Figure 5*) or use a spreadsheet program such as *Microsoft Excel*.

(d) Use the R_f values of the unknown sample polypeptides to estimate their molecular weights by interpolation.

5.6.3 Molecular weight estimation using gradient gels

A typical separation using a gradient gel is shown in *Figure 6*. The procedure of Lambin *et al.* (48, 49) is adopted for the determination of molecular weight from linear gradient gels.

(a) Swell the gel in 10% acetic acid for 1 h after destaining.

(b) Place the gel on a plastic plate and record the length of the resolving gel. Measure the distance between the top edge of the gradient gel and the centre of each band for the standard and unknown polypeptides.

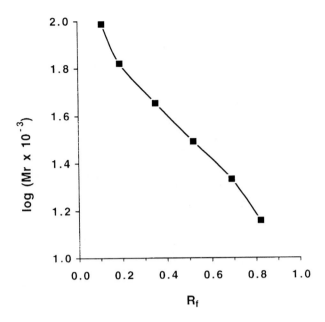

Figure 5. Plot of R_f versus log M_r for standard proteins ranging from 14.4 kDa to 97.4 kDa (Bio-Rad low range, see *Table 10*) separated on a 12.5% T, 2.6% C uniform concentration SDS–PAGE gel using a Bio-Rad minigel apparatus.

(c) Assuming a linear gradient, calculate the per cent gel concentration (%T) reached by each polypeptide by the formula:

%T = %T_L + (distance of protein migration/gel length) × (%T_H – %T_L)

where %T_L is the lowest %T of the gradient and %T_H is the highest %T of the gradient.

(d) Plot log M_r against log (%T) for the standard polypeptides (*Figure 7*), using either double-log graph paper or a spreadsheet program such as *Microsoft Excel* .

(e) From the %T values reached by the unknown polypeptides, derive their molecular weight by interpolation.

5.6.4 Problems with molecular weight estimation

It is known that high ionic strength reduces the amount of SDS bound to proteins (46), and may therefore interfere with molecular weight determination (50). However, See and Jackowski (51) found that the molecular weight of proteins in the range M_r 20000–66000 can be determined by SDS–PAGE even in the presence of NaCl in samples up to 0.8 M. Samples containing either up to 0.5 M KCl or ammonium sulfate up to 10% saturation do not disturb electrophoresis (52).

In the vast majority of cases, the molecular weights of sample polypeptides

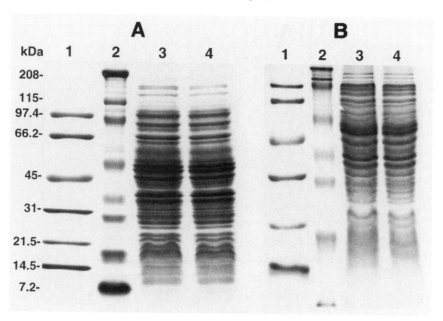

Figure 6. Comparison of *E. coli* lysate patterns obtained from (A) a gradient SDS–polyacrylamide minigel (5–20%T, 2.6%C) and (B) a uniform SDS–polyacrylamide minigel (12.5%T, 2.6%C). The same set of samples were loaded on both gels and the gels were stained with Coomassie brilliant blue R-250. Lane 1, Bio-Rad low range molecular weight standard; lane 2, Bio-Rad pre-stained broad range molecular weight standard; lanes 3 and 4, total *E. coli* proteins. Notice the differences in the sharpness of the protein bands and also the resolution above the 97.4 kDa marker and below the 21.5 kDa marker. Clearly, the 5–20% gradient gel is superior in resolving power to the 12.5% uniform concentration gel.

estimated by SDS–PAGE can be relied upon as reasonably accurate. However, in a minority of cases, for extreme sets of proteins, problems can be experienced and are described below.

(a) Glycoproteins often exhibit abnormal migration during SDS–PAGE, giving different apparent molecular weights when determined in different gel concentrations. This is caused by their hydrophilic glycan moiety which reduces the hydrophobic interactions between the protein and SDS, thus preventing the correct binding of SDS. A more accurate procedure appears to be the use of an SDS gradient gel (53).

(b) Proteins with high acidic residue content, such as caldesmon and tropomyosin, also migrate anomalously in SDS–polyacrylamide gels (54). This may be due to the repulsion of the negatively charged SDS by the acidic residues. Neutralization of the negative charge of the acidic residues restores normal electrophoretic mobility of these proteins during SDS–PAGE (55).

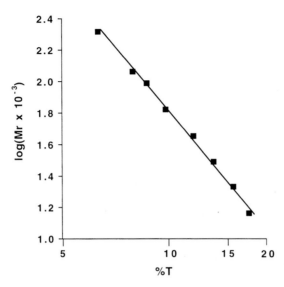

Figure 7. Plot of log ($M_r \times 10^{-3}$) versus log (%T) for 5–20% SDS–polyacrylamide gradient gel. The molecular weights of the standard proteins range from 14.4 kDa to 200 kDa.

(c) Highly basic proteins like histones and troponin I typically give abnormally large molecular weights in SDS–PAGE. The lower electrophoretic mobility of basic proteins is presumably due to the reduction of the charge/mass ratio of the SDS:polypeptide complex as a result of the high proportion of basic amino acids.

(d) Proteins with high proline content and other unusual amino acid sequences such as ventricular myosin light chain 1 and collagenous polypeptides may also have abnormally high molecular weights determined by SDS–PAGE. The anomaly may be due to the alteration of conformation of the SDS:protein complex.

5.7 Troubleshooting

(a) Failure of the samples to remain in the well. This indicates that the density of the sample solution is lower relative to that of the electrode buffer, usually caused by the accidental omission of glycerol or sucrose from the sample buffer.

(b) Yellow sample colour. This indicates that the sample has an acidic pH since bromophenol blue will turn yellow at lower pH values. Adjust the pH with Tris until the sample turns blue. However, note that excessive addition of Tris may lead to problems during electrophoresis.

(c) Inability of tracking dye to enter the gel. Check that current is flowing. If no bubbles are being formed on the electrodes, a poor connection between the electrode and power supply is suggested. If the current reading is normal, check that the polarities of the electrodes have not been reversed inadvertently.

(d) Inability of protein to enter the resolving gel. Most of the time this is the result of protein aggregation. Some proteins are very sensitive to increased concentration and aggregate due to the concentrating effect of the stacking gel. If this is the case, one would be advised to use less concentrated samples with a continuous buffer system. Other proteins may precipitate in the presence of SDS. If this is the reason, replace the SDS with another type of detergent (see Section 7.5). Some proteins tend to aggregate at pH 6.8 (stacking gel pH value). Again, the use of a continuous gel system may solve the problem.

(e) Distorted bands. During electrophoresis, heat is generated due to electric resistance of the gel. Since the mobility of migrating ions is increased as the temperature rises, a temperature gradient from the centre of the gel to the gel surface formed by uneven heat loss will cause distorted bands. This can always be corrected by proper cooling such as filling the lower buffer chamber with buffer all the way to the level of the sample well or reducing the power applied. Distorted bands can also result from poorly polymerized gels, insoluble material or bubbles in the gel, and high salt content of the sample.

(f) Protein bands observed in unloaded tracks of a slab gel. This is usually caused by:

- Overloading of wells leading to sample overflowing to adjacent wells.
- Poorly polymerized stacking gel resulting in partial teeth formation and hence leakage between wells.
- Contaminated running buffer.

For (i) and (ii) the unexpected bands show the same size of the adjacent sample, whereas for (iii), protein staining can usually be seen in all tracks.

(g) Unexpected bands observed only in loaded tracks of a slab gel. This could be caused by contamination of sample buffer. If this is true, the same band should appear in all loaded tracks. If the unexpected band has identical size with the adjacent protein sample, this could be the result of leakage between wells or sample overflowing.

(h) Poorer resolution than expected. The following reasons can cause decreased gel resolution:

- Aged gels, due to gradual diffusion of buffers between stacking and resolving gels.
- Impurities in the reagents.
- Sample overloading.

- Incorrect pH of buffers.
- Poorly polymerized gels.

(i) High background staining along individual tracks. Samples with extensive proteolysis or that have been overloaded give high background staining. Impurities of sample buffer components could also result in background staining.

(j) Protein streaking. This could be caused by protein precipitation followed by dissolution of the precipitates during electrophoresis. Be sure to centrifuge samples after denaturation to remove insoluble materials. Decreasing the amount of the sample to be loaded is recommended. Some proteins tend to aggregate at very high concentrations reached in the stacking gel. This problem can sometimes be overcome by using a continuous buffer system since this avoids concentration of sample proteins.

(k) Protein dimer or double band formation. This could be the result of using aged sample buffer in which not enough thiols are present or due to the oxidative power of the polyacrylamide gel caused by residual persulfate and its reaction products (56). The former can be corrected by using fresh sample buffer and the latter by adding thiols into the electrode buffer or instead using a photopolymerization system to prepare the poly-acrylamide gels.

(l) Protein smears. Accidental omission of SDS in the running buffer causes severe protein smearing and total loss of resolution possibly due to gradual dissociation of SDS from proteins.

(m) Abnormal run time. If the run time is too long, the running buffer may be too dilute or the gel buffer may be too concentrated. If the run time is too short, the running buffer may be too concentrated or the gel buffer may be too dilute.

6. Non-denaturing polyacrylamide gel electrophoresis

6.1 Introduction

Although SDS–PAGE is the most commonly used gel electrophoresis system for analysing proteins, it cannot be used to analyse intact protein complexes and proteins whose biological activity need to be retained for subsequent functional testing. In these situations, it is necessary to use a non-denaturing system.

A further limitation of SDS–PAGE is that different proteins with the same size are unlikely to be resolved by this technique. Any single band may be composed of multiple components of identical molecular mass. Since non-denaturing gel electrophoresis separates proteins on the basis of both size and charge, the analysis of native proteins by this procedure may reveal the

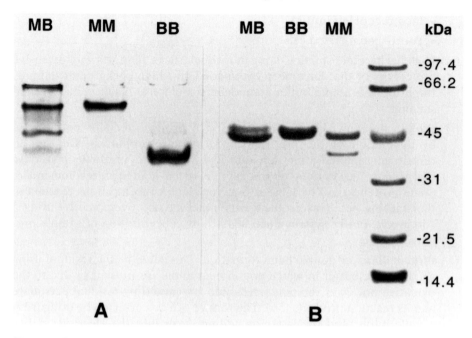

Figure 8. Comparison of denaturing and non-denaturing polyacrylamide gel electrophoresis of recombinant creatine kinase (CK) isoforms expressed in *E. coli* cells. Creatine kinase is a dimer composed of two possible types of subunit, M and B, thus three cytoplasmic isoforms exist: MM (muscle-type with a pI 6.5), BB (brain-type with a pI 5.0), and the heterodimeric form, MB (pI 5.8), found mainly in adult mammalian heart. (A) 7% native polyacrylamide gel using Tris–HCl buffer pH 8.8. (B) SDS–PAGE with a 10% uniform polyacrylamide gel. 5 μg of each isoform were analysed. SDS–PAGE analysis of the three isoforms suggest that: (i) the MM preparation is contaminated with a smaller protein about 40 kDa; (ii) the MB preparation has two bands; (iii) all three isoforms, when denatured, show similar size (43–45 kDa). Native PAGE analyses of the three isoforms indicate that: (i) both MM and BB are purified and BB has a higher mobility (due to its lower pI value); (ii) the MB preparation is not pure and it contains more than 50% of MM and a small amount of BB; (iii) there is protein aggregation in MB sample (the appearance of a high molecular weight band).

presence of the different polypeptides. For example, because they have similar sizes, it is difficult to distinguish the three cytoplasmic isoforms of creatine kinase by SDS–PAGE (*Figure 8B*). However, they are readily distinguishable by non-denaturing gels (*Figure 8A*) since the two creatine kinase subunits have different pI values. In addition, because the native protein structure is preserved, the enzymatic activities of creatine kinase can be measured following non-denaturing gel electrophoresis, an analysis which is not possible following SDS–PAGE.

Unfortunately there is no universal buffer system ideal for the electrophoretic separation of all native proteins. For example, cardiac troponin I has a pI of 10.31 and troponin C has a pI of 3.87. If a buffer system with a pH between 4–10 is employed, these two proteins will migrate in opposite directions. Only by carrying out electrophoresis at one or other pH extremes would these proteins have the same charge. However this is not feasible since some protein hydrolytic reactions occur at the extremes of pH. Furthermore extreme pH may denature the protein and thus result in the loss of physiological activities. Therefore, it is necessary to choose conditions for non-denaturing gel electrophoresis which provide optimal resolution of the proteins in the particular sample to be analysed.

6.2 Choice of buffers and polymerization catalyst

Proteins differ widely in their charge, solubility, and stability at different pH values, their sensitivity to ionic strength and type of ionic species, and their cofactor requirements. Therefore, choosing a buffer in a native gel system is critical compared to the SDS–PAGE system and the buffer chosen will depend entirely on the proteins under study. In choosing a suitable buffer the following points have to be met:

- the proteins must be charged during electrophoresis
- the proteins must be stable
- the proteins must remain soluble.

The further the pH of the electrophoresis buffer from the isoelectric points of the proteins to be separated, the higher the charge on the proteins. This leads to shorter times required for electrophoretic separation and hence reduced band diffusion. On the other hand, the closer the pH to the isoelectric points of the proteins the greater the charge differences between proteins, thus increasing the chance of separation.

Just like the two buffer systems described for SDS–PAGE, both discontinuous and continuous gel systems can be applied to native gel electrophoresis. Of these, the discontinuous buffer systems give the higher resolution of sample proteins. Over 4000 buffer systems for such moving boundary electrophoresis have been published (57, 58), and 19 of these, which operate at various pH values, have been selected by Chrambach and Jovin (59) as particularly suitable. Two widely used buffer systems, a high pH system resolving proteins at pH 9.5 (60) and a low pH system resolving proteins at pH 3.8 (61), are described in the next section.

Some native proteins aggregate and may precipitate at the very high protein concentrations reached in the sharply stacked zones of the discontinuous buffer systems and then either fail to enter the resolving gel or cause 'streaking'. The problem may be overcome by using a continuous buffer system. In general, almost any buffer between pH 3–10 with a concentration from about 0.01–0.1 M may be used. McLellan (62) has devised ten useful continuous buffer systems for electrophoresis at pH values ranging from 3.8–10.2. Typical buffer systems which have been used are Tris–glycine (pH range 8.3–9.5),

Tris–borate (pH range 7.0–8.5), Tris–acetate (pH range 7.2–8.5), Tris–citrate (pH range 7.0–8.5), and β-alanine–acetate (pH range 4–5).

Some protein complexes may need the presence of other components to retain their activity (for example, the need of Ca^{2+} for the troponin complex) so that these components must also be added to the electrophoresis buffer. Other proteins may require the presence of a reducing agent to retain activity. In this case, 1 mM dithiothreitol should be used since 2-mercaptoethanol inhibits acrylamide polymerization. Finally, certain classes of proteins such as histones, nuclear non-histone proteins, and membrane proteins are not soluble in the usual non-denaturing buffers, and additional agents such as urea and non-ionic detergents are required to ensure their continued solubility.

The choice of a suitable polymerization initiator system is dependent on what buffer pH has been selected since different initiators have different polymerization efficiencies at a given pH value (8). The optimum pH ranges for the three polymerization systems (*Table 4*) are presented in Section 2.2.1. In addition, different catalysts may have different sensitivities to gel additives. Therefore one should choose a polymerization system which is not inhibited by any component of the gel system.

6.3 Preparation of non-denaturing slab gels

Two discontinuous systems are presented here; a high pH system that has been widely used for many types of sample protein and a low pH system that is useful for basic proteins.

6.3.1 High pH discontinuous system

All stock solutions used in this system are very similar to those used for the Laemmli (1) system but lack SDS (see *Table 12*). Before using this buffer system, one should check whether the proteins under study have a pI close to or greater than pH 8.3 (stacking at pH 8.3). If so, this buffer system may not be the best choice since the proteins may precipitate or migrate very slowly.

Table 12. Stock solutions for high pH non-denaturing discontinuous buffer systems

Stacking gel buffer (see Section 5.2.4)

Resolving gel buffer (see Section 5.2.4)

Acrylamide:bisacrylamide mixture (30%T, 2.6%C) (see *Table 3* and Section 5.2.1)

TEMED (see Section 5.2.3)

10% (w/v) ammonium persulfate: prepare daily in distilled water (see Section 5.2.3)

5 × running buffer: to 15 g Tris base add 72 g glycine, add distilled water to 1 litre final volume. The pH should be pH 8.3 without adjustment. Store at room temperature; dilute to 1 × before use.

2 × sample buffer: mix 2 ml stacking buffer (0.5 M Tris–HCl pH 6.8), 1.6 ml glycerol, 4 ml water, and 0.4 ml of 0.1% (w/v) bromophenol blue

---

Prepare the non-dissociating stacking and resolving gel mixtures as described in *Tables 5* and *6* but replacing the 10% SDS with an equivalent volume of water. Cast the slab gel as described in *Protocol 1*.

6.3.2 Low pH discontinuous system

This system stacks proteins at pH 5 and separates at pH 3.8. Since the persulfate/ TEMED polymerization system is efficient only in the pH 7–10 range, a photo-polymerization system is recommended (see *Table 4* and Section 5.2.3). In addition, the usual bromophenol blue used in the sample buffer has to be replaced with a dye with a positive charge at the required pH such as pyronin Y, methylene green, or methylene blue. The stock solutions required are listed in *Table 13*. If the methylene blue photopolymerization system is used for the low pH discontinuous system, the final concentration of methylene blue, sodium toluene sulfinate, and diphenyliodonium chloride should be 30 μM, 500 μM, and 20 μM, respectively.

Cast the slab gel as described in *Protocol 1*. Photopolymerization should be carried out for at least 1 h as described in *Protocol 7*, step 1 for each gel layer (resolving and stacking). Completion of reaction is evident by photobleaching.

6.4 Sample preparation and electrophoresis of native proteins

The sample buffer for native proteins should contain 10% glycerol (or sucrose), suitable tracking dye (bromophenol blue for the high pH system, pyronin Y, methylene green, or methylene blue for the low pH system), and 1/4 to 1/8 diluted buffer stock used for the sample well gel. Mix the sample with one volume of sample buffer and load directly into the sample well for electrophoresis. If the low pH buffer system is employed to analyse basic proteins, make sure that the upper electrode is connected to the anode ($+$) of the power pack. If the sample is too dilute, any non-denaturing method listed in Section 5.4.2 can be used to concentrate the sample proteins. To increase protein solubility and reduce aggregation, mild detergents such as Triton X-100 may be included in the sample buffer and gel buffers. If the stacking gel

---

Table 13. Stock solutions for low pH non-denaturing discontinuous buffer systems

8 × stacking gel buffer (acetic acid–KOH pH 6.8): mix 48 ml of 1 M KOH, 2.9 ml glacial acetic acid, and 49.1 ml water

8 × resolving gel buffer (acetic acid–KOH pH 4.3): mix 48 ml of 1 M KOH, 17.2 ml glacial acetic acid, and 34.8 ml water

Photopolymerization catalyst solution (see *Table 4* and Section 5.2.3)

1 × running buffer (acetic acid–β-alanine pH 6.8): mix 8 ml glacial acetic acid, 31.2 g β-alanine, and add water to 1 litre final volume

2 × sample buffer: mix 2 ml stacking gel buffer (acetic acid–KOH pH 6.8), 1.6 ml glycerol, 4 ml water, and 0.4 ml of 0.4% (w/v in water) pyronin Y

---

causes aggregation of the sample proteins, use of a resolving gel without a stacking gel may solve the problem. Other problems associated with polyacrylamide gel electrophoresis and potential solutions are described in Section 5.7.

All steps involved in the preparation of native protein samples should be performed at 4°C to minimize attack by any protease in the sample. It is also important that the gel does not heat up excessively, since this could denature the protein in gel. This can be done either by running the gel in the cold room or by lowering the current. Prepared samples may be stored at 4°C, or frozen if the proteins are stable under these conditions.

7. Variations of standard polyacrylamide gel electrophoresis

7.1 Blue native polyacrylamide gel electrophoresis

The analysis of native proteins without knowing their isoelectric points makes the selection of buffer pH difficult. Blue native polyacrylamide gel electrophoresis (BN–PAGE) is a charge-shift method originally developed by Schagger and Jagow (63) which overcomes this problem. A negatively charged dye (Coomassie blue G-250) is added to the native proteins and induces a charge-shift as it binds to the surface hydrophobic domains of the proteins. The electrophoretic mobility of the proteins is then mainly determined by the negative charges of bound Coomassie dye and even basic proteins now migrate to the anode at pH 7.5. The drawback of this method is that not all native proteins bind to the Coomassie dye, although the majority of them do (64).

BN–PAGE differs from SDS–PAGE in two respects; Coomassie dye binds to the protein surface (not the interior), and the ratio of bound Coomassie: protein is variable, so that the final charge:mass ratio of different proteins is variable. The two consequences are that the dye does not denature the proteins and does not allow accurate molecular mass determination in uniform concentration gels. However Schagger demonstrated that using BN–PAGE with polyacrylamide gradient gels does allow a reasonably accurate molecular mass determination of native proteins (64). This is due to the fact that the pore size gradient of such gels determines the end-position of migration of proteins based on their size (see also Section 5.3.3).

Other features of BN–PAGE include:

(a) Increased solubility and reduced aggregation of native proteins by the presence of negative charges of bound Coomassie dye on the surface of the proteins.

(b) A working pH near pH 7.5 which minimizes protein denaturation and is specially suitable for membrane proteins.

(c) The fact that the native proteins are stained makes them visible during electrophoresis.

In addition, the oligomeric states of native proteins can be determined if one combines BN–PAGE in a first-dimensional separation with SDS–PAGE in the second dimension. This novel electrophoretic technique (*Protocol 6*) appears to be a sensitive, high-resolution method for analysis of molecular mass, oligomeric state, and homogeneity of native proteins. A detailed description of the procedures involved and some applications of the method are also described in refs 63–65.

Protocol 6. Blue native polyacrylamide gradient gel electrophoresis[a]

Equipment and reagents

- Mini slab gel apparatus (e.g. Bio-Rad Mini-Protean II Cell)
- Power pack (see Section 3.2)
- Acrylamide:bisacrylamide: 50%T, 2.6%C (see *Table 3*)
- TEMED (see Section 5.2.3)
- 10% (w/v) ammonium persulfate: prepare daily in distilled water (see Section 5.2.3)
- Anode buffer:[b] 50 mM bisTris–HCl pH 7.0

- 4 × resolving gel buffer:[b] 2 M 6-amino-caproic acid, 200 mM bisTris–HCl pH 7.0
- Cathode buffer: 50 mM Tricine, 15 mM bisTris–HCl, 0.02% Coomassie blue G-250 pH 7.0
- Sample buffer:[b] 750 mM aminocaproic acid, 50 mM bisTris–HCl, 1.25% dodecyl maltoside, and 0.35% Coomassie blue G-250 pH 7.0

Method

1. For the resolving gel, follow *Protocol 3*, steps 1–6 to prepare a 5%T–17.5%T gradient minigel.

2. Prepare the sample gel mixture[c] (4%T) by mixing:
   - acrylamide:bisacrylamide (50%T, 2.6%C)   0.4 ml
   - 4 × resolving gel buffer   1.25 ml
   - 10% (w/v) ammonium persulfate   40 μl
   - TEMED   4 μl
   - water   3.306 ml

3. Overlay the resolving gel with the sample gel mixture.

4. Insert the sample comb, leaving about 5 mm between the top of the resolving gel and the bottom of the comb. Allow the sample gel to polymerize for 2 h before running the gel.

5. Fill the anode and the cathode reservoirs with the respective buffers.

6. Load the protein samples dissolved in sample buffer.

7. Start electrophoresis at 4–7 °C from cathode (−) to anode (+) at 60 V as the samples migrate through the sample gel. Then increase the voltage to 250 V as the samples migrate through the resolving gel.

[a] Based on the method given in ref. 63, originally designed for the separation of membrane protein complexes.
[b] Store at 4 °C.
[c] Enough for two 1.5 mm thick Bio-Rad minigels.

7.2 Acid–urea polyacrylamide gel electrophoresis

In most cases, SDS–PAGE is able to resolve all histone types on the basis of their molecular weight. However, the system is less effective for separating histone variants or their modified forms resulting from post-translational modifications such as acetylation, methylation, and phosphorylation, since these have very similar molecular weights. In order to analyse such modifications, systems in which the charge will play a role in the separation must be used. For this reason, acid–urea polyacrylamide gel electrophoresis is commonly employed for the analysis of histones. The method can also be applied to other basic proteins (66). Urea is usually added to the gel to denature the proteins. This results in increased protein solubility and at the same time decreases mobility due to protein unfolding. Triton X-100 may also be included in the system to improve resolution since some histone variants have a greater affinity for Triton X-100 than others, which slows down their migration. Detergent binding is reduced by urea so that the electrophoretic patterns obtained vary markedly depending on the exact concentrations of urea and detergent.

Polymerization of acidic gels with the standard persulfate/TEMED system is more difficult since this system is less effective at pHs lower than 7 (8). Therefore, high concentrations of the initiators are needed leading to high residual concentrations of persulfate in the gel which could oxidize proteins unless removed before use (67). To eliminate the residual persulfate, the gel must be pre-run before the samples are loaded.

Recently, Rabilloud *et al.* (6) have described the successful application of the newly introduced methylene blue photopolymerization system (2) to the preparation of an acid–urea–Triton gel. This new system exhibits several advantages compared to the standard persulfate/TEMED system:

- faster polymerization rate
- non-oxidative feature (thus no pre-running is needed)
- lower backgrounds after silver staining (silver staining is described in Chapter 2)

A procedure for acid–urea continuous PAGE is given in *Protocol 7*. Although the method described is designed mainly for histones, it can be optimized to separate other basic proteins by altering the pH of the buffer, the urea concentration (2–8 M), and the concentration of Triton X-100 (6–8 mM).

7.3 Transverse gradient gel electrophoresis

7.3.1 Transverse pore gradient gels

A slab gel in which there is a continuous, linear gradient of acrylamide, perpendicular to the direction of protein migration, is called a transverse pore gradient gel. Proteins are applied uniformly across the top of the gel and electrophoresed transversely across the gradient of polyacrylamide. Each

---

Protocol 7. Acid–urea continuous polyacrylamide gel electrophoresis[a]

Equipment and reagents

- Mini slab gel apparatus (e.g. Bio-Rad Mini-Protean II Cell)
- Power pack (see Section 3.2)
- Light source (see Section 3.4)
- Acrylamide:bisacrylamide: 50%T, 1.5%C
- Glacial acetic acid
- Urea

- 2 mM methylene blue[b]
- 20 mM sodium toluene sulfinate[b]
- 1 mM diphenyliodonium chloride[b]
- Reservoir buffer: 0.9 M acetic acid, prepare as 5.4% glacial acetic acid in water
- Sample buffer: 2.5 M urea, 0.9 M acetic acid, and 0.01% pyronin Y

Method

1. Prepare the gel mixture (15% acrylamide, 2.5 M urea, 0.9 M acetic acid pH 2.7) by mixing:

   - acrylamide:bisacrylamide (50%T, 1.5%C)   3.0 ml
   - glacial acetic acid   0.54 ml
   - urea   1.5 g
   - 2 mM methylene blue   0.15 ml
   - 20 mM sodium toluene sulfinate   0.25 ml
   - 1 mM diphenyliodonium chloride   0.2 ml
   - water   10 ml final volume

2. Pour the gel into the mini slab mould. Photopolymerize for 2 h with constant illumination. Pre-electrophoresis is not necessary.

3. Fill the buffer reservoirs with the reservoir buffer.

4. Load the protein samples dissolved in sample buffer.

5. Start electrophoresis from anode (+) to cathode (−). Use 130 V constant voltage if the Bio-Rad minigel is used.

[a] Based on the method given in refs 6 and 68.
[b] See Section 5.2 for stock solution preparation.

---

protein will generate a continuous curve of mobility versus acrylamide concentration (*Figure 9A*). Since an experimental Ferguson plot (69) can be produced by this technique, it becomes an important tool for obtaining non-linear Ferguson plots [log (mobility) versus gel concentration] (70). This permits rapid determination of the %T which optimally separates the proteins, thus avoiding repeated experiments that would otherwise be used using uniform concentration gels. The procedure has also been used to determine the optimal polyacrylamide concentration required for high efficiency semi-dry transfer of proteins to membranes (71).

Transverse pore gradient gels can also be generated by varying the proportion of cross-linker (%C) rather than the total monomer concentration (%T).

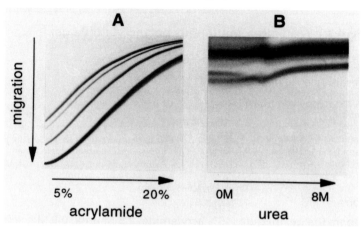

Figure 9. (A) Transverse pore gradient gel using a dissociating buffer system. Proteins analysed were (from top to bottom): creatine kinase MM (43 kDa, the first and the second band, see also *Figure 8B*); recombinant troponin I (29 kDa); recombinant myoglobin (20 kDa). (B) Transverse urea gradient gel using a 7.5% polyacrylamide gel with Tris–HCl buffer pH 8.8. Recombinant creatine kinase MB preparation from *Figure 8* was analysed. The mobilities of the three isoforms (from top to bottom: MM, MB, and BB) did not change until the urea concentration reached about 2 M. At about 2.5 M urea, dimeric MB disappeared indicating the dissociation of M and B subunits. As the urea concentration continues to increase, all isoforms dissociated and unfolding of M (the upper band) and B (the lower band) subunits resulted in two main bands with decreased mobility.

This approach allows one to test the effects of various cross-linkers and their proportions on the mobilities of macromolecules of interest (72).

7.3.2 Transverse urea gradient gels

The electrophoretic mobility of a protein depends upon not only its size and charge but also its conformation. Therefore protein conformational changes may be detected by electrophoresis. Transverse urea gradient (0–8 M) gels are used for such a purpose. At low urea concentrations, the protein is in the native conformation, while at higher urea concentrations the unfolded form of the protein, which migrates more slowly, becomes predominant. A continuous, graphic picture of variation in electrophoretic mobility with urea concentration will be generated (*Figure 9B*), and the approximate urea concentration at which unfolding occurs can be deduced (for review, see ref. 73). However, one should note that urea accelerates persulfate/TEMED initiated polymerization (thus inducing smaller pore size) so that homogeneous gels may not be obtained with a transverse urea gradient.

A transverse pore or urea gradient gel can be prepared by using a Bio-Rad minigel apparatus as described in *Protocol 8* or the Hoefer SE 600/400 systems as described by Smejkal (71). Polyacrylamide gradient gels prefabricated for the PhastSystem of Pharmacia have also been used for transverse pore gradient gel electrophoresis (74).

Protocol 8. Casting transverse gradient minigels

Equipment and reagents
- Bio-Rad minigel apparatus
- Gradient maker (Bio-Rad)
- Stacking gel mixture: composition depends upon the application but also see *Table 6*.

- Two resolving gel mixtures to form the required gradient gel: the composition of these depend upon the particular experiment but see *Table 5*.

Method

1. Assemble and orient the gel cassette as shown in *Figure 10*. Glue a small piece of rubber to the bottom clamp to level the assembly. Insert a single well comb (which spans almost the entire width of the gel cassette) against the lower spacer leaving an opening between the comb and upper spacer through which the gel can be poured.

2. Prepare the desired gel mixtures and pour these into the two chambers of the gradient maker. Pour the gradient until the gel mixture reaches the top edge of the comb. If the gel leaks, immerse the entire assembly immediately in water with the water level just above the gel level. The polymerization rate can be adjusted by altering the water temperature.

3. After the acrylamide monomers have polymerized (usually 30 min), prepare 5 ml of stacking gel mixture, omitting SDS in the stacking mixture if a urea transverse gradient or native pore gradient gel is being prepared. Turn the assembly at a 90 degree angle from which the gradient is formed. Remove any water that has entered the gel cassette. Re-position the comb leaving a space (about 5 mm) between the resolving gel and the comb. Transfer the stacking gel mixture into the space between the comb and one of the spacers using a Pasteur pipette. The remaining space at the side of the gel (which was the top of the gel when the gradient gel was being poured) will also be filled with stacking gel mixture.

4. After the stacking gel has polymerized, remove the comb[a] and load the protein sample into the single wide well.

5. Carry out electrophoresis (the exact conditions depend upon the type of gradient gel being used) and then detect the separated protein bands by staining (see Chapter 2).

[a] The gel may be stored at 4 °C for later use if placed in a sealed plastic bag with the comb left in place.

7.3.3 Transverse temperature gradient gel electrophoresis

Transverse temperature gradient gel electrophoresis (TGGE) is another type of transverse gradient technique. This approach allows the measurement of

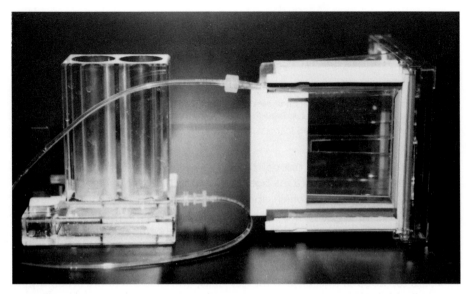

Figure 10. Assembly of a transverse gradient gel using the Bio-Rad Mini-Protean II apparatus.

the transition temperature of both heat- and cold-induced subtle conformational changes of proteins and detection of the different conformational states, native or denatured, at different temperatures in one experiment (75–77). In most cases, the gel is run horizontally on top of an electric insulated metal plate with two water-baths connected to provide a temperature gradient perpendicular to the electrophoretic migration. Apparatus for TGGE is commercially available from Diagen.

7.4 Tricine SDS–polyacrylamide gel electrophoresis

Proteins or oligopeptides with molecular mass below 14 kDa are not well resolved by standard SDS–PAGE using the Laemmli discontinuous buffer system (1). Even gradient gels in the 3–30% range cannot resolve oligopeptides smaller than 10 kDa. However, two successful approaches for gel electrophoresis of oligopeptides have been described. First, oligopeptides with molecular mass as low as 2.5 kDa can be resolved by inclusion of 8 M urea in a continuous 0.1 M Tris–phosphate pH 6.8 buffer system, (78). Secondly, a discontinuous buffer system has been described by Schagger and von Jagow (16) in which the tracking ion in the Laemmli system, glycine, is replaced with Tricine (*Protocol 9*). At the usual pH values, Tricine migrates much faster than glycine in a stacking gel and thus the stacking limit is shifted to the low molecular mass range so that small SDS:polypeptide complexes separate well from SDS. By using this superior method, a separation of proteins in the range from 1–100 kDa can be achieved at acrylamide concentrations as low as 10%. Alternatively, the resolving range can be shifted to 3.5–200 kDa by tailoring the ionic concentration of both stacking and separating phases (79). A further advantage is that the omission of glycine and urea in this method prevents problems which may occur in subsequent amino acid analysis and sequencing. The methodology of Tricine SDS–PAGE is described in *Protocol 9*. A comparison of results obtained from the Laemmli and Tricine SDS–PAGE systems is shown in *Figure 11*.

Protocol 9. Tricine SDS–polyacrylamide gel electrophoresis[a]

Equipment and reagents

- Power pack (see Section 3.2)
- Resolving gel mixture: prepare 10% acrylamide mixture following the recipes given in *Table 5*
- Stacking gel mixture: prepare 4% acrylamide mixture following the recipes given in *Table 6*
- Mini slab gel apparatus (e.g. Bio-Rad Mini-Protean II Cell)
- Upper reservoir buffer: 0.1 M Tris, 0.1 M Tricine, 0.1% (w/v) SDS, the pH will be about pH 8.25
- Lower reservoir buffer: 0.2 M Tris–HCl pH 8.9

Method

1. Cast the discontinuous gel following the steps in *Protocol 1*.

2. Fill the upper and the lower buffer reservoirs with the corresponding buffer.

3. Load the samples (prepared as for SDS–PAGE; see Section 5.4).

4. Start electrophoresis from cathode (−) to anode (+) at a constant current of 25 mA/gel.

[a] Based on the method given in ref. 16.

7.5 Use of cationic detergents for PAGE

Detergents are most commonly classified on the basis of their charge and thus they may be neutral, anionic, cationic, or zwitterionic detergents. Many types of detergents have been used for electrophoresis. The anionic detergent, SDS, is the most widely used due to the popularity of SDS–PAGE. Neutral detergents such as Triton X-100 and Nonidet P-40, and zwitterionic detergents such as CHAPS are often used for isoelectric focusing and two-dimensional electrophoresis since they do not alter the charge of proteins and help protein solubilization. Although SDS is usually the choice for one-dimensional electrophoresis, it still has some drawbacks. For example, SDS forms crystals at lower temperature and so in these situations detergents such as lithium dodecyl

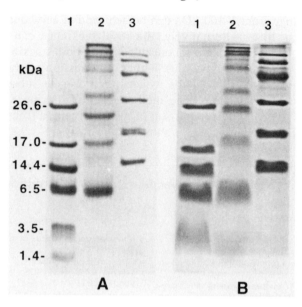

Figure 11. Comparison of (A) Tris–Tricine SDS–PAGE and (B) regular Tris–glycine SDS–PAGE (Laemmli). Both gels (16.5%T, 3.3%C) were prepared with a Bio-Rad minigel apparatus and stained with Coomassie brilliant blue G-250. Lane 1, Bio-Rad low range molecular weight standard (see *Table 10*); lane 2, Bio-Rad pre-stained broad range molecular weight standard; lane 3, Bio-Rad polypeptide molecular weight standard.

sulfate may have to be employed (80). In other cases, SDS causes protein aggregation (81) or precipitation. Some proteins are not well resolved or migrate abnormally in SDS gels. The last problem led to the development of cationic detergent solubilization and electrophoresis (82–84).

Most recently, Akins *et al.* (85) has described discontinuous polyacrylamide–agarose gel electrophoresis using the cationic detergent cetyltrimethylammonium bromide (CTAB). This system allows the separation of proteins based on molecular weight and is carried out at neutral pH. If samples are prepared without boiling and without the addition of a reducing agent, some proteins retain their native activities yet still migrate as a function of molecular weight. The binding of CTAB with proteins makes all proteins positively charged and thus migrate in the opposite direction compared to SDS–PAGE. The procedure is described in *Protocol 10*.

The fact that electrophoresis in the presence of CTAB offers the same degree of reliability in analysis of most proteins makes this system a useful alternative to SDS–PAGE. However, CTAB gel electrophoresis does have some drawbacks:

(a) Samples have to be prepared at room temperature immediately before running the gel since precipitation occurs at temperatures below 15°C.

(b) The lack of good staining protocols since cationic detergents interfere with protein dyes during staining.

(c) Some proteins exhibit anomalous migration (83).

Protocol 10. CTAB–polyacrylamide gel electrophoresis[a]

Equipment and reagents

- Power pack (see Section 3.2)
- Resolving gel mixture: 6% acrylamide, 375 mM Tricine–NaOH pH 8. Prepare this by mixing 9.4 ml water, 2.4 ml acrylamide: bisacrylamide monomer (40%T, 2.67%C), 4 ml 1.5 M Tricine–NaOH pH 8, 0.16 ml 10% ammonium persulfate, and 16 μl TEMED. Add the TEMED just prior to pouring the gel. This volume of gel mixture is sufficient to pour two minigels, 1.5 mm thick.
- 5 × running buffer: mix 22.4 g Tricine, 5 g CTAB, 75 ml 1 M arginine, and add water to 1 litre final volume. The pH should be about pH 8.2.

- Mini slab gel apparatus (e.g. Bio-Rad Mini-Protean II Cell)
- Stacking gel mixture: 0.7% agarose, 0.1% CTAB, 125 mM Tricine–NaOH pH 10. Mix 35 mg agarose, 1.25 ml 0.5 M Tricine–NaOH pH 10, 50 μl 10% CTAB, and add water to 5 ml. Melt the agarose by boiling or microwaving prior to pouring the gel. This gel mixture is sufficient to pour stacking gels for two minigels, 1.5 mm thick.
- Sample buffer: 10 mM Tricine–NaOH pH 8.8, containing 1% CTAB, 10% glycerol, and 10 μl/ml of a saturated aqueous solution of crystal violet

Method

1. Prepare the gels as described in *Protocol 1*, but without reference to *Table 5* and *6*; use the resolving gel mixture and stacking gel mixture described above instead.

2. Fill the upper buffer reservoir with 120 ml of 1 × running buffer and the lower buffer reservoir with 200–600 ml of the same buffer.

3. Prepare the protein samples by dissolving them in sample buffer at room temperature. If retaining protein activity is not a concern, heat the samples for 3 min in a boiling water-bath in the presence of 2% 2-mercaptoethanol.

4. Load the samples and start electrophoresis from anode (+) to cathode (−) at 100 V as the samples migrate through the stacking gel. Then increase the voltage to 150 V as the samples migrate through the resolving gel.

[a] Based on the method given in ref. 85.

Acknowledgements

We wish to thank our colleagues Jianying Yang for supplying us *Figures 5, 8, 9*, and *11*, and Tracy Yang for careful reading of the manuscript. This chapter is partly based on the earlier version (2nd edn) of the same chapter written by Dr B. David Hames.

References

1. Laemmli, U. K. (1970). *Nature*, **227**, 680.
2. Lyubimova, T., Cagilo, S., Gelfi, C., Righetti, P. G., and Rabilloud, T. (1993). *Electrophoresis*, **14**, 40.
3. Caglio, S., Chiari, M., and Righetti, P. G. (1994). *Electrophoresis*, **15**, 209.
4. Righetti, P. G. (1995). *J. Chromatogr.*, **698**, 3.
5. Rabilloud, T., Vincon, M., and Garin, J. (1995). *Electrophoresis*, **16**, 1414.
6. Rabilloud, T., Girardot, V., and Lawrence, J. J. (1996). *Electrophoresis*, **17**, 67.
7. Righetti, P. G., Gelfi, C., and Bianchi-Bosisio, A. (1981). *Electrophoresis*, **2**, 291.
8. Caglio, S. and Righetti, P. G. (1993). *Electrophoresis*, **14**, 554.
9. Gelfi, C. and Righetti, P. G. (1981). *Electrophoresis*, **2**, 213.
10. Righettti, P. G. and Caglio, S. (1993). *Electrophoresis*, **14**, 573.
11. Lyubimova, T. and Righetti, P. G. (1993). *Electrophoresis*, **14**, 191.
12. Fawcett, J. S. and Morris, C. J. O. R. (1966). *Sep. Stud.*, **1**, 9.
13. Righettti, P. G., Caglio, S., Saracchi, M., and Quaroni, S. (1992). *Electrophoresis*, **13**, 587.
14. Righetti, P. G., Bossi, A., Giglio, M., Vailati, A., Lyubimova, T., and Briskman, A. V. (1994). *Electrophoresis*, **15**, 1005.
15. Hjerten, S. (1962). *Arch. Biochem. Biophys.*, **99**, 466.
16. Schagger, H. and Jagow, G. V. (1987). *Anal. Biochem.*, **166**, 368.
17. Makowski, G. S. and Ramsby, M. L. (1993). *Anal. Biochem.*, **212**, 283.
18. Merle, P. and Kadenbach, B. (1980). *Eur. J. Biochem.*, **105**, 449.
19. Swank, R. T. and Munkres, K. D. (1971). *Anal. Biochem.*, **39**, 462.
20. Pennings, S., Meersseman, G., and Bradbury, E. M. (1992). *Nucleic Acids Res.*, **20**, 6667.
21. Rossini, K., Rizzi, C., Sandri, M., Bruson, A., and Carraro, U. (1995). *Electrophoresis*, **16**, 101.
22. Kuba, K. (1993). *Anal. Biochem.*, **213**, 200.
23. Sobieszek, A. (1994). *Electrophoresis*, **15**, 1014.
24. Gersten, D. M., Kimball, H., and Bijwaard, K. E. (1991). *Anal. Biochem.*, **197**, 59.
25. Gersten, D. M. and Bijwaard, K. E. (1992). *Electrophoresis*, **13**, 282.
26. Gersten, D. M. and Bijwaard, K. E. (1992). *Electrophoresis*, **13**, 399.
27. Studier, F. W. (1973). *J. Mol. Biol.*, **79**, 237.
28. Tyagi, R. K., Babu, B. R., and Datta, K. (1993). *Electrophoresis*, **14**, 826.
29. Weber, K. and Osborn, M. (1969). *J. Biol. Chem.*, **244**, 4406.
30. Fritz, J. D., Swartz, D. R., and Greaser, M. L. (1989). *Anal. Biochem.*, **10**, 2606.
31. Lambin, P. and Fine, J. M. (1979). *Anal. Biochem.*, **98**, 160.
32. Lambin, P., Herance, N., and Fine, J. M. (1986). *Electrophoresis*, **7**, 342.
33. Somerville, L. and Wang, K. (1981). *Biochem. Biophys. Res. Commun.*, **102**, 53.
34. Granzier, H. L. M. and Wang, K. (1993). *Electrophoresis*, **14**, 56.
35. Marshall, T. and Williams, K. M. (1992). *Electrophoresis*, **13**, 887.
36. Marshall, T., Abbot, N. J., Fox, P., and Williams, K. M. (1995). *Electrophoresis*, **16**, 28.
37. Sauve, D. M., Ho, D. T., and Roberge, M. (1995). *Anal. Biochem.*, **226**, 382.
38. Hyman, M. R. and Arp, D. J. (1993). *Electrophoresis*, **14**, 619.
39. Wilson, D., Hall, M. E., Stone, G. C., and Rubin, R. W. (1977). *Anal. Biochem.*, **83**, 33.

40. Rabilloud, T. (1996). *Electrophoresis*, **17**, 813.
41. Kumer, M. A. and Davidson, V. L. (1992). *BioTechniques*, **12**, 198.
42. Guttman, A. and Nolan, J. (1994). *Anal. Biochem.*, **221**, 285.
43. Shapiro, A. L., Vinuela, E., and Maizel, J. V. (1967). *Biochem. Biophys. Res. Commun.*, **28**, 815.
44. Pitt-Rivers, R. and Impiombato, F. S. A. (1968). *Biochem. J.*, **109**, 825.
45. Reynolds, J. A. and Tanford, C. (1970). *Biol. Chem.*, **245**, 5161.
46. Reynolds, J. A. and Tanford, C. (1970). *Proc. Natl. Acad. Sci. USA*, **66**, 1002.
47. Robinson, W. C. and Tanford, C. (1975). *Biochemistry*, **14**, 369.
48. Lambin, P., Rocher, D., and Fine, J. M. (1976). *Anal. Biochem.*, **74**, 567.
49. Lambin, P. (1978). *Anal. Biochem.*, **85**, 114.
50. Fish, W. W. (1975). *Methods Membrane Biol.*, **4**, 189.
51. See, Y. P., Olley, M. P., and Jackowski, G. (1985). *Electrophoresis*, **6**, 382.
52. Weber, K. and Osborn, M. (1975). In *The proteins* (ed. K. Neurath and R. L. Hill), Vol. I, p. 179. Academic Press, NY.
53. Poduslo, J. (1981). *Anal. Biochem.*, **114**, 131.
54. Bryan, J. (1989). *J. Muscle Res. Cell. Motil.*, **10**, 95.
55. Graceffa, P., Jancso, A., and Mabuchi, K. (1992). *Arch. Biochem. Biophys.*, **297**, 46.
56. Chiari, M., Micheletti, C., and Righetti, P. G. (1992). *J. Chromatogr.*, **598**, 287.
57. Chrambach, A. (1980). *J. Mol. Cell. Biochem.*, **29**, 23.
58. Jovin, T. M., Dante, M. L., and Chrambach, A. (1970). *Multiphasic buffer systems output*. Public Board Numbers 196085 to 196091, 203016, 259309 to 259312. National Technical Information Service, Springfield, VA, USA.
59. Chrambach, A. and Jovin, T. M. (1983). *Electrophoresis*, **4**, 190.
60. Davis, B. J. (1964). *Ann. N. Y. Acad. Sci.*, **121**, 404.
61. Reisfeld, R. A., Lewis, U. J., and Williams, D. E. (1962). *Nature*, **195**, 281.
62. McLellan, T. (1982). *Anal. Biochem.*, **13**, 336.
63. Schagger, H. and von Jagow, G. (1991). *Anal. Biochem.*, **199**, 223.
64. Schagger, H., Cramer, W. A., and von Jagow, G. (1994). *Anal. Biochem.*, **217**, 220.
65. Schagger, H. (1995). *Electrophoresis*, **16**, 763.
66. Harwig, S. S. L., Chen, N. P., Park, A. S. K., and Lehrer, R. I. (1993). *Anal. Biochem.*, **208**, 382.
67. Klarskov, K., Roecklin, D., Bouchon, B., Sabatie, J., Van Dorssalaer, J., and Bischoff, R. (1994). *Anal. Biochem.*, **216**, 127.
68. Panyim, S. and Chalkley, R. (1969). *Arch. Biochem. Biophys.*, **130**, 337.
69. Ferguson, A. O. (1964). *Metabolism*, **13**, 985.
70. Chrambach, A. and Wheeler, D. L. (1994). *Electrophoresis*, **15**, 1021.
71. Smejkal, G. and Gallagher, S. (1994). *BioTechniques*, **16**, 196.
72. Smejkal, G. B. and Hoff, H. F. (1992). *Electrophoresis*, **13**, 102.
73. Goldenberg, D. P. and Creighton, T. E. (1984). *Anal. Biochem.*, **138**, 1.
74. Buzas, Z., Wheeler, D. L., Garner, M. M., Tietz, D., and Chrambach, A. (1994). *Electrophoresis*, **15**, 1028.
75. Arakawa, T., Hung, L., Pan, V., Horan, T. P., Kolvenbach, C. G., and Narhi, L. O. (1992). *Anal. Biochem.*, **208**, 255.
76. Sattler, A. and Riesner, D. (1993). *Electrophoresis*, **14**, 782.
77. Curtil, C., Channac, L., Ebel, C., and Masson, P. (1994). *Biochim. Biophys. Acta*, **1208**, 1.

78. Swank, R. W. and Munkres, K. D. (1971). *Anal. Biochem.*, **39**, 462.
79. Khalkhali-Ellis, Z. (1945). *Prep. Biochem.*, **25**, 1.
80. Kubo, K. and Takagi, T. (1986). *Anal. Biochem.*, **156**, 11.
81. Bayer, E. A., Ehrlich-Rogozinski, S., and Wilchek, M. (1996). *Electrophoresis*, **17**, 1319.
82. Mocz, G. and Balint, M. (1984). *Anal. Biochem.*, **143**, 283.
83. Akin, D. T., Shapira, R., and Kinkade, J. M. (1985). *Anal. Biochem.*, **145**, 170.
84. MacFarlane, D. E. (1989). *Anal. Biochem.*, **176**, 457.
85. Akins, R. E., Levin, P. M., and Tuan, R. S. (1992). *Anal. Biochem.*, **202**, 172.

2

Protein detection methods

CARL R. MERRIL and KAREN M. WASHART

1. Detection of proteins in gels

1.1 Introduction

Advances in detection methods for proteins have paralleled the development of new separation techniques over the last three decades. Selection of the proper protein visualization procedure is as important as the separation technique employed in the analysis of proteins from complex mixtures. The ease of use, speed of detection, level of sensitivity, and integrity of the proteins following detection (i.e. to allow further processing of the proteins if necessary) are important criteria for determining the appropriate detection method. Suitable methods also need to maintain a linear and stoichiometric relationship between protein detection signal and protein concentration.

Until 1937, the detection of protein separated by electrophoresis of solutions and colloidal suspensions was restricted to the visualization of proteins coated onto microspheres or investigations of naturally coloured proteins such as haemoglobin, myoglobin, or ferritin (1–4). In 1937, Tiselius demonstrated the use of ultraviolet light for detection of non-coloured proteins, such as ovalbumin, serum globin fractions, and Bence Jones proteins (5). Tiselius also discovered that protein position and concentration in electrophoretic systems could be detected by analysing the shadows, or 'schlieren', that form based on the boundaries created between regions with different refractive indices (5). These detection methods were satisfactory for protein detection in liquid-based electrophoretic systems, but the advent of solid support electrophoresis (ranging from moist filter paper to polyacrylamide gels) allowed for the development of protein staining techniques that were simpler to use and provided greater sensitivity.

Organic stains, generally derived from dyes used in the textile industry, provided for the visualization of proteins separated on solid support electrophoretic mediums with detection sensitivities in the microgram range. Some of the commonly used organic dyes include Amido black, Fast green FCF, and Coomassie blue stains. *Table 1* compares detection sensitivities for these stains, as well as other detection methods for electrophoretic gels. The

Table 1. Comparison of staining sensitivities for *in situ* gel detection

Staining method	Protein (ng)	Reference
Amido black	30 000	166
Coomassie blue R-250	38	10
Fast green FCF	15 000	166
Silver stain	0.02	21
Negative stains	5–1500	31–34
MTA negative stain	0.5	35
Pre-electrophoretic fluorescent stains	6–10	167
Post-electrophoretic fluorescent stains	100	38
SYPRO red and orange fluorescent stains	0.5–1.0	40

Coomassie stains/dyes were named to commemorate the British victory over the Ashanti capital of Kumasi or 'Coomassie', which is now in Ghana. They were originally developed near the end of the 19th century as acid wool dyes and were reported to be intense protein stains by Fazekas de St. Groth and his collaborators in 1963 (6). Coomassie blue detects as little as 38 ng of protein (6). However, some fluorescent protein stains can detect as little as one nanogram of protein (7). The fluorescent stains were first introduced by Talbot and Yphantis in 1971 (8). Most fluorescent stains are covalently bound to the sample proteins prior to electrophoresis; this increases the proteins' molecular weights only slightly and so still permits their use in SDS–PAGE, which separates proteins on the basis of molecular weight. However, covalently bound fluorescent stains may alter the charge of certain proteins, resulting in altered isoelectric focusing patterns and altered two-dimensional (2D) electrophoretograms, since 2D PAGE also relies on the separation of proteins by charge in the first dimension.

In 1979, Merril *et al.* (9) introduced silver staining techniques for protein detection in polyacrylamide gels. The first of these techniques was adapted from a histological silver stain. The introduction of silver staining increased the sensitivity of protein detection by 2000-fold over Coomassie blue staining, the most commonly used organic stain at that time, from tenths of a microgram to tenths of a nanogram (9, 10).

In addition to staining methods for the detection of proteins, physical methods are also available for protein visualization. The introduction of auto-radiographic methods by Becquerel and Curie, while engaged in the discovery of radioactivity, provided the fundamental techniques for the use of radio-active isotopes (11). Detection of strong β-emitters, such as ^{14}C and ^{32}P, can be achieved using autoradiography, but weak β-emitters, such as ^{3}H, require enhancement by fluorographic techniques (12). The use of radioactive label-ling provides for highly sensitive detection of proteins but the difficulties encountered with radioactive isotope use, such as waste disposal, have recently encouraged the use of non-radioactive labelling techniques, such as biotin–streptavidin labelling (13).

Specific proteins may be detected by a number of methods. Post-translationally modified proteins may be detected by reactions directed at the modifying groups. Immunoglobulins specific for unique protein antigens may be coupled with fluorescent or other visualization systems to identify certain proteins. Proteins may also be detected on the basis of their enzymatic activity.

Many of the protein detection methods in current use were designed for specific applications. For this reason it is critical to review the advantages and disadvantages associated with each method prior to committing to the use of a specific method. Most of the methods can be used both qualitatively and quantitatively. However, many of them depend on reactions with specific structures, such as free amino groups, within each protein. For this reason quantitative comparisons should be performed in a protein-specific manner. That is, albumin on one electrophoretic gel should be compared with albumin on another gel but not with some other protein.

1.2 Organic dyes

Organic dyes provide for some of the simplest methods for protein detection but the sensitivity of detection varies from stain to stain and from protein to protein. Most of these organic dyes are believed to be electrostatically attracted to charged groups on the protein, forming strong dye:protein complexes that are further augmented by Van der Waals forces, hydrogen bonding, and hydrophobic bonding.

Some of the most common organic dyes utilized are Amido black, Fast green FCF, Coomassie blue R-250 (the letter 'R' stands for a reddish hue while the number '250' is a dye strength indicator), and Coomassie blue G-250 ('G' indicates that this stain has a greenish hue). Other Coomassie stains, Coomassie violet R-150 and Serva violet 49 (Serva violet 49 differs from Coomassie violet 150 by the substitution of a diethylamine group for a dimethylamine group) have gained some favour because, while they stain proteins rapidly, they do not stain carrier ampholytes used in isoelectric focusing (14, 15).

Amido black stains rapidly, but its sensitivity is surpassed by the Coomassie blue stains (6). 1 mg of protein can bind 0.17 mg of Amido black, 0.23 mg of Fast green, or 1.2 mg of Coomassie blue R-250 (CBB-R) and 1.4 mg of Coomassie blue G-250 (CBB-G) (16). This relatively high staining intensity of the Coomassie blue stains, compared to other organic dyes, is apparently due to dye:dye interactions with Coomassie blue dye molecules that are ionically bound to, or in hydrophobic association with, the protein molecules (16). Basic amino acids appear to be particularly important for the interaction between the Coomassie dye molecules and proteins as shown by correlations between the intensity of Coomassie blue staining and the number of lysine, histidine, and arginine residues in the protein (17). Further evidence comes from the observation that polypeptides rich in lysine and arginine are aggregated by Coomassie G dye molecules (18).

Staining with Coomassie blue-R is usually performed in methanol or ethanol–acetic acid solutions (see *Protocol 1*). An acidic solution is required to facilitate the electrostatic attractions between the dye molecules and the amino groups of the proteins. These ionic interactions, along with Van der Waals forces, bind the dye:protein complex together. Once staining has occurred, excess dye is removed from the polyacrylamide gel matrix by de-staining (16) (see *Protocol 1*). Coomassie blue-G has a diminished solubility in 12% trichloroacetic acid (TCA), allowing it to be used as a colloidal disper-sion that does not penetrate gels. This prevents undesired background from forming and allows for more rapid staining of protein bands in the gel.

A stain that has been reported to be equivalent to Coomassie blue-R in terms of sensitivity but requires less time for staining procedures than the customary techniques utilizing Coomassie blue-R, was introduced by Bikar and Reid (19). This stain employs copper phthalocyanine 3,4',4'',4'''-tetra-

---

Protocol 1. Coomassie blue staining method

Reagents

- Staining solution: 50% (v/v) methanol, 10% (v/v) acetic acid, 0.25% (w/v) Coomassie blue
- Destaining solution: 5% (v/v) acetic acid, 10% methanol

Method

1. Filter the staining solution with a Whatman No. 1 filter prior to use.

2. Immediately following electrophoresis, place the gels in clean glass trays (one gel per tray) containing sufficient staining solution to cover the gel.

3. Place the trays on a rocker/shaker and stain by shaking with gentle agitation for 3 h.[a]

4. To destain, decant the staining solution and add sufficient destaining solution to each tray. The destaining solution must be changed re-peatedly as the dye is removed from the gel. Alternatively, destaining solution may be conserved by placing a piece of wool felt in the destaining tray. The felt will absorb the stain which is leaching out of the gel.

5. When the bands appear to be distinct and the background staining is sufficiently reduced, remove the gels from the destaining solution.

6. Remove excess moisture from the gels and store them moist in sealed clear plastic bags.

[a] Exact staining times and solution volumes vary based on the size and thickness of the gel.

---

sulfonic acid tetrasodium salt (CPTS) for protein detection in polyacrylamide and agarose gels. However, the protocol for this stain requires removal of SDS and/or urea from the gels prior to staining and fixation of the proteins in the gel.

1.3 Silver stains

1.3.1 Introduction

All silver stain methods depend on the reduction of ionic to metallic silver to provide metallic silver images (20). Currently the three main silver staining methods used for detection of proteins separated on polyacrylamide gel are:

- diamine or ammoniacal stains
- non-diamine silver nitrate stains
- silver stains based on photodevelopment

Selective reduction of silver ions to metallic silver at gel sites occupied by proteins depends on differences in the oxidation–reduction potentials in the sites occupied by the proteins in comparison with adjacent sites in the gel that do not contain proteins. These relative oxidation–reduction potential differences may be altered to achieve either negatively or positively stained gels by changing the staining procedures (21). The signal-to-noise ratio of protein detection may be enhanced by reducing the background staining. Since general background staining has been demonstrated to be due in part to the chemistry of the polyacrylamide gels, a reduction of background staining has been achieved for certain diamine stains by the use of special cross-linking agents in the formation of the polyacrylamide gels (22).

The silver stains can be used qualitatively as well as quantitatively. If the silver stains are employed quantitatively it is important to note that the relationship between the density of silver staining and protein concentration is protein-specific. Such protein-specific reactions are not limited to the silver detection methods, as they have been observed with Coomassie blue staining and even with the commonly used Lowry protein assay (23, 24). In both Coomassie blue staining and silver staining, a correlation has been observed between the staining curves of denatured proteins and their mole per cent basic amino acids, particularly histidine and lysine (17, 24).

Most silver stains produce monochromatic brown or black colours. However, other colours may be produced. Lipoproteins may stain with a bluish hue, while glycoproteins may stain yellow, brown, or red (25, 26). This colour effect is due to the diffractive scattering of light by the microscopic silver grains (27). The colour produced depends on the size of the silver grains, the refractive index of the gels, and the distribution of the silver grains in the gel. In general larger silver grains produce black images while smaller grains (less than 0.2 μm in diameter) produce yellow to reddish images.

It should be noted that silver staining might interfere with the subsequent detection of radioactively labelled proteins. Diamine stains can cause as much as a 50% decrease in the autoradiography image density of ^{14}C-labelled proteins and all types of silver stains severely quench the detection of ^{3}H-labelled proteins (28). Destaining can restore the capability of detecting ^{14}C-labelled proteins and some of the capability to detect ^{3}H-labelled proteins fluorographically, providing that the initial staining was performed with a non-diamine silver stain.

Silver stained gels may be destained to enhance subsequent detection of radioactively labelled proteins or to correct a gel that was over-stained. Such destaining can be accomplished by the use of a two-solution photographic reducer (10). The two solutions (A and B) are fairly stable and can be stored for months. Solution A is made by dissolving 37 g of sodium chloride and 37 g of cupric sulfate in 850 ml of deionized water followed by the addition of concentrated ammonium hydroxide. The ammonia hydroxide is added until the precipitate that first appears is completely dissolved. Solution B contains 436 g of sodium thiosulfate dissolved in 1 litre of deionized water. These solutions are mixed in equal volumes just prior to use. The resulting mixture can be used in either a concentrated or diluted form depending on the amount of silver to be removed from the gel. Once they are mixed together they should be used within 30 minutes. If destained gels are to be restained, they should first be extensively washed with water containing 10% ethanol or methanol.

1.3.2 Diamine silver stains

In the diamine stains, silver nitrate is mixed with ammonium hydroxide to form silver diamine. Image development in gels soaking in ammoniacal silver solution containing silver diamine is initiated by acidifying the solution, usually by using citric acid in the presence of a reducing agent such as formaldehyde (see *Protocol 2*). Citric acid lowers the concentration of free ammonium ions, thereby liberating silver ions to a level where their reduction, by formaldehyde, to metallic silver is possible. Citric acid also participates in the reduction reaction of free silver ions to produce metallic silver (29).

Protocol 2. Diamine silver stain (for gels greater than 1 mm in thickness)[a]

Reagents[b]

- Fixation solution: 40% (v/v) ethanol, 10% (v/v) acetic acid
- Rehydration solution: 5% (v/v) ethanol, 5% (v/v) acetic acid
- Glutaraldehyde solution: 1% (w/v) glutaraldehyde[c]

- Silver diamine solution: 0.047 M silver nitrate, 0.2 M ammonium hydroxide, 0.02 M sodium hydroxide
- Reducing solution: 0.5 mM citric acid, 7 mM formaldehyde
- Stop-bath solution: 5% (v/v) acetic acid

Method

NB: Perform all staining steps in glass trays on a rocker or shaker at a very gentle speed. Perform all steps at room temperature. From the development step onward, use clean trays for each solution. For each solution, use volumes sufficient to fully immerse the gels.

1. After electrophoresis, place the gels into a tray containing 500 ml of fixation solution. Soak the gels in this solution for approx. 1 h.

2. Transfer the gels from the fixation solution into 500 ml of rehydration solution. Soak the gels in this solution for a minimum of 3 h. Gels may be left in this solution for extended periods of time, i.e. several days.

3. Transfer gels from the rehydration solution into 500 ml of deionized water. Wash the gels for 5 min.

4. Discard the wash solution and add 500 ml of the glutaraldehyde solution to each tray. Soak the gels in this solution for 30 min.

5. Discard the glutaraldehyde solution.

6. Wash the gels three times with 500 ml of deionized water for 10 min each wash.

7. Wash the gels four times with 500 ml of deionized water for 30 min each wash.

8. Transfer the gels to the silver diamine solution and soak the gels in this solution for 5–30 min.

9. Pour off the silver diamine solution and rinse the gels twice with 500 ml of deionized water for 15 sec each time, followed by three rinses for 5–10 min each time. This rinsing schedule may be varied, but two factors must be kept in mind: inadequate rinsing may leave a metallic sheen on the surface of the gels, but rinsing for more than 30 min may result in the gels sticking to the bottom of the trays.

10. Develop the image by adding the reducing solution and soaking the gels until the proteins are sufficiently stained. Proteins normally become visible in 1–2 min.

11. Stop the development by replacing the reducing solution with 500 ml of 5% (v/v) acetic acid. Soak the gels in this solution for 15 min.

12. Wash the gels three times with 500 ml of deionized water for 5 min each wash.

13. Store the gels moist in sealed clear plastic bags.

[a] Modified from Merril *et al.* (9), Switzer *et al.* (10), and Hochstrasser *et al.* (22).
[b] Volumes are given for the staining of 16 cm × 16 cm gels.
[c] Glutaraldehyde is a hazardous material and should not be used in solutions greater than 2% because the vapour pressure of stronger solutions. Avoid skin contact, and particularly avoid breathing vapours. Therefore, only make up the volume needed to fix the number of gels being stained. It is advisable to carry out preparation of this solution in a fume-hood.

1.3.3 Non-diamine silver nitrate stains

In these stains, silver ions released from silver nitrate under acidic conditions are reduced to metallic silver when the gels are placed in an alkaline solution containing a reducing agent (see *Protocol 3*). Sodium carbonate and/or sodium hydroxide are generally used to maintain an alkaline pH and formaldehyde is utilized as a reducing agent. In the reduction reaction formaldehyde is oxidized to formic acid which is buffered by the sodium carbonate.

Protocol 3. Non-diamine silver stain (for gels less than 1 mm in thickness)[a]

Reagents[b]

- Fixation solution: 50% (v/v) methanol, 10% (v/v) acetic acid
- Rehydration solution: 10% (v/v) methanol, 5% (v/v) acetic acid
- Glutaraldehyde solution: 1% (w/v) glutaraldehyde[c]
- Stop-bath solution: 3% (v/v) acetic acid
- Dichromate solution: 34 mM potassium dichromate, 32 mM nitric acid
- Staining solution: 0.118 M silver nitrate
- Reducing solution: 0.283 M sodium carbonate, 7 mM formaldehyde
- 10% methanol or ethanol

Method

NB: Perform all steps in glass trays, staining one gel per tray. Place the trays on a rocker or shaker at a very gentle speed. Perform all steps at room temperature.

1. After electrophoresis, place each gel into 500 ml of fixation solution. Soak the gels in this solution for 1 h. Gels can be stained 1 h after fixation; however, the gels may remain in this solution overnight if necessary.

2. Transfer each gel from the fixation solution into 500 ml of rehydration solution. Soak the gels in this solution for 10 min.

3. Discard the rehydration solution and add 200 ml of glutaraldehyde solution to each tray. Soak the gels in this solution for 30 min.

4. Discard the glutaraldehyde solution. Wash each gel with 500 ml of deionized water for 15 min.

5. Discard the wash and add to each gel 200 ml of dichromate solution. Soak the gels in this solution for 5 min.

6. Discard the dichromate solution and add to each gel 200 ml of staining solution. Soak the gels in this solution for 25 min.

7. Discard the staining solution.

8. Wash each gel briefly (approx. 15 sec) with 50 ml of the reducing solution, and discard this wash. Follow this initial rinse with 500 ml of reducing solution. If colloidal particles collect in the solution prior to

---

full image development, replace the reducing solution with fresh solution. Continue the development in this solution until the proteins are sufficiently stained. Proteins should become visible within 1–3 min.

9. To stop image development, discard the reducing solution and add 500 ml of stop-bath solution. Soak the gels in this solution for 5 min.

10. Discard the stop-bath solution. Wash each gel twice with 500 ml of deionized water with 10% methanol or ethanol for 10 min each wash.

11. Store the gels moist in sealed clear plastic bags.

[a] Modified from Merril *et al.* (29).
[b] Volumes are given for the staining of 16 cm × 16 cm gels.
[c] Glutaraldehyde is a hazardous material and should not be used in solutions greater than 2% because of the vapour pressure of stronger solutions. Avoid skin contact, and particularly avoid breathing vapours. Therefore, only make up the volume needed to fix the number of gels being stained. It is advisable to prepare the solution in a fume-hood.

1.3.4 Photodevelopment silver stains

The energy from light photons can also reduce ionic silver to metallic silver (30), a phenomenon that is utilized in the photodevelopment stains. In these stains, silver chloride, which is more light-sensitive than silver nitrate, is incorporated into the gel, and the presence of nucleic acids or proteins enhances the photoreduction procedure. Photodevelopment silver stains can be performed rapidly, but current versions sacrifice sensitivity and image preservation in comparison to the above silver stains. Since the silver chloride is dispersed throughout the gel, and depends on enhanced reduction at sites containing proteins, sensitivity is limited to the rate at which silver is reduced in regions devoid of protein. The widespread dispersal of silver chloride throughout the gel make it difficult to solubilize with thiosulfate and completely wash the gel free of silver chloride. For this reason, photography or other methods of image capture are recommended during development (30).

A relatively simple version of this type of stain can be performed by first fixing the gel for 5 min in a solution containing 50% (v/v) methanol, 10% (v/v) acetic acid, 2% (w/v) citric acid, 0.2% (w/v) sodium chloride in deionized water. Following fixation, the gel is rinsed rapidly with 200 ml of deionized water to remove surface chloride. The gel is then placed in 200 ml of a solution containing 50% (v/v) methanol, 10% (v/v) acetic acid, 2% (w/v) silver nitrate and illuminated with a uniform light source until an image appears.

1.4 Negative staining

Another series of methods for protein detection is the negative or reverse stains, which are generally based on the formation of insoluble metal salts, leaving protein bands unstained when viewed against a dark background. Most negative stains act rapidly (within 15 minutes), do not require any protein fixation within the gel matrix, and proteins stained in this manner are reported to be easily recovered from gels for further analysis. Some commonly used negative stains and their protein detection limits include:

- copper chloride (5 ng/mm) (31)
- zinc chloride (10–12 ng) (32)
- potassium acetate (0.12–1.5 μg) (33)
- sodium acetate (0.1 μg/mm^3) (34)

Detection limits for these negative stains fall between the sensitivity levels of Coomassie blue and silver stains. While most of the negative stains require the presence of detergent in the gel to achieve protein visualization, Candiano *et al.* (35) have described a negative staining technique for the presence of proteins in polyacrylamide gels in the presence or absence of SDS, based on the precipitation of methyl trichloroacetate (MTA). This two-step stain can be completed rapidly (see *Protocol 4*) and is reported to be able to detect as little as 0.5 ng of protein. In addition the staining is highly reversible, allowing for recovery of proteins for further processing (35).

Protocol 4. Negative staining of proteins with methyl trichloroacetate[a]

Reagents

- MTA precipitate: 30% (w/v) trichloroacetic acid (TCA) in 30% (v/v) methanol, 10% (v/v) acetic acid. Mix the above ingredients for 6 h at room temperature with gentle agitation. Precipitate the methyl trichloroacetate by adding 2 vol. of water. Wash the precipitate with water, four times for 10 min each wash. Decant the solution after each wash and allow the precipitate to air dry.

- Staining solution: prepare a 8% (w/v) solution of the MTA precipitate (prepared as described above), dissolved in 38% (v/v) isopropanol.

Method

1. Soak the gels in staining solution for 1 h. For optimal conditions, use 10 vol. of staining solution:1 vol. of gel volume.

2. Remove the staining solution and replace with distilled water.

3. Negative protein images should be visible within 1 min and last for 30 min.

[a] Modified from Candiano *et al.* (35).

1.5 Fluorescent stains

Proteins can also be detected by fluorescent labelling either prior to or following electrophoresis. Pre-electrophoretic stains allow for immediate visualization during and following electrophoresis. While these techniques are extremely sensitive, visualization requires ultraviolet light and direct quantitation requires sophisticated equipment, limiting the use of these procedures. Fluorescent labelling normally involves the covalent binding of a fluorescent dye to terminal amino groups of the proteins, altering the net overall charge (pI) of the labelled proteins. This effect is minimal for SDS–PAGE, which is based upon molecular weight separation, because the dye molecules are usually too small to have a significant effect (36) but may well affect electrophoretic separations based on protein charge.

Pre-electrophoretic fluorescent stains include dansyl chloride, fluorescamine, and 2-methoxy-2,4-diphenyl-3(2H)-furanone (MDPF). Dansyl chloride reacts with proteins in mildly alkaline SDS systems to produce dansylated proteins that can be analysed by SDS–PAGE without significantly altering apparent protein mass. Detection levels of dansylated proteins can be as low as 10 ng of protein (8). However, protein samples can be contaminated with fluorescent breakdown products when labelled with dansyl chloride. Fluorescamine and MDPF are not fluorescent themselves, nor are their hydrolysis products, so only labelled proteins will be detected by fluorescence. Fluorescamine has been shown to detect as little as 6 ng of myoglobin (37), and MDPF has been shown to exhibit similar properties to fluorescamine. However, MDPFs protein derivative fluoresces 2.5 times better than fluorescamine labelled proteins, and MDPFs fluorescence does not fade as readily. The labelling procedure is described in *Protocol 5*.

Protocol 5. MDPF or fluorescamine detection of proteins

Equipment and reagents

- Long wavelength UV fluorescent light box (e.g. UV Transilluminator, UVP, Inc.)
- 0.2 M borate buffer pH 9
- Protein sample (50–100 μg)

- MDPF stock solution: 2 mg MDPF in 1 ml of acetone
- Fluorescamine stock solution: 2 mg of fluorescamine in 1 ml of acetone

Method

1. Add 50 μl of 0.2 M borate buffer to 50–100 μg of protein solution.
2. While vortex mixing, add 30 μl of MDPF or fluorescamine stock solution.
3. Mix the solution for approx. 1 min to allow the proteins sufficient time to be labelled.
4. Air dry the labelled proteins to remove any acetone.
5. Dissolve the labelled proteins in sample buffer (composition depends on the gel system being used), load onto the gel, and electrophorese as normal.
6. During or after electrophoresis, visualize the proteins using the long wavelength UV fluorescent light box.

Most post-electrophoretic fluorescent stains react rapidly; staining is carried out in a slightly alkaline pH and induces minimal denaturation of the proteins. Fluorescamine and MDPF can also be used to stain proteins after electrophoresis. One-dimensional gels and two-dimensional gels are fixed overnight with 50% alcohol, 10% acetic acid; for 2D gels this removes any ampholyte used in isoelectrofocusing. The gels are then rinsed with 7% acetic acid followed by immersion in a DMSO solution for 1 h. The DMSO is then decanted and the gels submerged in a 0.04 M boric acid–DMSO pH 10 solution. After 2 h, an equal volume of 0.5 mg/ml fluorescamine in DMSO is added. Gels can be photographed under UV light after 8 h of gentle agitation, allowing for the detection of proteins in the range of 5 ng–7 μg (38). As discussed above, labelling with fluorescamine alters the pI of proteins and hence this affects separation in the first dimension (isoelectric focusing) of two-dimensional gel electrophoresis, limiting its use in 2D electrophoresis. However, using fluorescamine or MDPF post-electrophoretically requires excessive amounts of reagents, which can be costly. Anilinonaphthalene sulfonate (ANS) and 8-*p*-toluidino-1-naphthalenesulfonate (bisANS) have also been used for post-electrophoretic labelling. ANS can detect 100 μg of protein, while bisANS shows increased sensitivity over ANS as well as reduced background staining, producing a more reliable stain (39).

Recently, Steinberg, *et al.* (40) described two fluorescent protein stains, SYPRO red and SYPRO orange, that interact with SDS:protein complexes. They reported that these stains can produce sensitivity levels (0.5–1.0 ng/mm^2) that compare well with those achieved in silver staining. Staining time can be completed in 30–60 min, with no destaining required. In addition, SYPRO orange dye has been reported to stain only proteins, not nucleic acids, which could prove useful in some applications (41).

1.6 Direct detection of proteins

Under certain conditions, proteins can be detected directly in polyacrylamide gels without protein modification. These detection methods are useful when individual proteins need to be recovered on a preparative scale from SDS–polyacrylamide electrophoretic gels. Two of the earliest described methods include chilling the gels at 4 °C for 3–5 h to precipitate SDS:polypeptide complexes (42) and staining of the gel with pinacryptol yellow for 0.5 h (43). Pinacryptol yellow is an organic dye which forms a fluorescent complex with detergents, allowing SDS:protein complexes to be visualized under UV light. Precipitation of SDS:protein complexes can also be accomplished using potassium salts; the detection limits for this method have been reported to be less than 0.1 μg/mm^3 (33). Variations in SDS binding and SDS:protein complex precipitation for some proteins have been reported to cause some difficulties. Additionally, the bands formed following potassium salt precipitation are quite unstable and are visible only for approximately 20 minutes (34). Higgins and Dahmus (34) tested other salts to find one that produced a more stable banding pattern and discovered that sodium acetate could precipitate SDS that was not bound to protein, producing transparent protein bands within a white gel. Bands could be visualized against a dark background with side lighting, with a sensitivity comparable to that obtained with Coomassie blue-R.

1.7 Detection of radioactive proteins

Radioactive labelling of proteins either before or after electrophoresis is the most sensitive method available for protein detection to date. ^{3}H, ^{14}C, ^{32}P, ^{35}S, and ^{125}I are the most commonly used isotopes for radioactive labelling. The simplest detection method for these radioactively labelled proteins utilizes autoradiography (44), although fluorography (45, 46), liquid scintillation counting (47), and phosphorimaging (48, 49) can also be used.

1.7.1 Autoradiography

Autoradiography involves exposure of the polyacrylamide gels containing the radioactively labelled proteins to X-ray film. Depending on the isotope used to label the proteins, the radioactive emissions (β-emissions, γ-radiation, or X-rays) can create metallic silver latent images from the silver halides in photographic film emulsions. The latent images are then made visible by photographic development. Autoradiographic exposure times depend primarily on the radioisotope and the amount present in the gel. Additional factors that influence optimal autoradiographic exposure time include: temperature of exposure, whether the gels are hydrated or dehydrated, quenching by prior proteins stains (particularly silver stains), and the type of photographic film used. Generally photographic film sensitivity to ionizing radiation increases with increased temperature between $-100\,°C$ to $60\,°C$. If hydrated gels are to be autoradiographed, a barrier, such as thin plastic wrap, should be placed between the film and the gel to protect the film from the moisture in the gel. It should be noted that even thin plastic films can significantly impair weak β-emissions from exposing the photographic film. Dehydration of gels prior to autoradiography can eliminate this problem and dehydration also decreases the path length between the emitting isotope and the film.

   Marking the periphery of the gel with radioactive ink prior to autoradiography can facilitate alignment of a stained gel with its autoradiographic image. In addition, while single emulsion photographic films provide better image resolution, double emulsion films minimize exposure times, particularly for high energy β- or γ-emitters.

1.7.2 Fluorography

i. Detection of ^{3}H-labelled proteins

A direct autoradiography approach is not efficient for ^{3}H-labelled protein detections, as their β-emissions have insufficient energy to escape the polyacrylamide matrix efficiently (50). The sensitivity of detection for these weak β-emitters can be increased by incorporation of an organic scintillator, such as 2,5-diphenyloxazole (PPO), into the gel (see *Protocol 6*) to convert the β-energy to light within the gel matrix, which can then escape from the gel to expose the film (50). Similarly, intensifying screens may be used to convert the energy from strong β-emitters (i.e. ^{32}P) to light to expose the film (50). The use of two screens in a sample, screen, film, screen arrangement increases the sensitivity of detection of ^{32}P (51), but results in a loss of resolution (46). The sensitivity of detection for ^{125}I remains unchanged with the use of two screens.

Unlike autoradiography, in which a slight increase in temperature increases the sensitivity of detection (52), sensitivity levels can often be increased for fluorographic techniques by decreasing the exposure temperature. Two mechanisms have been suggested to account for this discrepancy. Luthi and Waser (53) proposed that an activated fluor incorporated into the gel can emit both light energy, which exposes the detection film, and thermal energy (due to intramolecular vibrations), which would go undetected by the film. As the temperature decreases, less intramolecular vibrations would occur, yielding greater energy levels available for light production. However, lowering the temperature was shown to increase light production only minimally for some fluors (54). For example, anthracene, the fluor employed in Luthi and Waser's study, did not generate any increase in light production when utilized at lower temperatures.

Alternatively, Randerath proposed that this paradoxical temperature effect between autoradiography and fluorography is based on differences in the interaction of light and elementary particles with the silver halide crystals of the film emulsion (54). Regardless of the temperature, elementary particles will darken a film at a rate proportional to the product of the intensity of the particle emissions and the exposure time. Light, however, will maintain a proportional relationship only at medium light intensity. At low light intensity, films no longer maintain a proportional relationship between light intensity and developed film density (this effect is known as reciprocity failure). It has been suggested that one photon of light can reduce one silver ion to metallic silver at low light intensity but four photons of light are needed to produce a stable latent image centre, consisting of four silver atoms (55). Anything less than that is unstable and decays rapidly. Lowering the temperature increases the lifetime of an unstable latent image centre to 4.8 msec (at 77 °K) from 1.7 μsec at room temperature (293 °K) (56). This increased lifetime allows sufficient photons from the radioactively-induced fluorescence to accumulate and form a stable latent image centre. Maintaining fluorographs at −70 °C to −80 °C during exposure increases sensitivity levels 12-fold for ^{3}H and ninefold for ^{14}C and ^{35}S.

Quantitation of fluorographic images is possible using microdensitometry or by simply cutting out spots from a dried fluorographed gel (49). Briefly, the gel is dried as usual (*Protocol 6*) and then spotted around the edges with radioactive ink. The resulting fluorograph (*Protocol 6*) is then placed over the dried gel and held up to a strong light to see the desired spots, which are then marked with pencil on the paper backing. The spots can be cut out and digested overnight at 37 °C in 1 ml of 95 parts of 30% H_2O_2 and 5 parts concentrated NH_4OH in plastic, capped scintillation vials. The radioactivity of each spot can be determined by addition of a water miscible scintillation solution and counting in a liquid scintillation counter.

Protocol 6. Fluorography for the detection of ^3H-labelled proteins in polyacrylamide gels[a]

Equipment and reagents

- Vacuum gel dryer
- X-ray film: Kodak X-Omat AR or equivalent (VWR Scientific)
- Photographic filter: Kodak Wratten 21 or 22 (orange)
- Electronic flash unit
- Whatman 3MM paper
- White bond paper
- Dimethyl sulfoxide (DMSO)[b]
- Scintillation solution: 22% (w/v) PPO in DMSO

Method

1. Place the hydrated gel in approx. 20 vol. of DMSO for 30 min.[c]

2. Soak the gel for an additional 30 min in fresh DMSO.

3. Submerge the gel in 4 vol. of scintillation solution for 3 h.

4. Soak the gel in excess water for 1 h to remove DMSO.[c]

5. Dry the gel under vacuum onto Whatman 3MM paper.

6. Attach the photographic filter to a flash unit and cover the flash unit with a single-ply sheet of white bond paper. To optimize exposure conditions, place a coin on a test strip of X-ray film and then expose the film with a single flash of light from the flash gun. Repeat this procedure using a series of film test strips, varying the intensity of the flash by adjusting the flash gun or varying the distance between the flash gun and the film, until the exposure is such that the coin is barely visible.

7. Expose the dried gel to the hypersensitized X-ray film overnight at −70 °C.

[a] Modified from Bonner and Laskey (57).
[b] DMSO can penetrate the skin; therefore, rubber gloves should be worn to prevent contact with the skin.
[c] Gels greater than 3 mm thick require longer incubation. The gels may also be soaked in 0.5% glycerol prior to drying; in some cases this reduces gel cracking during drying.

---

ii. Detection of dual-labelled proteins

Detection of proteins labelled with two isotopes is possible using X-ray film. The inability of ^3H to be detected by direct autoradiography allows dual-labelled proteins, with ^3H and ^{14}C or ^{35}S, to be detected by combining fluorography for detection of the ^{3}H-labelled molecules and direct auto-radiography for the detection of proteins labelled with ^{14}C or ^{35}S. The higher energy isotope emitter (in this case, ^{14}C) is easily detectable by placing a suitable film (e.g. Kodak No-Screen X-ray) directly in contact with the gel and allowing the β-particles emitted by the isotope to produce silver grains by direct interaction with the film emulsion (58). The weak β-emitter (^{3}H) can be detected only by fluorography (*Protocol 6*), impregnating the gel with a scintillator (e.g. PPO). Interaction of the β-particles with the scintillator in the gel produces light which is then captured on pre-sensitized film (e.g. Kodak XR5). This fluorogram will also record spots produced by the ^{14}C-labelled isotopes, which are indistinguishable from the ^{3}H-labelled spots. However, comparison of the two films (from direct autoradiography and fluorography) will identify spots unique to the fluorogram, which must be the ^{3}H-labelled proteins. The sensitivity of detection for both isotopes must be equal for this technique to be quantitatively accurate (58).

Alternatively, the gel may be marked with black ink following fluorography and exposure to X-ray film. The black ink prevents the light generated by the interaction of the β-emission from the tritium labelled proteins and the fluor-escent agent incorporated in the gel from reaching the film, leaving only ^{14}C-labelled proteins detectable during a subsequent exposure to X-ray film. Walton *et al.* (59) determined that a ^{3}H/^{14}C ratio exceeding 40:1 enabled only detection of ^{3}H on the fluorograph and only ^{14}C detection on the auto-radiograph.

Simultaneous exposure of the gel to two X-ray films was developed for detection of ^{32}P and ^{35}S double-labelled proteins (60). One side of the gel was wrapped in aluminium foil to prevent ^{35}S emissions from reaching the film, yielding only ^{32}P-labelled protein images. When the other side of the gel was exposed to film at –80^°C, only ^{35}S images were detectable.

Colour negative film has also been employed to differentiate the emissions from radioactive isotopes with colour film containing three emulsion layers: yellow, magenta, and cyan. Tritium exposes only the first, or yellow, layer while ^{14}C and/or ^{35}S can expose the first and second layers. The isotopes ^{32}P and ^{125}I can expose all three layers. This method has the potential advantage of detecting more than one isotope with a single film (61).

1.7.3 Storage phosphorimaging

Storage phosphorimaging was introduced in response to the shortcomings of autoradiography (i.e. general lack of sensitivity of X-ray films to β-radiation and the problems of linearity of response of film). Instead of being exposed to X-ray film, dried polyacrylamide gels containing radiolabelled proteins are exposed to an imaging plate coated with BaFBr:Eu^{2+} combined with an organic binder (48). The absorbed radiation induces excitation of the Eu^{2+} ions in the phosphor complex, storing the image. When subsequently scanned by a helium–neon laser, the stored energy is released as blue photons that can be quantitated based on the intensity of the resulting luminescence. Storage phosphorimaging has a 10–250-fold greater sensitivity to β-emissions than autoradiography and demonstrates an exposure range of more than five orders of magnitude (49, 62). This technology, built into equipment known as phosphorimagers, has rapidly become a common method of detecting and quantifying radioactive proteins after gel electrophoresis.

1.8 Detection of specific proteins

1.8.1 Glycoproteins

Detection of glycoproteins can be accomplished by utilizing any one of four reactions:

- thymol–sulfuric acid detection
- periodic acid–Schiff base detection.
- fluorescein isothiocyanate labelled (FITC) lectin or FITC antibody detection
- fluorophore labelled polysaccharide detection.

In addition, double-staining techniques utilizing Coomassie blue and silver staining have been reported for the detection of sialoglycoproteins (63, 64).

i. Thymol–sulfuric acid detection

Glycoproteins containing hexosyl, hexuronosyl, or pentosyl residues can be detected using the thymol–sulfuric acid method (*Protocol 7*). These residues react with sulfuric acid to form furfural derivatives, which then react with thymol to form a chromogen. Glycoproteins containing at least 50 ng of carbohydrate can be detected using this method (65).

Protocol 7. Specific detection of glycoproteins by thymol–H_2SO_4[a]

Reagents

- Fixative solution: 25% (v/v) isopropanol, 10% (v/v) acetic acid, 65% (v/v) H_2O
- Thymol solution: fixative solution containing 0.2% (w/v) thymol
- Sulfuric acid solution: 80% (v/v) concentrated H_2SO_4, 20% (v/v) absolute ethanol

Method

1. Wash the gel twice for at least 2 h each time with fixative solution.[b]
2. Incubate the gel for 2 h in thymol solution.
3. Decant the thymol solution and allow the gel to drain.
4. Add 10 ml of sulfuric acid solution per millilitre of gel at room temperature.
5. Gently agitate the gel for 2.5 h or until the opalescent appearance of the gel just disappears. Glycoproteins will stain red on a yellow background.

[a] Modified from ref. 66.
[b] Samples containing large concentrations of soluble carbohydrates may require additional washes.

---

ii. Periodic acid–Schiff base (PAS) detection

Addition of periodic acid to a polyacrylamide gel oxidizes and cleaves secondary alcohols of glycosyl residues to dialdehydes. Fuchsin (67), Alcian blue (68), or dansyl hydrazine (69) can then be added to react with aldehydes to form a Schiff base. Detection limits as low as 40 ng have been reported using dansyl hydrazine, while 2–3 μg of glycoprotein are needed for detection using Fuchsin or Alcian blue (69).

iii. Detection of glycoproteins using labelled lectin or antibody

Specific lectins or antibodies can be labelled with either fluorescent markers (108), enzymes (108), or radioactive isotopes (70) to facilitate detection of specific glycoproteins in SDS gels. If lectins with different carbohydrate binding specificities are used, specific information can be gleaned about the composition of the carbohydrate portion of the glycoprotein (70).

iv. Fluorophore-labelled detection of glycoproteins

Recent developments in glycoprotein detection have yielded techniques for fluorophore labelled detection of glycoproteins in polyacrylamide gel electrophoresis (71). Saccharides bound to proteins with free reducing end-groups may be labelled covalently with a fluorophore, and then detected following gel electrophoresis. Using this approach, subpicomolar quantities of individual saccharides have been detected (72). This technique is described in detail in Chapter 10.

1.8.2 Phosphoproteins

Phosphoproteins can be detected with some specificity following poly-acrylamide gel electrophoresis. Although there are direct staining techniques, such as 'Stains-all™' (73) and silver stains (74) which can detect phosphoproteins, a more specific phosphoprotein staining method depends on the entrapment of liberated phosphates (ELP) by staining with methyl green (75, 76) or rhodamine B (77). In the methyl green technique, the addition of calcium chloride initiates alkaline hydrolysis of susceptible protein phosphoester bonds producing insoluble calcium phosphate entrapped within the gel. The addition of 1% (w/v) ammonium molybdate in 1 M nitric acid (modified Fiske-Subba Row reagent) then results in an insoluble blue complex that can be detected as a bright green band following methyl green staining. 1 nmol of phosphate can be detected using this method (75). Debruyne (77) described a method based on rhodamine B. This involves alkaline hydrolysis, phosphate capture, and formation of insoluble rhodamine B–phosphomolybdate complexes. Staining with rhodamine B can be completed in 20–30 min and can detect 0.2 nmol of bound phosphate/mm^2.

1.9 Immunological detection methods

The use of antibodies specific for protein antigens provides detection methods of high sensitivity and specificity. Most immunological techniques for the detection of electrophoretically separated proteins require the transfer of the proteins from the gel to a blotting matrix, such as nitrocellulose. This technique is called Western blotting. Electrophoretic transfer of proteins from the gel to the blotting membrane (electroblotting) is one of the most efficient methods to achieve such transfers and is discussed in Section 2.

Immunological methods also exist for the detection of proteins directly in gels. Thus, following SDS–PAGE, fixed gels can be incubated with labelled antibody for detection of specific proteins. The antibody may be tagged by radiolabelling (i.e. ^{125}I) (70), labelling with a fluorescent marker (i.e. FITC) (78), or conjugating the antibody to an enzyme (i.e. peroxidase) (79). However, greater sensitivity is generally obtained by reacting the separated proteins with unlabelled specific antibody and then utilizing a secondary, labelled antibody against the primary antibody for actual visualization. Direct and indirect labelling techniques are discussed further in Section 2.

Antiserum overlays may also be used to detect specific proteins (80). In this approach, an agar or filter paper overlay containing antiserum is overlaid on the polyacrylamide gel following electrophoresis. After sufficient incubation time to allow diffusion of the antigens to the overlay, the overlay is removed from the gel and stained for immune precipitates.

While the use of antibodies can provide considerable specificity of detection, the experimental conditions must be controlled to prevent non-specific binding or cross-reactivity reactions from occurring. Non-specific binding can be avoided by:

- increasing the wash times, particularly between the first and second antibody incubations
- prevention of denaturation of the labelled antibody
- use of high titre antisera (70).

Cross-reactivity reactions can be minimized by:

- utilizing monoclonal antibodies
- adjusting the incubation time (81)
- adjusting the incubation temperature (82).

2. Western blots

2.1 Choice of blotting matrix

Some of the problems encountered when protein detection is performed in gels are eliminated by employing protein blotting techniques. Protein detection on blotted membranes can be generally performed in shorter times, with less reagents, and with multiple successive analyses if needed. The conditions utilized for protein transfer, particularly the type of membrane used, dictate the preferable detection method.

Nitrocellulose, which was introduced for use in protein blotting in 1967 (83), is the most commonly used blotting membrane at this time. Protein interaction with nitrocellulose is probably due in part to hydrophobic interactions, although the complete mechanism for the interactions is not well understood. However, nitrocellulose has a low affinity for some proteins, particularly those of low molecular weight (84). Given this problem, low molecular weight proteins may not be detectable using current nitrocellulose blotting techniques, although covalent fixation of the proteins to the nitrocellulose by glutaraldehyde (85) or *N*-hydroxysuccinimidyl-p-azidobenzoate (84) can help prevent the loss of low molecular weight proteins from cellulose nitrate membranes.

Other matrices have been developed such as diazobenzyloxymethyl (DBM) modified cellulose paper (86), and diazophenylthiother (DPT) paper (87). Negatively charged proteins interact electrostatically with the positively charged diazonium groups of these matrices, followed by irreversible covalent linkage via azo derivatives (86). DPT paper has been shown to be more stable than DBM paper, but equally efficient (87). Diazo paper is reported to show less resolution of separated proteins compared to nitrocellulose or nylon membranes (88). In addition, glycine may interfere with DBM protein blotting (84).

Nylon membranes are also used for protein blotting. Two commercially manufactured nylon membranes are GeneScreen™ (from NEN) (Boston, MA) and Zeta-bind™ from AMF/Cuno (Meriden, CT). Zeta-bind™ has a very high protein binding capacity (89) and requires extensive blocking to eliminate non-specific binding. Nylon membranes consistently provide a higher efficiency of protein transfer SDS gels than nitrocellulose (89, 90). However, one disadvantage of nylon matrices is that they bind the common anionic dyes such as Coomassie blue and Amido black which may result in background staining that can obscure the detection of transferred proteins.

Polyvinyl difluoride (PVDF) membranes can also be utilized for protein blotting (170). PVDF membranes are hydrophobic in nature and proteins bound to these membranes can be detected by most anionic dyes as well as by immunodetection protocols.

2.2 General detection of proteins on Western blots

2.2.1 Anionic protein stains

General staining of proteins transferred or blotted on membranes is needed to confirm that the overall size and geometry of the gel was maintained during the protein transfer. In addition, such staining is helpful if the blotted proteins are to be isolated from the gel for further processing, such as for protein sequencing. Many of the anionic stains utilized for protein detection in polyacrylamide gels can also be used to detect proteins bound to membranes following blotting; Amido black (90, 91), Coomassie blue R-250 (88), and Fast

Table 2. Staining sensitivities for detection of polypeptides on Western blots

Staining method	Protein (ng)	Reference
Amido black	50	168
Coomassie blue	100	88
Ponceau S	100	169
Fast green FCF	20	92
CPTS	10	19
India ink	1–2	95
Silver stain	1	24
Colloidal metal	0.5	100

green FCF (92, 93) are among the common stains utilized for this purpose (see *Table 2* for sensitivity of detection levels). However, most of the protein stains are less sensitive when used to detect proteins bound to membranes than they are when used for protein detection directly in gels. The detection of proteins on membranes generally requires at least 0.5 μg of protein (94).

The high binding affinity of most nylon membranes and diazo paper for the anionic protein stains induces high background staining. Because of this background staining problem, nitrocellulose is the matrix of choice when the blotted proteins are to be detected by the anionic protein stains. However, one of the anionic protein stains, Coomassie blue, may also produce significant background staining on nitrocellulose (95) and destaining of Coomassie blue stained nitrocellulose requires higher concentrations of methanol than does Amido black. Nitrocellulose is also affected by the alcoholic solutions used in the staining procedures. The membrane will disintegrate if left too long in methanol–acetic acid and may shrink in other alcoholic solutions. For these reasons and for rapid visualization, Amido black (0.1% Amido black, 25% isopropanol, 10% acetic acid) is often used (see *Protocol 8*), although Fast green FCF may be more sensitive (96).

Amido black and Coomassie blue both exhibit positive charges that may decrease the affinity of these dyes for protein and increase their non-specific binding to nitrocellulose membranes. The dye Ponceau S has no positive charge and a larger negative charge than Amido black or Coomassie blue, but it exhibits less sensitivity than these dyes, partly due to its weak staining colour. Copper phthalocyanine 3,4′,4′,4′′′-tetrasulfonic acid tetrasodium salt (CPTS), however, has no positive charge and can bind proteins tightly without high background staining on nitrocellulose, cellulose, or PVDF (19). Staining can be completed within 30 seconds, and bound proteins are detectable as turquoise blue bands on a pure white background. CPTS is about ten times more sensitive than Ponceau S, capable of detecting 10 ng of protein, only slightly less than silver staining or colloidal metal staining techniques (19).

Protocol 8. Amido black staining of Western blots

Reagents

- Staining solution: 0.1% (w/v) Amido black 10B[a]
- Destaining solution: 25% (v/v) isopropanol, 10% (v/v) acetic acid

Method

1. Rinse the membrane briefly with distilled water.
2. Place the membrane in 100 ml of staining solution in a clean glass dish for 1 min.
3. Transfer the membrane to 100 ml of destaining solution for 30 min. Alternatively, the membrane may be destained in two washes of destaining solution for 15 min each.
4. Allow the blot to air dry overnight.

[a] Staining solution may be used several times.

2.2.2 India ink staining

India ink has been used as a sensitive protein detection stain that can detect approximately 1 ng/mm of blotted protein on a nitrocellulose sheet (97). Although not as sensitive as silver staining (98), India ink is more sensitive than Amido black or Fast green (99).

2.2.3 Silver staining

The use of silver staining for detection of proteins following polyacrylamide gel electrophoresis has been proven to be more sensitive than most of the organic stains. Early attempts to use silver stain following protein blotting produced unacceptable background staining (30). However, in 1982, Yuen *et al.* (98) introduced a negative silver staining technique for transblotting proteins to nitrocellulose from SDS gels that could detect nanograms of proteins. Merril and Pratt later developed a membrane silver stain that produced positive protein images (24) (see *Protocol 9*). This stain employs photo and chemical development to positively stain transblotted proteins bound to nitrocellulose membranes. The staining procedure is rapid and it produces brownish to black images of the proteins with minimal background staining. The average sensitivity of detection is about 1 ng of protein.

2.2.4 Staining with colloidal metals

Colloidal metal staining has also been used to detect proteins bound to membranes. Colloidal gold particles, such as AuroDye (Janssen Life Sciences

Products) are the most frequently used of the colloidal metals for this purpose. The high affinity of nylon membranes for proteins transferred from SDS–PAGE gels limits the use of colloidal gold, which is negatively charged, to nitrocellulose membranes only (see *Protocol 10*) to prevent high background staining. However, colloidal iron particles, such as FerriDye (Janssen Life Sciences Products), are positively charged and can be used on both nitrocellulose and nylon membranes with only moderate background.

Protocol 9. Silver stain for detection of proteins on membranes[a]

Equipment and reagents

- Uniform light source (e.g. Aristo T-12 lamp with W45 daylight fluorescent lamp)
- 0.396 M cupric acetate solution
- Acetic acid solution: 1.74 M acetic acid, 34 mM NaCl, 0.104 M citric acid
- Fixing solution: 63 mM $Na_2S_2O_3$

- Silver nitrate solution: 1.74 M acetic acid, 0.118 M silver nitrate
- Developing solution: 0.182 M hydroquinone, 4% (v/v) concentrated formaldehyde

Method

1. Place the membrane containing the fixed protein samples into the cupric acetate solution for 1 min.

2. Transfer the membrane to amino acid solution solution for 1 min.

3. Transfer the membrane to silver nitrate solution. Irradiate for 1 min with the uniform fluorescent light source.

4. Return the membrane to the amino acid solution for 1 min.

5. Return the membrane to the silver nitrate solution, and irradiate it for 3 min with the uniform light source.

6. Place the irradiated membrane into the developing solution for 1 min, or until the membrane develops an even, dark brown colour.

7. Rinse the membrane in distilled water for 1 min.

8. Place the membrane in the fixing solution for 5 min, shaking gently every 30 sec.

9. Rinse the membrane with running distilled water, taking care to direct the flow of water parallel to the surface of the membrane to remove any unbound silver.

10. Photograph the membranes under incidental lighting at a 45° angle to the membrane.

[a] Modified from ref. 24.

Protocol 10. Protein staining with colloidal gold particles[a]

Reagents

- AuroDye®[b] (Janssen Life Sciences Products)
- PBS–Tween: phosphate-buffered saline
  pH 7.2, 0.3% (v/v) Tween 20

Method

1. Wash the membrane with PBS–Tween for 30 min at 37 °C.

2. Repeat the PBS–Tween washes three additional times for 15 min each time at room temperature.

3. Rinse the membrane in approx. 200 ml of distilled water for 3 min at room temperature.

4. Incubate the membrane in AuroDye at room temperature for at least 4 h, preferably overnight.

5. Wash the membrane several times with distilled water to remove excess gold particles and allow the membrane to air dry.

6. Photograph the membrane using incidental lighting at a 45° angle to the membrane.

[a] Modified from ref. 100.
[b] AuroDye® is a registered trademark of Janssen Life Sciences Products, a division of Janssen Pharmaceuticals.
[c] Do not reuse the AuroDye® stain. If the stain turns blue, replace with fresh stain solution.

In colloidal gold staining (*Protocol 10*), accumulation of the gold particles at protein banding sites results in red to dark red bands that can be detected in the subnanogram range/mm^2, a sensitivity comparable to that achieved with silver staining (100). The concentration of Tween 20 and the range of gold particle sizes (15–30 nm) can vary and still produce satisfactory results (101).

Iron binding of the FerriDye to negatively charged proteins following SDS–PAGE and blotting produces faint yellow/brown bands. Addition of potassium ferrocyanide (Perl's reaction) intensifies the FerriDye signal, yielding a deep blue colour. FerriDye is approximately eight times less sensitive than AuroDye (101) and can bind only SDS denatured proteins, but is the only general protein stain that can be used effectively with nylon membranes, including positively charged nylon membranes (*Protocol 11*).

2.2.5 Tagging of proteins

In this procedure, proteins are tagged with small hapten molecules and then detected by incubation with a chromogenic or enzymatic reagent. This provides a sensitive, easy-to-use procedure for protein detection. For example, Wojtkowiak *et al.* (102) incubated proteins bound to nitrocellulose sheets in 2,4-dinitrofluorobenzene (DNFB) to form 2,4-dinitrophenyl (DNP) labelled proteins. Detection of the DNP labelled proteins is accomplished by incubation with anti-DNP rabbit antiserum followed by reaction with peroxidase labelled goat anti-rabbit IgG and measurement of peroxidase activity. DNP tagging was determined to be 100-fold more sensitive than detection by Coomassie blue or Amido black staining (102).

Protocol 11.   Protein staining with colloidal iron particles[a]

Equipment and reagents

- Sealable plastic bags
- FerriDye solution: add, drop by drop, 10 ml of 0.5 M FeCl$_3$.6H$_2$O to 60 ml of boiling and stirring distilled water. Cool this to room temperature and add 1 vol. of it to 9 vol. of 0.1 M sodium cacodylate buffer pH 7.[b] Dilute this stock solution with sodium cacodylate (BDH) buffer to an OD$_{460\ nm}$ = 0.5. Add 1 ml 10% (v/v) Tween 20 per 49 ml of this diluted stock solution.

- Developing solution:[c] 1 vol. of 0.05 M K$_4$Fe(CN$_6$), 2 vol. of distilled water, 2 vol. of 1 M HCl

Method

1. Wash the protein blot with 200 ml of distilled water three times for 10 min each time in a large Petri dish.

2. Incubate the blot in 0.3 ml of FerriDye solution per square centimetre blotting membrane for 1 h in a sealed plastic bag.

3. Wash the blot thoroughly with 200 ml of distilled water, three times for 2 min each time.

4. Incubate the blot in developing solution at room temperature for 1 min.

5. Rinse the blot in a large excess of distilled water and allow to air dry.

[a] Adapted from ref. 100.
[b] Sodium cacodylate is toxic; use protective gloves and work in a fume-hood.
[c] Prepare solution in a fume-hood.

In another example, the proteins bound to the membrane are derivatized by pyridoxal 5'-phosphate (PLP) to produce 5'-phosphopyridoxyl (Ppxy) tagged proteins (103). The derivatized proteins are then incubated with an anti-Ppxy mouse monoclonal antibody and antibody–Ppxy derivatized protein complexes are detected using a peroxidase-linked goat anti-mouse F(ab')$_2$ antibody.

Immobilized proteins may also be detected by biotinylation followed by incubation with an avidin–horse-radish peroxidase conjugate (HRP) and an HRP chromogenic substrate. This procedure, which can be used with both nitrocellulose and nylon membranes, can detect as little as 10 ng of protein (104).

2.3 Detection of radioactively labelled proteins

Proteins labelled with radioactive isotopes that have been blotted to membranes can be detected by a number of methods. Direct autoradiography (see Section 1.7.1) will detect ^{35}S, ^{14}C, and even ^{32}P without the enhancement of fluorography as required for detection of ^{3}H on polyacrylamide gels (105). Nevertheless, the sensitivity of detection for radioactively labelled proteins can be augmented with fluorography (see Section 1.7.2). The degree of enhancement depends upon both the radioactive isotope and the fluorographic techniques utilized. For example, detection of ^{35}S- or ^{14}C-labelled proteins using PPO, an organic scintillator, improves sensitivity fourfold over autoradiography, while PPO increases ^{3}H detection 40-fold over autoradiography (105).

2.4 Detection of glycoproteins

Glycoproteins are normally detected by periodic acid–Schiff base (PAS) staining, but this is not a very efficient method for detection and the signal generated deteriorates with time. In addition, the procedure is time-consuming. Gershoni *et al.* (106) described a technique that combines the principles of periodate oxidation glycoprotein stains with enzymatic detection methods to stain glycoconjugates or selective glycoproteins. This enzyme–hydrazide technique (see *Protocol 12*) can be completed in less than 3 h and can detect nanograms of protein. Detection of specific glycoproteins can be achieved by utilizing specific enzyme oxidases (e.g. galactose oxidase) in the enzyme–hydrazide compounds. Gangliosides, lipopolysaccharides, and bacterial capsular polysaccharides can also be detected with this technique using 10 mM periodate (106).

Glycoproteins can also be radiolabelled for detection on membrane filters (89). Direct labelling of lectins (107) or the use of lectins then anti-lectin antibodies (108) on nitrocellulose also provide efficient means for the detection of glycoproteins. Moroi and Jung (107) described a technique for detection of platelet glycoproteins using peroxidase labelled lectins that improved upon methods previously introduced (109). Glass *et al.* (108) utilized anti-lectin antibodies to detect glycoproteins from cell fractions separated by PAGE and transferred to nitrocellulose. Glycoproteins were detected after incubation with anti-lectin antibodies and secondary detection using peroxidase–anti-peroxidase. Staining for peroxidase activity localized the lectin–antibody complexes. Comparison of glycoprotein content in various cell fractions can be performed using this technique.

2.5 Immunological detection of proteins

The development of protein immunoblotting techniques provides a powerful tool for the detection of specific proteins. Many of the antigenic properties of proteins are maintained following separation and blotting to membranes,

---

Protocol 12. Detection of glycoproteins using enzyme–
 hydrazides[a]

Reagents

- Enzyme–hydrazide: add 1 ml of 0.15 M NaCl to 0.45 ml of alkaline phosphatase (2500 U; 2.25 mg enzyme) and dialyse overnight against 0.15 M NaCl. Add 100 μl of 0.5 M adipic dihydrazide pH 5 to the dialysed enzyme. Add 0.15 M NaCl to bring the volume to 2 ml, followed by 0.5 ml of 0.1 M water soluble carbodiimide (WSC). Maintain pH 5 by adding 1 M HCl as needed. Incubate for 6 h at 25°C. Then dialyse first against 0.15 M NaCl, and then against 0.1 M Tris–HCl pH 8, 1 mM MgCl$_2$, 0.02% sodium azide.

- PBS (per litre): 0.144 g KH$_2$PO$_4$, 0.795 g Na$_2$HPO$_4$.7H$_2$O, 9 g NaCl pH 7.4 (e.g. from Boehringer)
- 2% (w/v) BSA in PBS
- 10 mM sodium periodate (made fresh each time)
- Alkaline phosphatase: 1 μg/ml enzyme
- Substrate solution: 10 mg naphthol–AS–MX phosphate (solubilized in 200 μl dimethylformamide) and 30 mg fast red in 100 ml of 100 mM Tris–HCl pH 8.4.

Method

1. Incubate the nitrocellulose or positively charged nylon membrane (i.e. Western blot) bearing the transferred glycoproteins in 2% (w/v) BSA for 1 h at 37°C.

2. Incubate the blot in 10 mM sodium periodate solution at room temperature for 30 min.

3. Rinse the blot briefly in PBS.

4. Incubate the blot for 1 h in 2% BSA in PBS containing the enzyme hydrazide.

5. Wash the blot twice for 10 min each time with PBS.

6. Incubate the blot with the substrate solution at room temperature. Glycoproteins will stain red within 30 min.

[a] Adapted from ref.106.

---

rendering antibodies an effective means for detection of specific proteins. As with any blotting procedure, the use of blocking agents is generally needed to saturate additional protein binding sites and so reduce non-specific binding. It is therefore important to avoid cross-reactivity of the primary or secondary antibodies with the blocking agent used.

Direct detection of proteins can be achieved by using antibodies labelled with radioactivity (91), fluorescence (91, 110, 111), enzymes (112, 113), or biotin (114). These labelling methods can also be used for indirect immuno-detection in which a primary antibody is directed against the desired protein and is then detected using a secondary, labelled antibody directed against the primary antibody. Alternatively, labelled *Staphyloccus aureus* protein A or protein G can be used in lieu of a secondary antibody.

Immunodetection can also increase the sensitivity of detection on blots after general protein stains. Thus staining of electroblots followed by immunodetection has been described for Coomassie blue (115), silver stains (24, 116), and colloidal metals (117, 118).

Additionally, Kaufmann *et al.* (119) have developed an efficient technique for the removal of primary and secondary antibodies without loss of immobilized proteins from nylon or nitrocellulose membranes, allowing for multiple probing of a single blot. Techniques have also been introduced for detection of different antigens on a single blot (120).

2.5.1 Antibody labelling

i. Radiolabelling

Antibodies are often radiolabelled for immunodetection purposes. Towbin *et al.* (91) initially described detection of blotted proteins by radioactively labelling antibodies using ^{125}I or ^{3}H. Typically, the antibody is labelled with ^{125}I via acylating reagents, i.e. Bolton–Hunter procedure (121), chemical oxidation of I⁻ with chloramine T (122), or enzymatic oxidation of I⁻ with lactoperoxidase (123).

ii. Fluorescent antibodies

The and Feltkamp (110) determined that a strong fluorescein/protein ratio could be established using relatively pure IgG conjugated to high quality fluorescein isothiocyanate (FITC). Towbin (91) demonstrated its use for detection of *E. coli* ribosomal proteins on nitrocellulose membranes by conjugating fluorescein to rabbit anti-goat IgG. MDPF, which has been demonstrated to be an effective fluorescent stain for proteins in gels, can be utilized in antibody labelling as well. With minimal background fluorescence, MDPF labelled proteins were stained with an intensity at least equivalent to that obtained with FITC labelled proteins (111).

iii. Enzyme-conjugated antibodies

Antibodies conjugated with alkaline phosphatase (AP) (112, 124), horseradish peroxidase (HRP) (125), and β-galactosidase from *E. coli* (β-gal) (126) are commonly used for visualization of proteins following blotting. Detection of a coloured precipitate formed by conversion of a soluble substrate by the enzyme localizes the specific proteins. Chromogenic reagents used for detection are 3,3'-diaminobenzidine (DAB), 4-chloro-1-naphthol, 3-amino-9-ethylcarbazole, 5-bromo-4-chloro-3-indolyl phosphate (BCIP), and nitroblue tetrazolium (NBT). Detection limits for HRP fall into the 100–500 pg range for proteins (94). AP can detect as little as 10–50 pg of protein when conjugated to BCIP or NBT (112). Note, however, that HRP labelled proteins will fade in colour when exposed to light, whereas AP labelled proteins are not light-sensitive.

Indirect peroxidase labelling involves detection of a primary antibody by a peroxidase labelled secondary antibody. A more sensitive technique is known as peroxidase–anti-peroxidase (PAP) labelling (127) (see *Figure 1*). In the PAP method, the secondary antibody is unlabelled and binds to the primary (protein-specific) antibody. However, it can also bind the peroxidase–antiperoxidase complex and this allows detection of the protein on the nitrocellulose.

iv. Biotin labelling
Biotin can also be coupled to peroxidase for protein detection on blotting matrices. A 68 kDa egg white glycoprotein, avidin, which can be labelled,

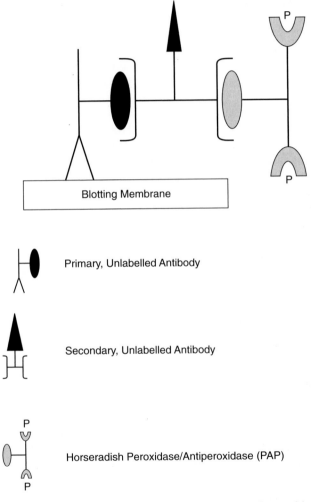

Blotting Membrane

Primary, Unlabelled Antibody

Secondary, Unlabelled Antibody

Horseradish Peroxidase/Antiperoxidase (PAP)

Figure 1. Detection method utilizing horse-radish peroxidase–anti-peroxidase (PAP) complexes. An unlabelled secondary antibody is used for detection of the primary antibody. The secondary antibody can then bind PAP to the immune complex of the primary antibody and the antigen on the nitrocellulose, providing greater levels of sensitivity than can be achieved using a peroxidase-labelled secondary antibody alone.

demonstrates a high affinity for the bound biotin, yielding a sensitive detection method. Biotin can also be bound by streptavidin, a slightly acidic non-glycosylated protein from *Streptomyces avidinii* (128). In this detection system, the secondary antibody is labelled with biotin and detected using peroxidase-conjugated streptavidin. This yields a lower background (decreased non-specific binding) when compared to the biotin–avidin detection.

2.5.2 Protein A and Protein G labelling

Staphyloccus aureus protein A is often used for secondary antibody labelling; it binds to the F_c region of various immunoglobulins. The affinity of protein A for the F_c portion of the primary antibody varies based on the species and antibody isotype (129). For instance, protein A binds well to rabbit IgG, but does not react well with murine or rat immunoglobulins (130). Protein A has four binding sites for antibodies, but only two of them can be used simultaneously (131).

Protein G, derived from group C or group G streptococci, is a protein with binding properties similar to protein A. However, protein G will bind IgG from species that protein A binds weakly, such as rat and mouse (132). Protein G and protein A binding exhibit different pH-dependence, suggesting that IgG binding occurs utilizing different molecular mechanisms for each of the proteins.

Protein A and protein G can be labelled with radioactivity or conjugated with enzymes such as alkaline phosphatase for visualization of proteins on a membrane. Briefly, an unlabelled antibody directed against the protein(s) of interest is added to the membrane and the immunocomplex formed is quantified by a subsequent incubation with labelled protein A or protein G. Alternatively, unlabelled protein A or protein G can be used to bridge two antibody molecules (i.e. the primary detection antibody from the immunocomplex and a labelled antibody) (131).

2.5.3 Immunogold staining

Colloidal metals have been used for immunodetection labelling. For example, Hsu demonstrated the use of gold particles conjugated to IgG for detection of proteins on nitrocellulose membranes (114). Detection here was achieved following dot blotting, in which proteins are directly applied to the nitrocellulose as spots. Antigen:antibody complexes were visualized as pink spots, and immunogold staining could detect proteins in the picogram range.

Surek and Latzko used a modified version of immunogold staining to demonstrate detection of proteins on nitrocellulose following SDS–PAGE (118). Colloidal gold particles were conjugated to *S. aureus* protein A and compared to methods of detection utilizing protein A–peroxidase techniques. The colloidal gold stain was at least twofold more sensitive than the protein A–peroxidase protocol. The gold stain was capable of detecting nanograms of protein as visualized by bright red bands on the nitrocellulose. Additionally, immunogold labelling complexes do not covalently bond the proteins, allowing the blots to be stripped and reused without reducing the antigenic properties of the protein for future probes.

2.5.4 Shift-Western assays

Interactions of proteins with DNA molecules have been studied extensively using mobility-shift (or gel-retardation) assays whereby binding of a protein to the DNA slows down the migration rate of the DNA band in a poly-acrylamide gel when compared to unbound DNA sequences. Differences in migration can be detected by labelling the DNA (133). Proteins can be isolated from specific protein:DNA complexes following gel electrophoresis and Western blotting using immunodetection methods (134).

Demczuk *et al.* (135) reported a technique known as Shift-Western blotting that can detect both the DNA and the protein from protein:DNA complexes separated by polyacrylamide gel electrophoresis by utilizing multiple membranes. Nitrocellulose, which will not bind double-stranded DNA, is used for immunodetection of the proteins, while the DNA is bound to an anion exchange membrane and detected by autoradiography. The protein blot can then be probed with specific antibodies, while the DNA blot is characterized using hybridization procedures.

3. Enzyme staining

Most of the techniques for detection of enzymatic activity in gels after electrophoretic separation are based on techniques originally described by histochemists. Electron transfer dyes, such as methyl thiazolyl tetrazolium which is reduced by electron donors to produce a dark blue insoluble formazan, are currently employed for most of the enzyme staining reactions. Detection of enzymes separated on gels requires limiting the diffusion of the enzyme and/or its reaction products for best resolution. In addition, it is important to perform pre-electrophoresis of the gels that are to be stained for enzyme activity to reduce the levels of unreacted monomers and persulfate ions that could otherwise inactivate the enzymes (143).

3.1 Categories of enzymes detected

Enzyme localization methods can be divided into categories based on functional enzyme properties. Slight alterations of these protocols can be used for the detection of enzymes exhibiting similar properties.

3.1.1 Oxidoreductases

Oxidoreductases, such as catalases, utilize NAD or NADP as a co-substrate. The gel is incubated in a tetrazolium salt solution that acts as a terminal electron acceptor and yields a coloured, insoluble formazan upon reduction. This protocol can be completed in a single incubation, although it must be conducted in the dark, since the tetrazolium salts are sensitive to light and oxygen (146). Commonly used tetrazolium salts include nitroblue tetrazolium (NBT) and methyl thiazolyl tetrazolium (MTT). Phenazine methosulfate (PMS) is occasionally utilized as a hydride ion carrier between the reduced enzyme and the tetrazolium salt. Controls, such as the omission of substrate, must be utilized with this protocol, since staining may occur in the absence of oxidoreductases (147).

3.1.2 Transferases and isomerases

The transfer of a group (such as methyl or glycosyl) from one compound to another is catalysed by transferases, such as glutathione *S*-transferase. Isomerases (i.e. triosephosphate isomerase) catalyse reactions that result in molecular rearrangements. Both classes of enzymes are detected by converting the primary reaction products to secondary product(s) in a coupling reaction. Formation of the secondary product is coupled to indicators, such as tetrazolium dyes, forming a coloured complex or a fluorogenic substance for easy visualization. However, the coupling enzymes penetrate polyacrylamide gels slowly, increasing the background staining. This can be prevented by sandwiching the polyacrylamide gel with agar or paper overlays containing the coupling enzymes. Ultrathin rehydratable gels have often been used for this purpose (148).

3.1.3 Hydrolases

In the presence of water, hydrolases catalyse cleavage of their substrate, and can be detected by fluorescence or chromogenic means. Detection can be accomplished using umbelliferyl (149, 150), phenolphthalein (151, 152), *p*-nitrophenyl (153), and other chromogenic derivatives. Specific enzymes that release inorganic phosphate, pyrophosphate, or carbon dioxide can also be detected by calcium ion precipitation (154).

3.2 Specific localization methods

Detection methods for the localization of specific enzymes in gels are almost as abundant as the enzymes themselves. Excellent reviews have been compiled containing reagent lists and protocols for the detection of enzymes separated on gels (136–138). A previous edition of this book (139) and general reviews on enzyme staining (140–142) provide additional information about specific enzyme detection methods. Appendix 2 of this book also includes references for specific enzyme detection methods. Recently, many enzyme localization techniques have been refined to decrease the use of potentially carcinogenic reagents. For example, eugenol (144) or tetrabase (145) can be used in place of benzidine and *o*-dianisidine in many protocols.

3.3 Protein renaturation assays

In 1973, Anfinsen received the Nobel Prize for his work showing that proteins could be renatured to regain their enzyme activities (155). Since then, techniques have been developed for the renaturation of proteins following electrophoresis, including high-resolution two-dimensional protein electrophoresis. Manrow and Dottin (156) described a technique for renaturation of proteins and detection of biological activity following the removal of urea or SDS. Scheele *et al.* (157) demonstrated the efficiency of a one-step elution and renaturation protocol for detection of enzymatic activity of human exocrine pancreatic proteins. However, a significant limitation is that multisubunit enzymes cannot be detected with these techniques.

4. Quantitative analysis of electrophoretically separated proteins

Quantitative intergel comparisons of proteins separated by protein electrophoresis present a number of problems. These problems may range from difficulties in sample handling during the loading of the gels, to protein loss during the transfer from the first to the second gel dimension in high-resolution two-dimensional electrophoresis, to variations in staining and/or variations in film development during autoradiography or fluorography in the case of radioactively labelled proteins. Proteins labelled with β-emitting isotopes ^{14}C, ^{35}S, or ^{3}H present special problems because autoradiographic detection results in a non-linear densitometric curves. However, radioactive step-tablets can be employed during the exposure of the film to calibrate the resulting autoradiogram (158–160). Radioactive step-tablets are calibrated wedges containing known concentrations of radioactivity increasing in stepwise fashion. Comparison of the autoradiographic images from radioactive step-tablets may be useful in comparing sample images made by varying the length of autoradiographic exposures, to obtain an expanded dynamic range. Photostimulated phosphorimaging systems also provide extended dynamic ranges for the detection of radioactive isotopes. Similarly, expanded ranges can be achieved for silver stained gels by sequentially photographing or scanning gels during the silver image development (161). It is difficult to incorporate protein concentration standards in the gel in a manner analogous to the use of radioactive step-tablets but it is possible to correct for variations introduced by scanning by including a standard calibrated photographic step wedge in the scanned image. Photographic step wedges contain standardized amounts of silver that are exposed to increasing amounts of light. The amount of silver that remains after light exposure can then be compared to the scanned image. The National Institute of Standards and Technology (NIST) has constructed a photographic step wedge containing known concentrations of silver, upon which other photographic step wedges are now calibrated. These step wedges are available commercially from photographic supply stores.

Scanning photographic step wedges and radioactive step-tablets cannot correct for other types of variations, such as variations in sample loading. To correct for these in 2D PAGE, Miller *et al.* (162) used a complex normalization function which is based on the use of multiplication factors to take into account the counts per minute per milligram of the total protein loaded on the gel, the specific activity of the incorporated label, and the number of counts found in each autoradiogram, as estimated by summing over all spots matched on each of the gels. Other researchers have utilized total gel densities, the sum of the densities of a group of well-defined spots present on each gel, a single well-defined middle-density spot, or spots showing low coefficients of variation in the database (163, 164). In addition some researchers normalize their gels by 'scaling' spots by fitting a linear, quadratic, or cubic function to correct the volumes of the spots on each gel being analysed (164).

Many of the normalization methods outlined above can be perturbed by the behaviour of abundant proteins, particularly if they saturate the detection system. A normalization algorithm has been developed which is based on known constitutive proteins (165). Unfortunately, given the current level of knowledge of the molecular genetics of most organisms, the number of proteins that fall in this class are very limited. However, the algorithm operates by identifying 'operationally constitutive proteins' by calculating the ratios of integrated densities for each possible pair of spots defined in each gel. These ratios are then compared across all gels and by iterative applications of the algorithm, and proteins that have minimal variations in their density ratios are defined as 'constitutive'. These 'operationally constitutive proteins' are then used for normalization between the gels in the study. In addition to its use as a normalization tool, this algorithm is useful as an aid in identifying proteins that may in fact be constitutive.

References

1. Porrett, R. (1816). *Ann. Philos.*, **July**, 78.
2. Abramson, H.A. (1934). In *Electrokinetic phenomena and their applications to biology and medicine*, p. 17. The Chemical Catalog Co. Inc., New York.
3. Picton, H. and Linder, S.E. (1892). *J. Chem. Soc.*, **61**, 148.
4. Davis, B.D. and Cohn, E.J. (1939). *Ann. N. Y. Acad. Sci.*, **39**, 209.
5. Tiselius, A. (1892). *Trans. Faraday Soc.*, **61**, 148.
6. Fazakas de St. Groth, S., Webster, R.G., and Datyner, A. (1963). *Biochim. Biophys. Acta*, **71**, 377.
7. Barger, B.O., White, F.C., Pace, J.L., Kemper, D.L., and Ragland, W.L. (1976). *Anal. Biochem.*, **70**, 327.
8. Talbot, D.N. and Yphantis, D.A. (1971). *Anal. Biochem.*, **44**, 246.
9. Merril, C.R., Switzer, R.C., and Van Keuren, M.L. (1979). *Proc. Natl. Acad. Sci. USA*, **76**, 4335.
10. Switzer, R.C., Merril, C.R., and Shifrin, S. (1979). *Anal. Biochem.*, **98**, 231.
11. Becquerel, A.H. (1896). *Comp. Rend. Acad. Sci. (Paris)*, **122**, 420.
12. Wilson, A.T. (1958). *Nature*, **182**, 524.

13. Kessler, C. (1992). In *Nonradioactive labeling and detection of biomolecules* (ed. C. Kessler), p. 2. Springer–Verlag, Berlin.
14. Frater, R. J. (1970). *Chromatography*, **50**, 469.
15. Radola, B.J. (1980). *Electrophoresis*, **1**, 43.
16. Wilson, C.M. (1979). *Anal. Biochem.*, **96**, 236.
17. Tal, M., Silberstein, A., and Nusser, E. (1985). *J. Biol. Chem.*, **260**, 9976.
18. Righetti, P.G. and Chillemi, F. (1978). *J. Chromatogr.*, **157**, 243.
19. Bickar, D. and Reid, P.D. (1992). *Anal. Biochem.*, **109**, 115.
20. Merril, C.R. (1987). *Adv. Electrophoresis*, **1**, 111.
21. Merril, C.R. and Goldman, D. (1984). In *Two-dimensional gel electrophoresis of proteins* (ed. J.E. Celis and R. Bravo), p. 93. Academic Press, New York.
22. Hochstrasser, D.F., Patchornik, A., and Merril, C.R. (1988). *Anal. Biochem.*, **173**, 412.
23. Nielsen, B.L. and Brown, L.R. (1984). *Anal. Biochem.*, **141**, 311.
24. Merril, C.R. and Pratt, M.E. (1986). *Anal. Biochem.*, **156**, 96.
25. Goldman, D., Merril, C.R., and Ebert, M.H. (1980). *Clin. Chem.*, **26**, 1317.
26. Sammons, D.W., Adams, L.D., and Nishizawa, E.E. (1981). *Electrophoresis*, **2**, 135.
27. Merril, C.R., Bisher, M.E., Harrington, M., and Steven, A.C. (1988). *Proc. Natl. Acad. Sci. USA*, **85**, 453.
28. Van Keuren, M.L., Goldman, D., and Merril, C.R. (1981). *Anal. Biochem.*, **116**, 248.
29. Merril, C.R., Goldman, D., Sedman, S.A., and Ebert, M.H. (1981). *Science*, **211**, 1437.
30. Merril, C.R., Harrington, M., and Alley, V. (1984). *Electrophoresis*, **5**, 289.
31. Lee, C., Levin, A., and Branton, D. (1987). *Anal. Biochem.*, **166**, 308.
32. Dzandu, J.K., Johnson, J.F., and Wise, G.F. (1988). *Anal. Biochem.*, **174**, 157.
33. Nells, L.P. and Bamburg, J.R. (1976). *Anal. Biochem.*, **73**, 522.
34. Higgins, R.C. and Dahmus, M.E. (1979). *Anal. Biochem.*, **93**, 257.
35. Candiano, G., Porotto, M., Lanciotti, M., and Ghiggeri, G.M. (1996). *Anal. Biochem.*, **243**, 245.
36. Pace, J.L., Kemper, D.L., and Ragland, W.L. (1974). *Biochem. Biophys. Res. Commun.*, **57**, 482.
37. Ragland, W.L., Pace, J.L., and Kemper, D.L. (1974). *Anal. Biochem.*, **59**, 24.
38. Jackowski, G. and Liew, C.C. (1980). *Anal. Biochem.*, **102**, 321.
39. Horowitz, P.M. and Bowman, S. (1987). *Anal. Biochem.*, **165**, 430.
40. Steinberg, T.H., Jones, L.J., Haugland, R.P., and Singer, V.L. (1996). *Anal. Biochem.*, **239**, 223.
41. Steinberg, T.H., Haughland, R.P., and Singer, V.L. (1996). *Anal. Biochem.*, **239**, 238.
42. Wallace, R.W., Yu, P.H., Dieckart, J.P., and Dieckart, J.W. (1974). *Anal. Biochem.*, **61**, 86.
43. Stoklosa, J.T. and Latz, H.W. (1974). *Biochem. Biophys. Res. Commun.*, **20**, 393.
44. Hahn, E.J. (1983). *Am. Lab.*, **15**, 64.
45. Bonner, W.M. (1984). In *Methods in enzymology* (ed. W.B. Jakoby), Vol. 104, p. 460. Academic Press, New York.
46. Bonner, W.M. (1983). In *Methods in enzymology* (ed. S. Fleischer and B. Fleischer), Vol. 96, p. 215. Academic Press, New York.

47. Andrews, A.T. (1986). In *Electrophoresis: theory, techniques and biochemical and clinical applications*, 2nd edn. Oxford University Press, Oxford.
48. Amemiya, Y. and Miyahura, J. (1988). *Nature*, **336**, 89.
49. Johnston, R.F., Pickett, S.C., and Barker, D.L. (1990). *Electrophoresis*, **11**, 355.
50. Laskey, R.A. (1980). In *Methods in enzymology* (ed. L. Grossman and K. Moldave), Vol. 65, p. 363. Academic Press, New York.
51. Swanstrom, R. and Shanks, P.R. (1978). *Anal. Biochem.*, **86**, 184.
52. Mees, K.C.E. (1952). In *The theory of the photographic process*, 1st edn, p. 563. Macmillan Press, New York.
53. Luthi, U. and Waser, P.G. (1965). *Nature*, **205**, 1190.
54. Randerath, K. (1970). *Anal. Biochem.*, **34**, 188.
55. Hamilton, J.F. and Logel, P.C. (1974). *Photogr. Sci. Eng.*, **18**, 507.
56. Kellogg, L.M. (1977). In *The theory of the photographic process* (ed. T.H. James), 4th edn, p. 404. Macmillan, New York.
57. Bonner, W.M. and Laskey, R.A. (1974). *Eur. J. Biochem.*, **64**, 147.
58. McConkey, E.H. (1979). *Anal. Biochem.*, **96**, 39.
59. Walton, K.E., Stryer, D., and Gruenstein, E. (1979). *J. Biol. Chem.*, **254**, 795.
60. Cooper, P.C. and Burgess, A.W. (1982). *Anal. Biochem.*, **126**, 301.
61. Kronenberg, L.H. (1979). *Anal. Biochem.*, **93**, 189.
62. Johnston, R.F., Pickett, S.C., and Barker, D.L. (1991). In *Methods: a companion to methods in enzymology*, Vol. 3, p. 128. Academic Press, New York.
63. Dzandu, J.K., Deh, M.E., Barratt, D.L., and Wise, G.E. (1984). *Proc. Natl. Acad. Sci. USA*, **81**, 1733.
64. Dzandu, J.K. (1989). *Appl. Theor. Electrophoresis*, **1**, 137.
65. Racusen, D. (1979). *Anal. Biochem.*, **99**, 474.
66. Gander, J.E. (1984). In *Methods in enzymology* (ed. W.B. Jakoby), Vol. 104, p. 447. Academic Press, New York.
67. Zaccharia, R.M., Zell, T.E., Morrison, J.H., and Woodlock, J.J. (1969). *Anal. Biochem.*, **30**, 148.
68. Wardi, A.H. and Miehos, G.A. (1972). *Anal. Biochem.*, **49**, 607.
69. Eckhardt, A.E., Hayes, C.E., and Goldstein, I.E. (1976). *Anal. Biochem.*, **73**, 192.
70. Burridge, K. (1978). In *Methods in enzymology* (ed. V. Ginsburg), Vol. 50, p. 54. Academic Press, New York.
71. Jackson, P. and Jackson, P. (1994). *Anal. Biochem.*, **216**, 243.
72. Jackson, P. (1996). *Mol. Biotechnol.*, **5**, 101.
73. Green, M.R., Pastewka, J.V., and Peacock, A.C. (1973). *Anal. Biochem.*, **56**, 43.
74. Saroh, K. and Busch, H. (1981). *Cell Biol. Int. Rep.*, **5**, 857.
75. Cutting, J.A. and Roth, T.F. (1973). *Anal. Biochem.*, **54**, 386.
76. Cutting, J.A. (1984). In *Methods in enzymology* (ed. W.B. Jakoby), Vol. 104, p. 451. Academic Press, New York.
77. Debruyne, I. (1983). *Anal. Biochem.*, **133**, 110.
78. Stumph, W.E., Elgin, S.C.R., and Hood, L. (1974). *J. Immunol.*, **113**, 1752.
79. Olden, K. and Yamada, K.M. (1977). *Anal. Biochem.*, **78**, 483.
80. Showe, M.K., Isobe, E., and Onorato, L. (1976). *J. Mol. Biol.*, **107**, 55.
81. Vining, R.F., Compton, P., and McGinley, R. (1981). *Clin. Chem.*, **27**, 910.
82. Miller, J.J. and Levinson, S.S. (1996). In *Immunoassay* (ed. E.P. Diamandis and T.K. Christopoulos), p. 165. Academic Press, New York.
83. Kuno, H. and Kihara, H.K. (1967). *Nature (London)*, **215**, 974.

84. Kakita, K., O'Connell, K., and Permutt, M.A. (1982). *Diabetes*, **31**, 648.
85. Kay, M.M.B., Goodman, S.R., Sorensen, K., Whitfield, C.F., Wong, P., Zaki, L., *et al.* (1983). *Proc. Natl. Acad. Sci. USA*, **80**, 1631.
86. Alwine, J.C., Kemp, D.J., Parker, B.A., Reiser, J., Renart, J., Stark, G.R., *et al.* (1979). In *Methods in enzymology* (ed. R. Wu), Vol. 68, p. 220. Academic Press, New York.
87. Reiser, J. and Wardale, J. (1981). *Eur. J. Biochem.*, **114**, 569.
88. Burnette, W.N. (1981). *Anal. Biochem.*, **112**, 195.
89. Gershoni, J.M. and Palade, G.E. (1982). *Anal. Biochem.*, **124**, 396.
90. Gershoni, J.M. and Palade, G.E. (1983). *Anal. Biochem.*, **131**, 1.
91. Towbin, H., Staehelin, T., and Gordon, J. (1979). *Proc. Natl. Acad. Sci. USA*, **76**, 4350.
92. Reinhart, M.P. and Malmud, D. (1982). *Anal. Biochem.*, **123**, 229.
93. Parchment, R.E., Ewing, C.M., and Shaper, J.H. (1986). *Anal. Biochem.*, **154**, 460.
94. Hames, B.D. (1990). In *Gel electrophoresis of proteins: a practical approach* (ed. B.D. Hames and D. Rickwood), 2nd edn, p. 85. IRL Press, Oxford.
95. Hancock, K. and Tsang, V.C.W. (1983). *Anal. Biochem.*, **133**, 157.
96. Moeremans, M., Daneels, G., and DeMey, J. (1985). *Anal. Biochem.*, **145**, 315.
97. Tsang, V.C.W., Hancock, K., Maddison, S.E., Beatty, A.L., and Moss, D.M. (1984). *J. Immunol.*, **132**, 2607.
98. Yuen, L.K.C., Johnson, T.K., Denell, R.E., and Consigli, R.A. (1982). *Anal. Biochem.*, **126**, 398.
99. Hancock, K. and Tsang, V.C.W. (1988). In *Handbook of immunoblotting of proteins* (ed. O.J. Bjerrum and N.H.H. Heegaard), Vol. 1, p. 127. CRC Press, Inc., Boca Raton, FL.
100. Moeremans, M., Daneels, G., De Raeymaeker, M., and De Mey, J. (1988). In *Handbook of immunoblotting of proteins* (ed. O.J. Bjerrum and N.H.H. Heegaard), Vol. 1, p. 137. CRC Press, Inc., Boca Raton, FL.
101. Moeremans, M., Daneels, G., and DeMey, J. (1986). *Anal. Biochem.*, **153**, 18.
102. Wojtkowiak, Z., Briggs, R.C., and Hnilica, L.S. (1983). *Anal. Biochem.*, **129**, 486.
103. Kittler, J.M., Meisler, N.T., Viceps-Madore, D., Cidlowski, J.A., and Thanassi, J.W. (1984). *Anal. Biochem.*, **137**, 210.
104. Bio Radiations. (1985). From Bio-Rad Laboratories, Ltd., No. 56 EG.
105. Roberts, P.L. (1985). *Anal. Biochem.*, **147**, 521.
106. Gershoni, J.M., Bayer, E.A., and Wilchek, M. (1985). *Anal. Biochem.*, **146**, 59.
107. Moroi, M. and Jung, S.M. (1984). *Biochim. Biophys. Acta*, **798**, 295.
108. Glass, W.F., Briggs, R.C., and Hnilica, L.S. (1981). *Anal. Biochem.*, **115**, 219.
109. Phillips, D.R. and Agin, P.P. (1977). *J. Biol. Chem.*, **252**, 2121.
110. The, T.H. and Feltkamp, T.E.W. (1970). *Immunology*, **18**, 865.
111. Weigele, M., De Bernado, S., Leimgruber, W., Cleeland, R., and Grunber, E. (1973). *Biochem. Biophys. Res. Commun.*, **54**, 899.
112. Blake, M.S., Johnson, K.H., Russel-Jones, G.J., and Gotschlich, E.C. (1984). *Anal. Biochem.*, **136**, 175.
113. Knecht, D.A. and Dimond, R.L. (1984). *Anal. Biochem.*, **136**, 180.
114. Hsu, S.M., Raine, L., and Fanger, H. (1981). *Am. J. Clin. Pathol.*, **75**, 816.
115. Jackson, P. and Thompson, R.J. (1984). *Electrophoresis*, **5**, 35.
116. Yuen, K.C.C., Johnson, T.K., Dennell, R.E., and Consigli, R.A. (1982). *Anal. Biochem.*, **126**, 398.

117. Hsu, Y. (1984). *Anal. Biochem.*, **142**, 221.
118. Surek, B. and Latzko, E. (1984). *Biochem. Biophys. Res. Commun.*, **121**, 284.
119. Kaufmann, S.H., Ewing, C.M., and Shaper, J.H. (1987). *Anal. Biochem.*, **161**, 89.
120. Steffen, W. and Linck, R.W. (1989). *Electrophoresis*, **10**, 714.
121. Bolton, A.E. and Hunter, W.M. (1973). *Biochem. J.*, **133**, 529.
122. McConahey, P.J. and Dixon, F.J. (1980). In *Methods in enzymology* (ed. H. Van Vunakis and J.L. Langone), Vol. 70, p. 210. Academic Press, New York.
123. Morrison, M. (1980). In *Methods in enzymology* (ed. H. Van Vunakis and J.L. Langone), Vol. 70, p. 214. Academic Press, New York.
124. Ishikawa, E., Imagawa, M., Hashida, S., Yoshitake, S., Hamaguchi, Y., and Ueno, T. (1983). *J. Immunoassay*, **4**, 209.
125. Wilson, M.B. and Nakane, P.K. (1978). In *Immunofluorescence and related staining techniques* (ed. W. Knapp, K. Holubar, and G. Wick), p. 215. Holland Biomedical Press, New York, Amsterdam.
126. Inoue, S., Hashida, S., Tanaka, K., Imagawa, M., and Ishikawa, E. (1985). *Anal. Lett.*, **18**, 1331.
127. Sternberger, L.A., Hardy, P.H., Jr., Cuculis, J.J., and Meyer, H.G. (1970). *J. Histochem. Cytochem.*, **18**, 315.
128. Chaiet, L. and Wolf, F.J. (1964). *Arch. Biochem. Biophys.*, **106**, 1.
129. Richman, D.G., Cleveland, P.H., Oxman, M.N., and Johnson, K.M. (1982). *J. Immunol.*, **128**, 2300.
130. Langone, J.J. (1982). In *Advances in immunology* (ed. D.J. Dixon and H.G. Kunkel), Vol. 32. Academic Press, NY.
131. Christopoulos, T.K. and Diamandis, E.P. (1996). In *Immunoassay* (ed. E.P. Diamandis and T.K. Christopoulos), p. 234. Academic Press, New York.
132. Åkerstrom, B. and Björck, L. (1986). *J. Biol. Chem.*, **261**, 10240.
133. Garner, M.M. and Revzin, A. (1981). *Nucleic Acids Res.*, **9**, 3047.
134. Granger-Schnarr, M., Lloubes, R., De Murcia, G., and Schnarr, M. (1988). *Anal. Biochem.*, **174**, 235.
135. Demczuk, S., Harbers, M., and Vennström, B. (1993). *Proc. Natl. Acad. Sci. USA*, **90**, 2574.
136. Shaw. C.R. and Prasad, R. (1980). *Biochem.Genet.*, **4**, 297.
137. Siciliano, M.J. and Shaw, C.R. (1976). In *Chromatographic and electrophoretic techniques* (ed. I. Smith), Vol. 2, p. 185. Heinemann, London.
138. Harris, H. and Hopkinson, D.A. (1976). *Handbook of enzyme electrophoresis in human genetics*. North Holland Publishers, Amsterdam.
139. Hames, B.D. and Rickwood, D. (1990). *Gel electrophoresis of nucleic acids: a practical approach*. IRL Press, Oxford.
140. Ostrowski, W. (1983). *J. Chromatogr. Libr.*, **18B**, 287.
141. Heeb, M.J. and Gabriel, O. (1984). In *Methods in enzymology* (ed. W.B. Jakoby), Vol. 104, p. 416. Academic Press, New York.
142. Gabriel, O. (1971). In *Methods in enzymology* (ed. W.B. Jakoby), Vol. 22, p. 578. Academic Press, New York.
143. Bennick, A. (1968). *Anal. Biochem.*, **26**, 453.
144. Liu, E.H. and Gibson, D.M. (1977). *Anal. Biochem.*, **79**, 597.
145. Lomholt, B. (1975). *Anal. Biochem.*, **65**, 569.
146. Worsfold, W., Marshall, M.J., and Ellis, E.B. (1977). *Anal. Biochem.*, **79**, 152.

147. O'Conner, J.L., Edwards, D.P., and Bransome, E.D. (1977). *Anal. Biochem.*, **78**, 205.

148. Hofemann, M., Kittsteiner-Eberle, R., and Schrier, P. (1983). *Anal. Biochem.*, **128**, 217.

149. Chang, P.L., Ballantyne, S.R., and Davidson, R.G. (1979). *Anal. Biochem.*, **97**, 36.

150. Coates, P.M., Mestriner, M.A., and Hopkinson, D.A. (1975). *Ann. Hum. Genet.*, **39**, 1.

151. Huggins, C. and Talalay, P. (1945). *J. Biol. Chem.*, **159**, 339.

152. Hopkinson, D.A., Spencer, N., and Harris, H. (1964). *Am. J. Hum. Genet.*, **16**, 141.

153. Hodes, M.E., Crisp, M., and Gelb, E. (1977). *Anal. Biochem.*, **80**, 239.

154. Nimmo, H.G. and Nimmo, G.A. (1982). *Anal. Biochem.*, **121**, 17.

155. Anfinsen, C.B. (1973). *Science*, **181**, 223.

156. Manrow, R.E. and Dottin, R.P. (1980). *Proc. Natl. Acad. Sci. USA*, **77**, 730.

157. Scheele, G., Pash, J., and Bieger, W. (1981). *Anal. Biochem.*, **112**, 304.

158. Anderson, N.L., Taylor, J., Scandora, A.E., Coulter, B.P., and Anderson, N.G. (1981). *Clin. Chem.*, **27**, 1807.

159. Garrels, J.I. (1983). In *Methods in enzymology* (ed. R. Wu, L. Grossman, and K. Moldave) Vol. 100, p. 411. Academic Press, New York.

160. Olson, A.D. and Miller, M.J. (1988). *Anal. Biochem.*, **169**, 49.

161. Merril, C.R., Goldman, D., and Van Keuren, M.L. (1982). *Electrophoresis*, **3**, 17.

162. Miller, M.J., Vo, P.K., Nielsen, C., Geiduschek, E.P., and Xuong, N.H. (1982). *Clin. Chem.*, **28**, 867.

163. Lester, E.P., Lemkin, P., and Lipkin, L. (1982). *Clin. Chem.*, **28**, 828.

164. Taylor, J., Anderson, N.L., Scandora, A.E., Jr., Willard, K.E., and Anderson, N.G. (1982). *Clin. Chem.*, **28**, 861.

165. Merril, C.R., Creed, G.J., Joy, J., and Olson, A.D. (1993). *Appl. Theor. Electrophoresis*, **3**, 329.

166. Wilson, C.M. (1983). In *Methods in enzymology* (ed. C.H.W. Hirs), Vol. 91, p. 236. Academic Press, New York.

167. Ragland, W.L., Benton, T.L., Pace, J.L., Beach, F.G., and Wade, A.E. (1978). In *Electrophoresis '78* (ed. N. Catsimpoolas), Vol. 2, p. 217. Elsevier, North Holland, Amsterdam.

168. Schaffner, W. and Wiessmann, C. (1973). *Anal. Biochem.*, **56**, 502.

169. Aebersold, R., Leavitt, J., Saavedra, R., Hood, L., and Kent, S. (1987). *Proc. Natl. Acad. Sci. USA*, **84**, 6970.

170. Christiansen, J. and Houen, G. (1992). *Electrophoresis*, **13**, 179.

3

Preparative gel electrophoresis

KELVIN H. LEE and MICHAEL G. HARRINGTON

1. Introduction

Protein analysis has long served an important role in the biological and biomedical sciences. The most commonly used methods for separating protein mixtures are chromatography and gel electrophoresis. The essence of these two methods are the same because, in both, a sample mixture is separated into components by various molecular interactions while passing in a 'liquid phase' though a 'solid phase' matrix. The difference lies in the force used to drive the liquid phase through the solid phase: pressure in chromatography and an electric field in electrophoresis. Among its many applications, analytical protein electrophoresis can be used to determine biochemical characteristics such as the isoelectric point and mass of a protein, to catalogue protein content in complex mixtures with two-dimensional electrophoresis, and to identify the relative quantity of a particular protein in a mixture.

There are several preparative approaches available to the protein biochemist. Among the methods which can prepare larger quantities of material are different chromatographic techniques, electrophoresis, and immunoprecipitation. Chromatographic preparation of proteins is often used when a specific affinity interaction is available for purification or when a continuous system is preferred. Because chromatography and electrophoresis are technologies with different advantages and disadvantages, it is important for the reader to use preparative gel electrophoresis in the context of the overall experimental agenda. For example, many of the electrophoretic purification techniques rely on the use of urea, ionic (e.g. SDS), or non-ionic (e.g. Nonidet P-40) detergents which can result in a loss of enzymatic activity. Typical preparative gel electrophoresis experiments will involve at least 1 mg of protein for 'micropreparative' studies and up to several grams of material in liquid-based preparative separations.

It should be noted that there are tradeoffs between liquid- and gel-type separations. Liquid-type procedures, such as those used in chromatography or liquid isoelectric focusing, give protein yields superior to methods that require recovery from gel matrices (such as SDS–PAGE). However, separations performed in gel matrices have greater resolving capabilities than liquid phase separations. An emerging technology which offers both excellent recovery as well as resolution is capillary electrophoresis. While capillary electrophoresis separations are not performed at the true preparative scale, they can be used as a precursor to some biochemical characterization techniques such as mass spectrometry.

Preparative scale (> 1 mg protein) gel electrophoresis has been used to obtain sufficient quantities of material for further biochemical or functional studies. In this role, preparative gel electrophoresis is often a precursor to other analytical techniques such as amino acid composition analysis, amino acid sequence analysis, or protease digestion, followed by peptide mass finger-printing. Because of recent developments in the arena of genomics and in-formation technology, the ability to obtain significant quantities of pure proteins from complex mixtures by preparative electrophoresis has become a tool complementary to chromatography for characterizing both individual proteins and complex biological systems alike. The preparative electrophoresis step is typically performed to separate a component from a mixture of other proteins or to clean up a sample which contains trace amounts of contamin-ants in order to perform functional or structural studies with the purified molecule.

The three most widely used analytical electrophoresis techniques—isoelectric focusing (IEF), sodium dodecyl sulfate–polyacrylamide gel electrophoresis (SDS–PAGE), and high-resolution two-dimensional electrophoresis (2DE)— can all be scaled-up or otherwise adapted to the preparative level. For IEF and SDS–PAGE, preparative scale experiments typically involve the purifica-tion of at least milligram quantities of protein and often quantities on the order of several grams. For 2DE, the quantities are much smaller due to limi-tations in current technology. Here, up to milligram quantities of protein can be loaded onto gels in order to obtain suitably resolved material for amino acid sequencing and other microchemical characterization techniques. Because these three methods are the most commonly used and because commercial instruments are available for preparative IEF and preparative SDS–PAGE, this chapter covers each of these preparative techniques individually. Differ-ent protocols and options are available for each of these separation methods and are described in the following sections. More general aspects of these techniques are covered in other sections of this book.

There are several practical aspects that can significantly influence the suc-cessful separation and recovery of desired proteins (particularly if they are to be further analysed by amino acid sequencing, mass spectrometry, etc.). In particular, the use of high quality reagents is *essential* for optimal results and this includes a good source of deionized water. Moreover, contamination by extraneous proteins should be minimized by the use of disposable gloves and laboratory coats.

2. Preparative isoelectric focusing

2.1 Technology options—immobilized pH gradient versus carrier ampholyte isoelectric focusing

Preparative scale isoelectric focusing can be simply and reliably performed by taking advantage of several commercially available instruments. As in analytical isoelectric focusing, it is natural to categorize these instruments based on the technology for forming the pH gradient: immobilized pH gradient or carrier ampholyte isoelectric focusing. Hoefer–Pharmacia–Amersham sells the IsoPrime unit for preparative scale isoelectric focusing (1). PrIME (preparative isoelectric membrane electrophoresis) technology is based on Pharmacia Immobiline technology and uses acrylamido buffers which are covalently bound to membranes. Several membranes with different pH characteristics are used to create a pH gradient wherein proteins can migrate (across membranes) to the appropriate isoelectric point in a liquid matrix. At equilibrium, a protein is in a chamber bounded by two membranes of different pH values and can be retrieved into a reservoir. The IsoPrime technology does not require the use of carrier ampholytes and can separate proteins which differ by fractions of a pH unit in isoelectric point. It is useful for purifying from hundreds of micrograms to hundreds of milligrams of a single protein in a single experiment.

Bio-Rad Laboratories sells the Rotofor for preparative scale isoelectric focusing (see *Figure 1*). Like the Hoefer–Pharmacia–Amersham system, the Rotofor is also a liquid-based system; however, the Rotofor is based on

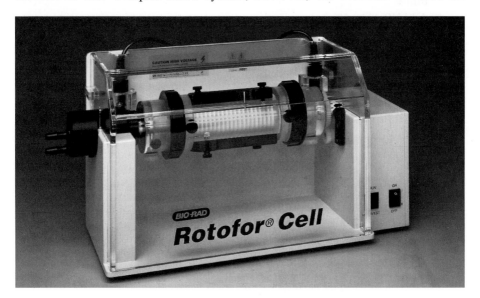

Figure 1. The Bio-Rad Rotofor for preparative isoelectric focusing.

carrier ampholyte technology. The separation range can be tuned by the selection of different ampholytic ranges. The instrument has interchangeable focusing chambers (mini and standard) which can be used to focus from microgram to gram quantities of protein and 500-fold purifications can be obtained in three hours by collecting one of the 20 focused zones.

2.2 Preparative liquid isoelectric focusing

Both the Rotofor and the IsoPrime systems are suitable for general use and the sample preparation techniques are similar for both instruments. *Protocol 1* has been optimized for use with the Rotofor.

Protocol 1. Preparative isoelectric focusing using the Rotofor

Equipment and reagents
- Rotofor system with standard accessories (Bio-Rad)
- pH 3–10 carrier ampholytes (Bio-Rad)
- 0.1 M NaOH
- 0.1 M H$_3$PO$_4$
- NP-40 (Nonidet P-40)
- 2% CHAPS [3-([3-cholamidopropyl)dimethyl-ammonio]-1-propanesulfonate], 2% 2-mercaptoethanol
- Urea

Method
1. Prepare the protein sample for isoelectric focusing by sonicating, extracting, precipitating, etc. as required by standard procedures.
2. Dissolve the protein sample in 2% CHAPS, 2% 2-mercaptoethanol.
3. Heat the sample to 95°C for 5 min and then let it cool to room temperature.
4. Add NP-40 to 1%.[a]
5. Pellet the insoluble debris and discard. Add urea to 8 M final concentration to the supernatant.[b]
6. Add pH 3–10 carrier ampholytes to 1% to the supernatant.[c]
7. Cool the Rotofor apparatus to 4°C and load the sample. Use 0.1 M H$_3$PO$_4$ and 0.1 M NaOH as electrolyte solutions.
8. Run the apparatus at 10 W (constant power) until focusing is achieved.[d]
9. Collect the fractions and analyse them.[e]
10. The sample should be run through the Rotofor again for maximum resolving capability. The carrier ampholytes present in the collected fraction can be used to further separate and resolve the desired product.

[a] Up to 0.2% SDS may also be added to help solubilize protein.
[b] The approximate total volume at this point may be 50–60 ml.
[c] The selection of carrier ampholyte pH range and their concentration will depend on the nature of the sample and the isoelectric point of the protein of interest. This is freely variable and should be adjusted as appropriate.
[d] The current will drop and become stable after approx. 3 h.
[e] The samples may be analysed by protein assay, functional assay, or other simple electrophoresis technique (e.g. the Phast system from Hoefer–Pharmacia–Amersham).

---

3. Preparative SDS–PAGE

3.1 Technology options—the Bio-Rad Prep Cell versus scale-up

As with preparative isoelectric focusing, there are two main options for performing SDS–PAGE at the preparative level. The most direct approach is to scale-up the analytical SDS–PAGE procedures. One can simply load milligram quantities of protein and perform SDS–PAGE in large format (e.g. 16 × 20 cm) gels. With partially purified material even larger quantities of protein can be analysed and purified. The drawback with this simple approach is a significant loss in resolving power which may result in insufficiently purified material.

An alternate and more robust system for preparative SDS–PAGE is the Model 491 Prep Cell from Bio-Rad. This apparatus can tolerate a sample capacity up to 0.5 g of protein in volumes from 500 μl to 15 ml. The procedure is based on continuous elution electrophoresis (*Figure 2*). Samples are loaded onto an upper gel (polyacrylamide) surface and the molecules are electrophoresed through glass tubes (inner diameter can be 28 mm or 37 mm and the length is 14 cm). As the bands migrate off the bottom of the gel, they pass directly into an elution frit and a dialysis membrane traps these bands in place. Elution buffer flows from an elution chamber and as the individual bands migrate off the bottom of the gel, they are drawn to the centre of the elution frit. A peristaltic pump drives the collection of individual bands in elution buffer, up the centre of the frit and into a fraction collector. The overall temperature at which the separation takes place can be controlled by an external recirculating pump. With this apparatus, proteins differing by as little as 2% in molecular weight can be separated in approximately six hours.

3.2 Preparative SDS–PAGE using the Bio-Rad Prep Cell

The detailed procedure is described in *Protocol 2*.

Protocol 2. Preparative SDS–PAGE

Equipment and reagents

- Model 491 Prep Cell plus associated equipment (Bio-Rad)
- Reagents for polymerizing polyacrylamide gels
- 10 × running buffer: mix 30.3 g Tris base, 144 g glycine, 10 g SDS, and add deionized water to a volume of 1 litre

- 2 × SDS–PAGE sample buffer (8 ml): mix 3.8 ml deionized water, 1.6 ml 10% (w/v) SDS, 1 ml 0.5 M Tris–HCl pH 6.8, 0.8 ml glycerol, and 0.4 ml 0.5% bromophenol blue—just prior to use, add 400 μl 2-mercaptoethanol

Method

1. Prepare the proteins for SDS–PAGE by sonicating, extracting, precipitating, etc. as required by standard procedures.

Protocol 2. *Continued*

2. Mix the protein sample with an equal volume of 2 × SDS–PAGE sample buffer.

3. Heat the sample to 95°C for 5 min and then let it cool to room temperature.

4. Prepare 4%T stacking/12%T separating polyacrylamide gels (adjust the pore size as appropriate for the desired experiment) in either 28 mm diameter tubes or 37 mm tubes as dictated by the amount of protein to be loaded.

5. Assemble the apparatus according to the manufacturer's instructions.

6. Dilute the 10 × running buffer to a working concentration using deionized water.

7. Load the sample and set the elution rate for 1 ml/min.

8. Run at 12 W constant power. The elution time for the dye front will be between 2–3 h for the smaller (28 mm) tubes and 3–5 h for the larger (37 mm) tubes. The proteins will elute after the dye front.

9. Collect the fractions and analyse them by protein assay, functional assay, or other simple electrophoresis technique (e.g. on the Phast system from Hoefer–Pharmacia–Amersham).

4. Micropreparative two-dimensional electrophoresis

The main limitation in performing micropreparative two-dimensional electrophoresis (2DE) is that the isoelectric focusing gels can tolerate only a limited protein load. The low percentage polyacrylamide tube gels often used in carrier ampholyte first-dimension separations can crumble under higher protein loads (close to 1 mg). This leads to a loss of protein and poor reproducibility. In addition, larger amounts of protein disturb carrier ampholyte pH gradients to an unacceptable level. On the other hand, immobilized pH gradient gels (where gradients are not disturbed by large quantities of protein) may suffer from poor sample solubility at protein loads much higher than 1 mg. Nevertheless, substantially higher protein loads can be used in immobilized pH gradient isoelectric focusing after optimizing the sample preparation and application details of the procedure.

Often the most direct approach for obtaining suitable quantities of material from a 2DE separation is to load the maximum protein tolerable by the protocol of choice. The two standard first-dimension technologies for use in 2DE are tube gel carrier ampholyte-based separations and immobilized pH gradient-based separations. The former has the advantage of better sample entry into the gel while the second, by virtue of a support medium (Pharmacia GelBond) for the polyacrylamide gel, offers better physical and pH gradient

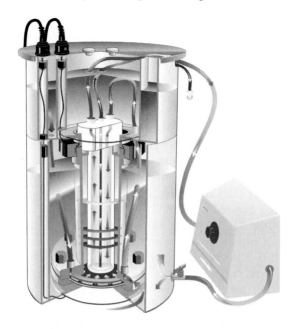

(a)

(b)

Figure 2. (a) Schematic depiction of the use of the Bio-Rad Model 491 Prep Cell for preparative electrophoresis. (b) The Model 491 Prep Cell.

stability and can tolerate higher proteins loads without gel crumbling. The use of a narrower range of ampholytic species during carrier ampholyte isoelectric focusing or immobilized pH gradient isoelectric focusing enables those proteins that are not in the desired pH range to migrate off the gels and into the electrolyte solutions. This is a useful approach since it allows the protein load to be increased significantly during isoelectric focusing of micropreparative two-dimensional electrophoresis.

The ideal isoelectric focusing medium for use in micropreparative 2DE would be chemically and physically stable, capable of tolerating high protein loads, and would be reusable. While no such medium has found widespread use thus far, several examples offer exciting potential toward this end (2–4).

Another tactic for alleviating this sample load bottleneck is to pre-enrich the sample for the desired protein. This can be done in any of a number of ways: by ammonium sulfate precipitation, by filtering using an Amicon filter, by performing an initial chromatography step, or by performing an initial electrophoresis purification, among others. The choice of method is largely up to the researcher and depends on the characteristics of the protein of interest.

Micropreparative scale isoelectric focusing can also be performed using Immobiline technology. Hochstrasser and colleagues (5) have described some simple modifications to the Immobiline dry strips which are often used in 2DE. The modifications which include the use of a larger sample cup and a larger gel surface area at the basic end of the pH gradient, allow the loading of up to 1 ml of sample solution. Another modification to the immobilized pH gradient experimental protocol to increase the protein loading capacity involves reswelling the gels in the standard buffer solution and the sample simultaneously (6). This change in the sample application mode for isoelectric focusing enables micropreparative two-dimensional electrophoresis to tolerate up to 5 mg of protein in 500 µl solution while maintaining resolution at a level compatible with most applications.

Because many different laboratories use micropreparative 2DE and the techniques and technology are ever-evolving, no one protocol nor approach can provide all users with the desired results. In general, however, using a pre-enrichment step and focusing in narrower pH ranges are often worthwhile approaches. The reader is encouraged to investigate refs 5–8 for helpful ideas.

5. Protein recovery from gels

Protein recovery is the final, and perhaps most important step. From one-dimensional gels, the protein of interest is most easily recovered by excising the relevant region from an unstained gel by running with appropriate prestained markers. For two-dimensional gels, the protein is most simply recovered from blotted membranes. These two alternatives, gel elution and blotting, are the two main options available for protein recovery. Recent developments (9) in protein sequencing protocols allow the sequencing of proteins directly from the blots. Some devices, such as the Bio-Rad Model 491 Prep Cell, have liquid fraction collector methodologies inherent in the instrument and will not be discussed further.

5.1 Elution of proteins from blotting membranes

Electrotransfer of proteins onto blotting membranes (e.g. nitrocellulose or polyvinylidene fluoride—PVDF) has become routine in most laboratories. Protein recovery from the membranes is sometimes preferred over gel elution because contaminant salts and detergents can be easily removed by membrane washing. Elution of the protein from the membrane is described in *Protocol 3*. The composition of the elution buffer is chosen to counteract the binding of the protein to the membrane and this is sequence- and membrane-dependent. Recovery is often erratic and poor for proteins, but is improved for small peptides, such as those derived from partially digested proteins. The elution buffer may be volatile or non-volatile (see *Procotol 3*). A volatile buffer has the advantage of being easily removed by evaporation without the a resulting concentration of residual buffer salts. This is often important in the recovery of dilute proteins which are to be subjected to analysis requiring small volumes and low salt.

Protocol 3. Protein recovery from nitrocellulose/PVDF membranes

Equipment and reagents

- Standard blotting apparatus (e.g. the Bio-Rad Transblot cell or a semi-dry apparatus such as is available from Bio-Rad) and associated equipment[a]
- PVDF or nitrocellulose membrane[b]
- Blotting buffer: 192 mM glycine, 20% methanol, 25 mM Tris base in deionized water.

- Ponceau S staining solution: 0.1% (w/v) Ponceau S in 1% (v/v) acetic acid
- Fresh elution buffer: non-volatile elution buffer (e.g. 50 mM Tris pH 9, 2% (w/v) SDS, 1% (v/v) Triton X-100) (10), or a volatile elution buffer (30% trifluoroacetic acid and 40% acetonitrile) (11).

Method

1. After running the gel, blot the proteins to the desired membrane using blotting buffer by following the manufacturer's instructions.[c]

2. Stain the membrane with Ponceau S staining solution for 30 sec and then destain it in water for 2 sec.

3. Excise the smallest area of the membrane corresponding to the protein of interest and place it in an Eppendorf tube.[d]

4. Add elution buffer in the smallest volume sufficient to just cover the entire membrane fragment (20–50 μl).

5. Rotate the tube at room temperature for 1 h.

6. Centrifuge at 10 000 *g* for 10 min. Remove and collect the supernatant.

Protocol 3. *Continued*

7. Rinse the tube and the membrane with the smallest volume of fresh elution buffer sufficient to just cover the entire membrane fragment (20–50 μl).

8. Centrifuge at 10000 *g* for 5 min and collect the remaining supernatant.

9. Pool the supernatants from steps 6 and 8 and analyse.

[a] It has been our experience that substantially greater recovery of sequenceable protein is obtained when gel separated protein is transferred using the wet rather than the semi-dry apparatus.

[b] We prefer the use of nitrocellulose membrane for greater recovery. Generally it has been found that elution is more efficient from nitrocellulose than PVDF membranes and that smaller proteins elute faster than large ones.

[c] When using PVDF membrane, the membrane must be pre-wet with 100% methanol and soaked in blotting buffer before use.

[d] At this stage, proteins can be digested on the membrane (12) with an enzyme (commonly Lys–C or trypsin). Protein digestion allows for internal sequencing of a protein when the protein is N-terminally blocked thus preventing standard N-terminal sequencing.

5.2 Elution of proteins from gels

For one-dimensional gels, proteins are often eluted directly from the gels, without membrane blotting. This is effectively achieved with the aid of commercially available instruments such as the Model 422 Electroeluter from Bio-Rad (*Figure 3*) or the Schleicher and Schuell Elutrap (11). Either instrument

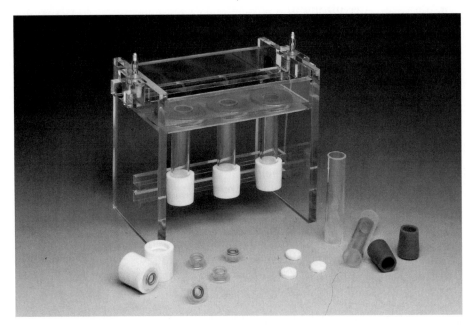

Figure 3. The Bio-Rad Model 422 Electroeluter for elution of proteins from electrophoresis gels.

can elute proteins from single or multiple gel pieces and can elute several different samples in parallel. The gel slices are held in place between porous membranes and the proteins are electrophoresed out of the gel slice, through the frit/membrane, and collected in a membrane cap. The final volume for the eluted sample is approximately 500 μl. *Protocol 4* describes the use of the Bio-Rad electroeluter but it can also be easily adapted for use with the Elutrap.

Protocol 4. Elution from gels using the Bio-Rad Model 422 Electroeluter

Equipment and reagents
- Bio-Rad Model 422 Electroeluter and associated equipment
- Fresh elution buffer: 0.1% SDS, 50 mM NH_4HCO_3 pH 8.5

Method
1. Assemble the unit (*Figure 3*) prior to use according to the manufacturer's instructions. The membrane cap to be used should be soaked in elution buffer for 1 h.
2. Add elution buffer to the lower reservoir to cover the silicon adapter comfortably.
3. Stir vigorously with a magnetic stirrer to prevent bubbles from sticking to the dialysis membrane.
4. Cut the gel containing the protein of interest into 3 mm^2 pieces and place them in each tube. Do not fill each tube past the half-way point.
5. Cover the tubes using elution buffer and carry out the elution at 10 mA/tube for 4 h.
6. After the elution, remove the upper (elution) buffer and aspirate it from each glass tube down to the level of the frit.
7. Remove the adapter and cap and carefully recover the eluate by aspiration.
8. Rinse the cap with 200 μl of fresh elution buffer and add this to the eluate from step 7.
9. Analyse the eluate.

---

6. Conclusions

While preparative scale electrophoresis experiments do not have the same resolving capabilities as analytical scale electrophoresis experiments, preparative gel electrophoresis remains a powerful tool in protein analysis because it generally offers greater resolving power than other methods for the purification of specific proteins. Moreover, ongoing developments in electrophoresis technology (instrumentation, computer control, gel media, and sample preparation) are increasing the efficiency, throughput, and quantities of material which can be recovered while decreasing the separation times and cost. Further refinement of these associated technologies will inevitably result in more automated technology for preparative gel electrophoresis in the future.

References

1. Righetti, P.G., Wenisch, E., and Faupel, M. (1989). *J. Chromatogr.*, **475**, 293.
2. Harrington, M.G., Lee, K.H., Bailey, J.E., and Hood, L.E. (1994). *Electrophoresis*, **15**, 187.
3. Zewert, T.E. and Harrington, M.G. (1992). *Electrophoresis*, **13**, 817.
4. Zewert, T.E. and Harrington, M.G. (1992). *Electrophoresis*, **13**, 824.
5. Bjellqvist, B., Sanchez, J.C., Pasquali, C., Ravier, F., Paquet, N., Frutiger, S., *et al.* (1993). *Electrophoresis*, **14**, 1375.
6. Rabilloud, T., Valette, C., and Lawrence, J.J. (1994). *Electrophoresis*, **15**, 1552.
7. Harrington, M.G., Gudeman, D., Zewert, T., Yun, M., and Hood, L. (1991). *Methods: a companion to methods in enzymology*, **3**, 98.
8. Hanash, S.M., Strahler, J.R., Neel, J.V., Hailat, N., Malhem, R., Keim, D., *et al.* (1991). *Proc. Natl. Acad. Sci. USA*, **88**, 5709.
9. Patterson, S.D. (1994). *Anal. Biochem.*, **221**, 1.
10. Szewczyk, B. and Summers, D.F. (1988). *Anal. Biochem.*, **168**, 48.
11. Harrington, M.G. (1990). In *Methods in enzymology* (ed. M. Deutscher), Vol. 182, p. 488. Academic Press, San Diego.
12. Aebersold, R., Leavitt, J., Saavedra, R.A., Hood, L.E., and Kent, S.B.H. (1987). *Proc. Natl. Acad. Sci. USA*, **84**, 6970.

4

Polymer solution mediated size separation of proteins by capillary SDS electrophoresis

ANDRÁS GUTTMAN, PAUL SHIEH, and BARRY L. KARGER

1. Introduction

Researchers in the late 1960s found that the molecular weight of protein molecules can be easily derived from their electrophoretic mobility, measured by SDS–PAGE (1). The methodology was later significantly improved (2) and progressed as a standard molecular weight determination method for proteins in slab or rod gel electrophoresis methods (3, 4). The gel acts both as a molecular sieve, enabling separations based on the size (hydrodynamic volume) of the SDS:protein molecules, and as an anti-convective medium, minimizing the conductive transport and reducing diffusion of the analyte molecules during the electrophoretic separation. Classical slab and rod gel electrophoresis is still performed manually, requiring gel casting, manual sampling, and laborious staining/destaining procedures for detection of the separated proteins.

Capillary electrophoresis is rapidly becoming an important separation technique in analytical biochemistry and molecular biology (5–7). Using narrow bore fused silica capillaries filled with cross-linked gels or linear polymer networks, extremely high resolving power can be achieved in size separation of biopolymers (8). Polymer solution-mediated capillary electrophoresis offers the ability of multiple injections onto the same polymer solution-filled capillary column, on-line detection, and automation (9). There has been a considerable activity in the separation and characterization of protein and peptide molecules by capillary SDS electrophoresis, especially using replaceable polymer solutions (10, 11). Using this technique, one can rapidly define molecular mass, as well as protein purity. Early papers on the use of capillary gel electrophoresis for protein separations were published a decade ago by Hjerten (12) under non-dissociating conditions and Karger and co-workers (13, 14) under dissociating conditions (SDS). In this chapter we distinguish gels from entangled polymer solutions. The former is a viscoelastic medium with significantly greater rigidity than that of the latter.

2. Fundamentals of capillary gel electrophoresis

When a uniform electric field (E) is applied to an SDS:protein polyion with a net charge of Q, the electrical force (F_e) is given by:

$$F_e = QE. \tag{1}$$

In cross-linked polyacrylamide gel or entangled polymer solutions, the applied electric field results in electrophoretic migration, while a frictional force (F_f) acts in the opposite direction:

$$F_f = f(dx/dt) \tag{2}$$

where f is the translational friction coefficient and dx and dt are the distance and time increments, respectively. Differences in the size, shape, and overall charge of the solute molecules results in variances in electrophoretic mobilities, providing the basis of the electrophoretic separation of the SDS:protein complexes. Under steady state conditions, the above two forces are counterbalanced (*Equations 1* and *2*), and the solute will move with a steady state velocity (v):

$$v = dx/dt = EQ/f. \tag{3}$$

The electrophoretic mobility (μ) is defined as the velocity per unit field strength (4):

$$\mu = v/E. \tag{4}$$

The retardation of the SDS:protein molecules in polymer solution-mediated capillary electrophoresis is an exponential function of the separation polymer concentration (P) and the retardation coefficient (K_R):

$$\mu = \mu_0 \exp(-K_R P). \tag{5}$$

where μ is the apparent electrophoretic mobility, and μ_0 is the free solution mobility of the analyte (3, 4, 15–17).

When the average pore size of the matrix is in the same range as the hydrodynamic radius of the migrating analyte, actual sieving exists in the classical sense (the Ogston regime, *Figure 1*, see also ref. 18). In this instance, using constant polymer concentration, the retardation coefficient (K_R) is proportional to the molecular weight of the analyte. M_r (19) and the mobility of the solute is an exponential function of M_r:

$$\mu \sim \exp(-M_r). \tag{6}$$

This results in linear Ferguson plots (log μ versus P, see ref. 15), crossing each other at zero gel concentration.

The Ogston theory assumes that the migrating solute behaves as an unperturbed spherical object with a size comparable that of the pores of the gel. However, large DNA molecules, as well as SDS:protein complexes, which are

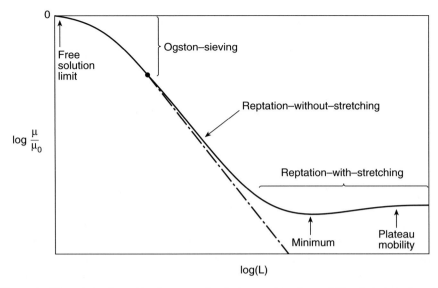

Figure 1. Plot of the logarithmic normalized electrophoretic mobility (μ/μ_0) as function of the molecular size (L) showing the Ogston, the reptation-without-stretching, and the reptation-with-stretching regimes. (From ref. 49 with permission.)

flexible chain biopolymer molecules, have been reported to migrate through polymer networks with pore sizes significantly smaller than the radius of gyration of the DNA molecule (20, 21). This phenomenon is explained by the reptation model (*Figure 1*, reptation-without-stretching regime) which explains the migration of the large biopolymer molecules as a 'head first, snake-like' motion through the much smaller size gel pores (20, 22–24). This reptation model suggests an inverse relationship between the size and mobility of the analyte (23):

$$\mu \sim 1/M_r. \qquad [7]$$

At high electric field strengths, this reptation model converts to reptation-with-stretching mode (*Figure 1*, reptation-with-stretching, i.e. charge alignment with the electric field) and the resulting mobility of the analyte is described by:

$$\mu \sim (1/M_r + bE^a) \qquad [8]$$

where b is a function of the mesh size of the sieving matrix, as well as the charge and the segment length of the migrating polyanions (19), and $1 < a < 2$.

Plotting log solute mobility as a function of log solute molecular weight is an easy method to identify the Ogston-, the reptation-, or the reptation-with-stretching regimes (22, 25, 26). When the slope values are close or equal to -1, pure reptation occurs (21). *Figure 2* depicts the log μ versus log solute M_r

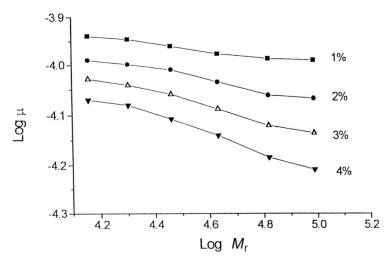

Figure 2. Double logarithmic plots of the electrophoretic mobility and protein molecular weight (M_r) for the protein test mixture α-lactalbumin (M_r 14 200); soybean typsin inhibitor (M_r 21 500); carbonic anhydrase (M_r 29 000); ovalbumin (M_r 45 000); bovine serum albumin (M_r 66 000); and phosphorylase B (M_r 97 400) with polyethylene oxide sieving matrix, M_r 100 000 PEO. The percentages on the right-hand side correspond to the actual sieving matrix concentration for each plot. (From ref. 27 with permission.)

plots for various SDS:standard protein complexes using polyethylene oxide (PEO, M_r 100 000) sieving matrix in 1%, 2%, 3%, and 4% concentration (27). The corresponding slopes are: −0.07, −0.1, −0.14, and −0.18, respectively. Similar behaviour was observed when different size (molecular weight) polyethylene oxide entangled polymers were applied in 1% concentration (data not shown, see ref. 27).

3. Operational variables

Classical SDS–PAGE uses cross-linked polyacrylamide gels. The three-dimensional structure of the gel creates a molecular sieve enabling size separation of SDS:protein complexes as they migrate through the gel medium. Unlike SDS–PAGE, in capillary electrophoresis of SDS:protein complexes, both high viscosity cross-linked gels and high or low viscosity non-cross-linked polymer solutions have been used. In classical slab gel electrophoresis, the gel is polymerized in the separation chamber (between two glass, or plastic plates) before use and then connected to the buffer tanks. In capillary electrophoresis, a narrow bore capillary is filled by the medium employed and the separation is then carried out at high voltage, providing fast separation with high efficiency. Depending on the separation medium used in the separation, the gel can be polymerized *in situ* inside the capillary, or filled and replaced by applying pressure on the polymer solution container which is connected to the separation capillary column. Alteration of the separation parameters, such as applied electric field, temperature, capillary dimensions (diameter and length), and the type and concentration of the gel, affects the separation of SDS: protein complexes (17, 28, 29). Since at basic pH the SDS:protein complexes are negatively charged, the cathode should be at the injection side (referred to as reversed polarity in CE) during electrophoresis.

3.1 Applied electric field and temperature

As *Equation 3* shows, the migration velocity is a linear function of the electric field strength; this relationship has been experimentally verified in SDS: protein separations (17, 28, 29). With an increase of the field strength, the necessary separation time is decreased, and simultaneously, both peak efficiency and resolution increased. The separation temperature is also an important parameter in polymer solution-mediated capillary SDS electrophoresis. Elevated temperatures can usually be used with non-cross-linked linear polymer matrices, although temperature changes may initiate bubble formation and discontinuity within the polymer-filled capillaries. As the temperature increases, the viscosity of the sieving matrix and the friction of the migrating SDS:protein complexes decreases. Temperature also changes the mean mesh size of any dynamic pore structure, modifying its sieving capability. Significant sieving differences were found at different temperatures on capillary electrophoresis separation of SDS:protein molecules when PEO and dextran matrices were employed (30). The temperature-dependent separation could not be rationalized simply by the friction–temperature relationship of the two polymer network systems. However, experimentally obtained activation energy data for the polymer network formation suggested that dynamic temperature-dependent formation and dissociation of oriented conformers exist in polymer solutions. The data suggested that more or less oriented arrangements (i.e. channel-like structures) are formed in concentrated polymer solutions, and these structures play a significant role in the sieving capability.

3.2 Capillary dimensions

The dimensions of the separation capillary also affect the resulting separation. Usually, longer capillaries result in higher separation power for a given electric field. Increase of the separation voltage also enhances the separation efficiency, resulting in higher resolution with similar selectivity values. Thus, an optimum exists between the applied field and column length that ensures appropriate separation with high efficiencies. The internal diameter (i.d.) of the capillary affects the detection limits and injection amounts. However, it is important to note that, with larger column diameters, column temperature will increase in a non-linear manner, due to excess Joule heat. The maximum applied power to a capillary is limited to approximately 1–2 W with passive cooling and 5 W with active (liquid) cooling.

3.3 Capillary coating

Uncoated and coated capillary columns have both been used for polymer solution-mediated capillary SDS electrophoresis. When bare fused silica capillary is used, a higher concentration or higher viscosity polymer is generally needed. An uncoated capillary usually possesses strong electro-osmotic character (EOF) and may have strong interaction with proteins in the samples. At high pH, the EOF can be so high that the negatively charged SDS:protein complexes do not even migrate towards the anode, and thus cannot be detected. Another drawback is that bare fused silica capillaries last only for a few runs with real biological samples before deteriorating due to irreversible adsorption of matrix or sample components.

Today, most polymer-mediated electrophoresis techniques use coated capillary columns for biopolymer separations. Both dynamically coated and covalently coated capillaries have been shown to be useful in polymer solution-mediated capillary SDS electrophoresis. With dynamic coating, a thin layer of polymer is adsorbed on the capillary wall before or during separation. This dynamic coating is easily applied on any bare fused silica capillary, and the coating can be regenerated whenever needed. It has been demonstrated that good resolution with fast separation can be obtained with dynamically coated capillaries (31, 32). However, dynamically coated capillaries require a long equilibration time before use and sometimes are not so efficient with biological samples. On the other hand, covalently coated capillaries offer high speed separations with little or no requirement for pre-separation equilibration. Several different capillary coatings have been reported in protein separations by polymer solution-mediated capillary SDS electrophoresis. Cohen and Karger (13) first suggested the use of a polyacrylamide coated capillary column for SDS capillary gel electrophoresis separation of a standard protein mixture. The coated capillary used had no interaction with SDS:protein complexes in the mixture and exhibited high resolution with sharp peaks. Ganzler *et al.* (10) reported a dextran coated capillary filled with dextran polymer for the separation of proteins by polymer solution-mediated capillary SDS electrophoresis. Their coated capillary was stable and had good run-to-run migration time reproducibility. Recently, Shieh *et al.* (31) developed a novel capillary coating which combines the advantages of both the dynamic and covalent coatings. A thin layer of polyacrylamide polymer is dynamically coated to the inner surface of the capillary followed by allylamine treatment (33). The capillary coated in this way can be used with biological samples for more than 200 injections, even with strong acid rinses (1 M HCl) in between runs.

Figure 3 shows the separation of a seven protein mixture with 100, 200, and 400 injections, employing polyethylene oxide sieving matrix solution in a coated capillary. Nakatani *et al.* (34) demonstrated a stable polyacrylamide coating through an Si–C linkage and used an entangled polymer solution for

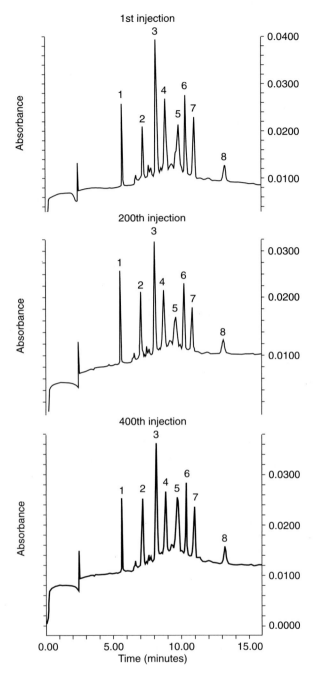

Figure 3. Migration time reproducibility of a standard protein test mixture over 100, 200, and 400 runs on a coated capillary column using PEO separation matrix. Note that the matrix was replaced after each run. Peaks: 1 = Orange G; 2 = α-lactalbumin; 3 = carbonic anhydrase; 4 = ovalbumin; 5 = bovine serum albumin; 6 = phosphorylase B; 7 = β-galactosidase; 8 ≐ myosin. (From ref. 31 with permission.)

111

more than a 100 injections, with a one month shelf-life at pH 8.7. Coatings with the Si–C linkage offer high pH stability, reduced electro-osmotic flow, and prevention of sample interaction with the capillary surface. With a stable coating, the capillary can be washed with strong acids and still be used at high pH for long periods of time. The coated capillary also provides good run-to-run reproducibility and excellent peak efficiency.

3.4 Gels and polymer solutions
3.4.1 Gels
Gels are formed of covalently cross-linked polymer networks, possessing very high viscosity and well-defined pore structure. The pore size is determined by the relative concentration of monomer and cross-linker used during polymerization (%T = total monomer concentration; %C = cross-linker concentration as a per cent of the total monomer and cross-linker concentration). Polyacrylamide is the most widely used gel material, usually cross-linked with N,N'-methylene-bisacrylamide (BIS). In the early work of capillary SDS gel electrophoresis, Cohen and Karger (13) used cross-linked polyacrylamide gels, filled in a narrow bore polyacrylamide coated capillary for the separation of proteins up to M_r 35 000. They demonstrated the separation power of this system by resolving the two chains of insulin in less than eight minutes, using a 7.5%T, 3.3%C polyacrylamide gel with 8 M urea and 0.1% SDS. Others also reported the separation of proteins by capillary SDS gel electrophoresis using polyacrylamide as a sieving medium (35).

Highly cross-linked polyacrylamide gels are very rigid and can not be replaced in the capillary after polymerization. Most gels are attached to capillary wall *via* a bifunctional reagent (36), and samples can only be introduced to the capillary by the non-quantitative electrokinetic injection method in which the sample migrates into the capillary by simply starting the electrophoresis from the sample vial. Gels usually have a short shelf-life due to sensitivity to changes in temperature, pH, and high voltage. During polymerization or separation, bubbles can be formed due to gel shrinkage (37). However, gels have well-defined pore sizes and provide high resolution, especially for the separation of low molecular weight biopolymers.

3.4.2 Polymer solutions
The other type of sieving media, polymer solutions, is widely used today in capillary electrophoresis (8, 11, 32). The pore sizes of these entangled polymer solutions are defined by the dynamic cross-linkage between the polymer chains, and can be varied by changing variables such as capillary temperature, separation voltage, salt concentration, or pH. Polymer solutions are not heat-sensitive, and even if a layer is attached to the capillary wall, the separation matrix can usually be replaced in the capillary simply by applying pressure to the end of the column. Thus, fresh matrix can be used for each analysis which

prevents any cross-contamination from previous samples and often leads to improved reproducibility of migration times. Two types of injection methods can be used with such matrices. The first is electrokinetic injection (see Section 3.4.1) which usually offers sharp peaks with sample pre-concentration if the separation and sample buffers are chosen appropriately (8). However, the method suffers from poor peak area (injection amount) reproducibility and this makes quantitative analysis difficult. The second method is the pressure injection technique, in which sample injection occurs by applying constant pressure for a short period of time in a sample vial connected to the capillary. The method offers excellent run-to-run peak area reproducibility and can be used for routine quantitative analysis.

Kenndler and co-workers (38) demonstrated that linear polyacrylamide matrices can be used for the separation of SDS:protein molecules in the molecular mass range of 17.8 kDa to 77 kDa, in a run time of only 60 minutes. Later, Regnier and co-workers (39) and others (40, 41) successfully applied linear polyacrylamide solutions for the separation of standard proteins in biological samples. More recently, low concentrations of entangled linear polymers, such as dextran, polyethylene oxide, hydroxy ethyl cellulose, pollulane, and polyvinyl alcohol have also been employed for SDS:protein separations.

It is important to note that polymer solutions offer several advantages over their gel counterparts, such as longer shelf-life, lower viscosity, and ease of manufacturing. These matrices are also much less sensitive to changes in physical variables, such as temperature, and can be used with pressure injection. Not surprisingly, therefore, this type of matrix is now extensively used in the separation of biopolymers by capillary electrophoresis.

4. Separations

In capillary polymer solution-mediated capillary SDS electrophoresis, the protein molecules are completely denatured by boiling in the presence of a reducing agent, such as 2-mercaptoethanol, and a sufficient amount of SDS (minimum of 0.1%, w/v). This procedure results in full reduction of the disulfide bonds of the protein, and the polypeptide chain forms a random coil shape. The SDS binds to the protein in a 1:1.4 weight ratio. The free solution electrophoretic mobility of different SDS:protein complexes is considered to be almost identical since their mass to charge ratios are practically equal. Therefore, the separation is based on sieving that occurs as the proteins migrate through the polymer matrix.

4.1 Cross-linked polyacrylamide gel

Protein separations by capillary electrophoresis were first carried out using cross-linked polyacrylamide gels adapted from classical slab gel SDS–PAGE. Hjerten was the first to show the separation of membrane proteins by

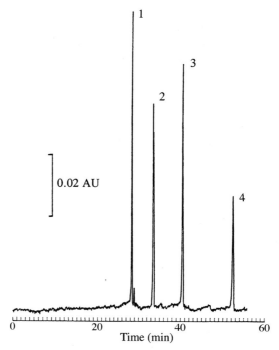

Figure 4. High-performance capillary SDS–PAGE separation of a four protein test mixture. Peaks: 1 = α-lactalbumin; 2 = β-lactoglobulin; 3 = trypsinogen; 4 = pepsin. Conditions: E = 400 V/cm, temperature = 27°C, cross-linked polyacrylamide gel (T = 10%, C = 3.3%). Separation buffer: 90 mM Tris–phosphate pH 8.6, 8 M urea, 0.1% SDS. (From ref. 13 with permission.) AU = absorbance unit.

capillary polyacrylamide gel electrophoresis (12). Cohen and Karger (13) provided a detailed investigation of cross-linked polyacrylamide gel for separation of SDS:protein complexes by capillary gel electrophoresis.

Figure 4 shows a separation of four standard proteins by Cohen and Karger (13) using a 10%T and 3.3%C polyacrylamide gel in 90 mM Tris–phosphate buffer pH 8.6, 0.1% SDS, 8 M urea, with detection at 280 nm wavelength. Following this work, Tsuji (35) added ethylene glycol to the gel to reduce bubble formation and extend the shelf-life of cross-linked polyacrylamide gel-filled capillaries. He silanized the inside surface of the capillary and filled it with 5.1%T and 2.6%C polyacrylamide in 375 mM Tris buffer pH 8.8, 3.2 mM SDS, 2.35 M ethylene glycol for the separation of a standard protein mixture ranging from 14 kDa to 98 kDa to evaluate the separation performance. It was apparent that the presence of ethylene glycol in 1.8–2.7 M concentration significantly improved the longevity of the gel-filled capillary. Capillaries made by this method were continuously used for more than two weeks, exceeding 300 injections.

4.2 Linear polyacrylamide solutions

Linear polyacrylamide matrices offer a number of advantages to cross-linked polyacrylamide gels, including the fact that the separation matrix can be easily removed and refilled in the capillary so that the polymer can be replaced after each run, and uncross-linked polymer solutions are not very sensitive to changes in the physical environment. The use of a coated capillary provides good separation in most instances, with longer shelf-life when compared to the use of an uncoated column. However, with higher polymer concentrations, uncoated capillaries still provide reasonable resolution of SDS:protein mixtures.

Figure 5 shows the separation of several human salivary proteins by polymer solution-mediated capillary SDS electrophoresis employing uncoated fused silica capillary filled with 4% linear polyacrylamide gel. Unfortunately, polyacrylamide (both the cross-linked gel and linear polymer matrix) has strong absorption at low UV wavelengths (< 230 nm) which interferes with the detection of the protein amide bond at 310 nm, reducing the sensitivity of

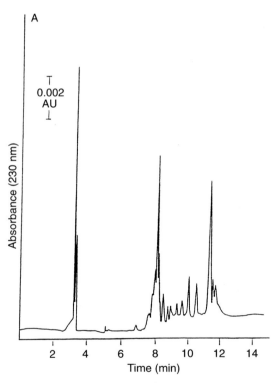

Figure 5. Separation of several human salivary proteins by capillary SDS gel electrophoresis employing an uncoated capillary and 4% linear polyacrylamide solution. (From ref. 39, with permission.) AU = absorbance unit.

115

detection. For this reason, other matrices are now often used for capillary SDS electrophoresis of proteins, as described below.

4.3 Other entangled polymer solutions

Non-polyacrylamide-based polymer solutions (also referred to as polymer networks) have been popular recently in capillary electrophoresis separations. These polymers may offer unique advantages including high UV transparency at lower wavelength (< 230 nm), stability at both high or low pH, non-toxicity, and usually low viscosity at the concentration necessary for separation. Their sieving capability is usually dependent on the polymer concentration being above the entanglement threshold, which is a function of the polymer chain length, temperature, pH, and salt concentration. Several well-characterized polymers have been employed for capillary SDS electrophoresis separation of protein molecules, such as dextran, polyethylene oxide, polyvinyl alcohol, and pollulane.

4.3.1 Dextran

Dextran is a widely used polymer in separation sciences, due to its high solubility in aqueous solutions. It is usually slightly branched as a natural polymer, and accessible in a wide range of molecular weights. Ganzler *et al.* (10) were the first to describe the separation of SDS:protein complexes using different dextran polymers of low to moderate viscosity range.

Figure 6 depicts the separation of a standard protein mixture, ranging from 17 kDa to 206 kDa, in a 18 cm coated capillary, filled with 10% (w/v) dextran (M_r 2 000 000) in a unique buffer composition of 0.06 M 2-amino-2-methyl-1,3-propanediol (AMPD) cacodylic acid (CACO) pH 8.8, containing 0.1% SDS. The relative mobility of the SDS:protein complexes were shown to be dependent on the chain length (M_r) of the dextran used for the separation. At constant polymer concentration, higher molecular weight dextran is more viscous and offers better resolving power, especially for high molecular weight samples.

4.3.2 Polyethylene oxide

Polyethylene oxide (PEO) is also frequently used in polymer solution-mediated capillary SDS electrophoresis of proteins. Polyethylene oxide is a synthetic, linear (non-branched) hydrophilic polymer. A broad range of different molecular weight products are commercially available ranging from several hundreds to several millions in degree of polymerization. *Figure 3* shows the separation of a standard protein test mixture ranging in molecular mass from 14.4 kDa to 98 kDa employing a 3% polyethylene oxide sieving matrix (31). The molecular weight estimation of proteins resolved by capillary electrophoresis using PEO is comparable to traditional slab gel electrophoresis (42).

Shieh *et al.* (31) reported on the stability and reproducibility of SDS:protein

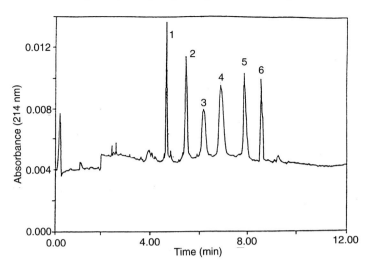

Figure 6. Capillary SDS gel electrophoresis separatioin of a standard protein test mixture with molecular mass range from 17 kDa to 206 kDa on a 18 cm coated capillary, filled with 10% (w/v) dextran (M_r 2 000 000). Buffer: 0.6 M 2-amino-2-methyl-1,3-propanediol (AMPD) cacodylic acid (CACO) pH 8.8., containing 0.1% SDS. (From ref. 10 with permission.)

separations using a mixture of polyethylene oxides as a sieving matrix. Seven standard proteins were separated on this polymer solution in less than 15 minutes. The linearity of log molecular weight versus mobility plot was greater than $R^2 = 0.998$ which enabled a good molecular mass estimation. The use of a coated capillary prevented any sample interaction with the glass surface and allowed acid wash cleaning steps between runs. Crude samples were injected directly into the capillary without any sample clean-up requirement. The run-to-run and day-to-day reproducibility was shown to be less than 2% relative standard deviation. Thus, this polymer can be applied for a routine daily use of rapid protein molecular mass estimation and purity check. A suitable procedure is described in *Protocol 1* and sample data are shown in *Figure 7*.

Protocol 1. Molecular weight determination by polymer solution-mediated capillary SDS electrophoresis

Equipment and reagents

- Capillary electrophoresis unit (Beckman Instruments, Model P/ACE 5000 or similar instrument)
- 2-mercaptoethanol (Sigma Chemicals)
- eCAP™ SDS coated capillary: 65 cm, 100 μm i.d. (Beckman Instruments)
- eCAP™ SDS 14–200 gel buffer (Beckman Instruments)
- eCAp™ SDS test mixtures: contains seven proteins, ranging from 14.2 kDa to 205 kDa (Beckman Instruments)
- eCAP™ SDS reference standard, Orange G (Beckman Instruments)
- eCAP sample buffer: 0.12 M Tris–HCl, 1% SDS pH 6.6 (Beckman Instruments)

117

Protocol 1. *Continued*

A. *Preparing the test mixture and sample*

1. Dissolve the protein test mixture in 750 μl of eCAP sample buffer. Once dissolved, add 750 μl deionized water and mix thoroughly. Set aside 200 μl aliquots of this solution and store promptly at −20 °C.

2. For each unknown sample and for a 200 μl aliquot of the protein test mixture, combine the following ingredients in the specified quantities in a 400 μl microcentrifuge tube:
   - 0.1–1.0 mg protein (200 μl when using the protein test mixture, which corresponds to 0.45 mg total protein)
   - 100 μl sample buffer (do not add to the prepared test mixture)
   - 10 μl reference standard, Orange G
   - 5 μl 2-mercaptoethanol
   - 85 μl deionized water filtered through 0.2 μm nylon membrane filter (add only to the unknown sample); add no additional water to the protein test mixture

3. Mix each solution on a vortex mixer for 2 min or until the proteins are totally dissolved.

4. Boil the mixtures in a water-bath at 100 °C for 10 min in closed micro-centrifuge tubes.

5. Place the tubes in an ice-bath to cool for 3 min before injection/

B. *Performing a test run*

1. Run the protein test mixture by capillary SDS gel electrophoresis following the instructions of the manufacturer of the capillary electrophoresis unit.

2. Then run the unknown samples.

3. Wash the column by rinsing with 1 M HCl for 1 min, followed by a 1 min water rinse, and a 2 min buffer rinse.

C. *Calculation of protein molecular weights*

1. Calculate the migration time of the standard proteins in the test mixture relative to the migration time of the reference standard (Orange G) or the relative migration time, RMT, using:

$$RMT = \frac{\text{migration time of protein}}{\text{migration time of reference standard (Orange G)}}$$

2. Obtain a standard curve by plotting the log of the molecular weights versus 1/RMT for each standard protein in the test mixture.

3. Calculate the RMT of each unknown protein and interpolate the log of its molecular weight from the plot. The molecular weight can then be determined by calculating the antilog.

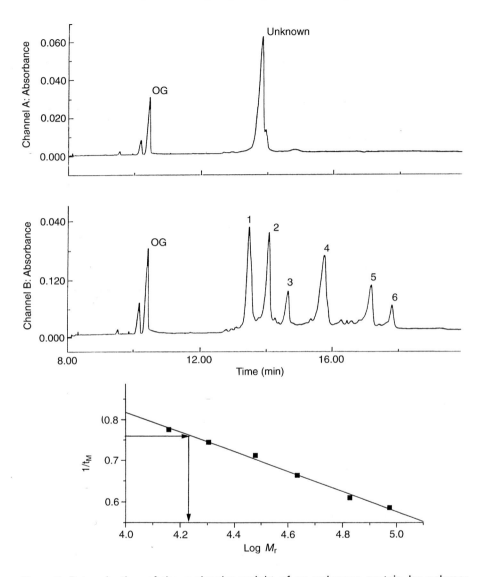

Figure 7. Determination of the molecular weight of an unknown protein by polymer solution-mediated capillary SDS electrophoresis as described in *Protocol 1*. Upper panel: the electropherogram of the unknown protein with the internal standard Orange G (OG). Middle panel: separation of the molecular weight standard test mixture. Peaks: OG = Orange G; 1 = α-lactalbumin; 2 = soybean trypsin inhibitor; 3 = carbonic anhydrase; 4 = ovalbumin; 5 = bovine serum albumin; 6 = phosphorylase B. Lower panel: the calibration plot of reciprocal migration time ($1/t_M$) versus log M_r. The determination of the molecular weight of the unknown (M_r 18000) is indicated by horizontal and vertical arrows.

4.3.3 Polyvinyl alcohol

Simo-Alfonso and co-workers (43) have studied the effect of different concentration of polyvinyl alcohol (PVA) as a separation matrix for SDS capillary electrophoresis of proteins. *Figure 8* shows the separation of a protein mixture at constant field with 3–8% of polyvinyl alcohol-filled in a 75 μm i.d. capillary column. The resolution of higher molecular weight proteins diminish using the polymer solution as separation matrix above 6% concentration. On the other hand, at low concentration (< 4%), resolution was lost for smaller proteins. These results suggest that PVA solutions in the 4–6% (w/v) range offer a useful dynamic matrix for separation of SDS:protein complexes by capillary electrophoresis. However, PVA is rarely used in capillary electrophoresis due to its poor solubility and higher viscosity when dissolved in aqueous solutions.

4.3.4 Pollulan

Several carbohydrate analogues, such as various celluloses and pollulan, have also been evaluated as separation matrices in capillary SDS electrophoresis of

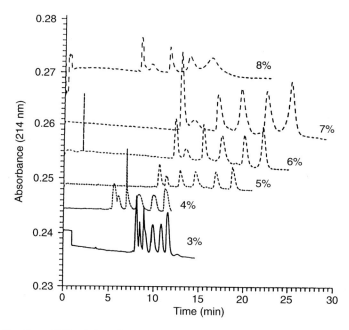

Figure 8. Capillary SDS electrophoresis separation of six protein standards; α-lactalbumin (M_r 14 400); trypsin inhibitor (M_r 20 100); carbonic anhydrase (M_r 30 000); ovalbumin (M_r 43 000); bovine serum albumin (M_r 67 000); phosphorylase B (M_r 94 000) at constant field with 3–8% polyvinyl alcohol polymer in a 75 μm i.d. coated capillary column. (From ref. 43 with permission.)

proteins. These polymers offer good stability and low UV absorption below 230 nm. Nakatani and co-workers (34) reported the separation of a standard protein mixture on 7% pollulan solution in 0.1 M Tris–Ches buffer pH 8.7 containing 0.1% SDS. Using a coated capillary, the relative standard deviation of the migration time of seven proteins over ten separations was found to be less than 0.5%.

4.4 The Ferguson method

The use of a standard calibration curve usually furnishes the user with the appropriate M_r value for an unknown protein by CE. However, this is not true for proteins that have covalent post-translational modifications, such as carbohydrate (glycoproteins), lipid (lipoproteins), or other prosthetic groups, which bind SDS differently (3), causing uneven SDS complexation and resulting in changes in the constant mass-to-charge ratio of the complex. In these instances, the Ferguson method is used to obtain more precise molecular weight estimates (15). The procedure is described in *Protocol 2*.

The Ferguson method involves capillary SDS electrophoresis separations employing various sieving polymer concentrations. The logarithmic reciprocal migration times of the individual proteins are plotted as a function of the polymer concentration. Linear regression analysis of these plots provide the slopes (the negative of the retardation coefficients: K_R, in *Equation 5*). Then, the logarithmic molecular weight data is plotted as a function of K_R, producing a universal calibration curve, the K_R plot (14, 40, 43). K_R is reported to be proportional to the effective molecular surface area (or to the radius of a spherical molecule with the same surface area) and not directly to the molecular weight (3). In SDS–PAGE (3, 4) and in polymer solution-mediated capillary SDS electrophoresis (40, 44) a linear relationship was found between K_R and log M_r. It is important to note that applying the Ferguson method to classical SDS–PAGE is very labour-intensive and time-

Table 1. Comparison of estimated molecular weight of glycoproteins by SDS CE using the standard curve method and Ferguson method[a]

Protein	Molecular Weight		
	Literature value	*Standard curve method*[b]	*Ferguson method*[c]
Amylase	56 500	68 200 (21%)[d]	54 900 (3%)
IgG light chain	23 000	4280 (87%)	27 200 (17%)
IgG heavy chain	55 000	68 500 (25%)	49 300 (10%)

[a] From ref. 50 with permission.
[b] The standard curve method was performed as described in *Protocol 1* using eCAP SDS 14–200 gel buffer.
[c] See *Protocol 2*.
[d] The figures in parentheses give the percentage difference between the estimated value from CE measurement and that cited in the literature.

consuming because of the necessity of pouring different gel concentrations in the slab format as well as the evaluation of the separated bands by regular staining–destaining procedures. However, the process is readily automated in polymer solution-mediated capillary SDS electrophoresis format (44) which is a major advantage.

Table 1 depicts the molecular weight determination of several glycoproteins by capillary SDS electrophoresis, comparing the regular calibration curve method (see *Protocol 1*) to the Ferguson method (44). As *Table 1* shows, the Ferguson method considerably increases the accuracy of the molecular weight assessment for these post-translationally modified (glycosylated) proteins.

Protocol 2. Performing the Ferguson method

Equipment and reagents

- See *Protocol 1*
- F75 (75% gel buffer): mix 15 ml SDS 14–200 gel buffer with 5 ml dilution buffer
- F67 (67% gel buffer): mix 14 ml SDS 14–200 gel buffer with 7 ml dilution buffer
- F50 (50% gel buffer): mix 10 ml SDS 14–200 gel buffer with 10 ml dilution buffer

- Dilution buffer: dissolve 1.21 g of electrophoresis grade Tris in 50 ml deionized water. Slowly add electrophoresis grade Ches [2-(cyclohexylamino)ethanesulfonic acid] until a pH of 8.8 is reached (approx. 1.5 g Ches). To this solution add 50 mg SDS and mix well.

Method

1. Perform capillary SDS gel electrophoresis of the standard protein test mixture and the unknown sample with eCAP™ SDS 14–200 gel buffer and diluted gel buffers, F75, F67, and F50.

2. Calculate the relative migration time of each of the proteins as described in *Protocol 1*, part C.

3. Plot the log of 1/RMT of the individual test proteins and unknown protein as a function of the gel buffer concentration.

4. Determine the slopes of each standard protein and the unknown protein.

5. Multiply the slope associated with each protein by -1 to convert the slope to a positive number.

$$\text{Retardation coefficient} = K_R = -1 \times \text{slope}.$$

6. Plot the log of the molecular weight of each standard protein versus the square root of the corresponding K_R value.

7. The K_R value of the unknown protein calculated in step 5 can then be used to read off the molecular weight of the unknown protein from the plot.

4.5 Detection

In capillary electrophoresis, on-column real time detection is used, mainly with ultraviolet absorption (UV) or laser-induced fluorescence (LIF) detection. Since proteins always have UV active groups, their ultraviolet absorption detection does not require any labelling methodology. On the other hand, derivatization with fluorescence dyes (pre-column labelling) would significantly increase detection sensitivity. Gump and Monning (45) used fluorescamine, naphthalene-2,3-dicarboxyaldehyde, and *o*-phthaldialdehyde to enhance detection sensitivity in both UV absorption and fluorescence detection modes. They found an enhancement factor of 22 with UV detection at 280 nm whereas with fluorescence detection an attomole detection limit was attained. Recently, the fluorescence characteristics of the immunoconjugate of a monoclonal antibody chimeric BGR96 and the anticancer drug doxorubicin have been utilized by Liu *et al.* (46) for monitoring CE separations.

4.6 Ultrafast separations

Ultrafast high-resolution separations by capillary SDS electrophoresis were introduced for fast purity checks and molecular weight determination of protein molecules. A standard six protein test mixture with molecular weights ranging from 14 400–97 400 was separated by Benedek and Guttman (47) in just 168 seconds (*Figure 9*) using polyethylene oxide matrix, without any significant loss in peak efficiency and resolution. Lausch *et al.* (48) have also shown rapid analysis of the IgG subunits by separating the light and heavy chains of IgG molecules in less than two minutes on a highly UV transparent dextran matrix (detection: UV 200 nm). These examples suggest that this method is a readily applicable rapid analysis technique for the biotechnology industry.

5. Conclusions

Polymer solution-mediated capillary SDS electrophoresis is a rapid automated separation and characterization technique for protein molecules. Size separation of SDS:protein complexes can easily be attained using coated capillaries filled with non-cross-linked polymer solutions. Although the separation mechanism was thought to be based on molecular sieving manifested by the protein sample passing through gel or polymer network (Ogston model), recent results have shown that in polymer solution-mediated capillary SDS electrophoresis, the separation occurs in the reptation regime. The technique has proven to be very useful in the rapid estimation of molecular weight and purity of recombinant proteins. If the detection sensitivity level of this method can be improved by one order of magnitude, polymer solution-mediated capillary SDS electrophoresis will become one of the most important separation tools available for the analysis of recombinant proteins in the biotechnology industry. Recently several research groups have employed fluorescent labelling and LIF detection to reach this higher sensitivity of detection (51; W. Nasbeh, personal communication).

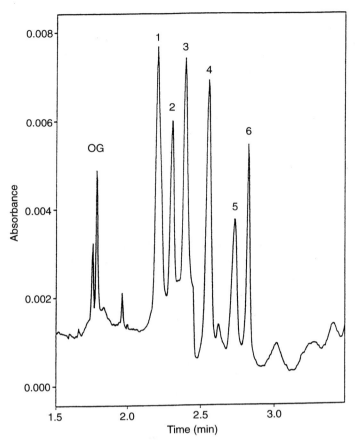

Figure 9. Ultrafast separation of SDS: protein complexes. Conditions: capillary length= 7 cm, electric field strength = 888 V/cm, detection = 214 nm UV. Peaks: OG = Orange G, 1 = α-lactalbumin; 2 = soybean trypsin inhibitor; 3 = carbonic anhydrase; 4 = ovalbumin; 5 = bovine serum albumin; 6 = phosphorylase B. (From ref. 47 with permission.)

Acknowledgements

The authors gratefully acknowledge Beckman Instruments for their support and the Beckman Research Library for their help. B. L. Karger gratefully acknowledges NIH for support of this work under GM15847. Contribution 695 from the Barnett Institute.

References

1. Shapiro, A., Vinuela, E., and Maizel, J. (1967). *Biochem. Biophys. Res. Commun.* **28**, 815.
2. Weber, K. and Osborn, M. (1969). *J. Biol. Chem.*, **244**, 4406.

3. Andrews, A. T. (1986). *Electrophoresis*, 2nd edn (ed. A. R. Peacock and W. F. Harrington), pp. 117–47. Clarendon Press, Oxford.
4. Chrambach, A. (1985). *The practice of quantitative gel electrophoresis* (ed. V. Neuhoff and A. Maelicke), pp. 177–86. VCH, Deerfield Beach, FL.
5. Li, S. F. Y. (1993). *Capillary electrophoresis*, pp. 173–83. Elsevier, Publishers, B.V., Amsterdam, The Netherlands.
6. Claire, R. L. St. (1996). *Anal. Chem.*, **68**, 569R.
7. Landers, J. P. (ed.) (1993). *CRC handbook of capillary electrophoresis: principles, methods, and applications.* CRC Press, Inc., Boca Raton, FL.
8. Karger, B. L., Chu, Y. H., and Foet, F. (1995). *Annu. Rev. Biophys. Biomol. Struct.*, **24**, 579.
9. Karger, B. L., Cohen, A. S., and Guttman, A. (1989). *J. Chromatogr.*, **492**, 585.
10. Ganzler, K., Greve, K. S., Cohen, A. S., Karger, B. L., Guttman, A., and Cooke, N. (1992). *Anal. Chem.*, **64**, 2665.
11. Guttman, A. (1996). *Electrophoresis*, **17**, 1333.
12. Hjerten, S. (1983). *J. Chromatogr.*, **270**, 1.
13. Cohem. A. S. and Karger, B. L. (1987). *J. Chromatogr.*, **397**, 409.
14. Karger, B. L., Paulus, A., and Cohen, A. S. (1987). *Chromatographia*, **24**, 15.
15. Ferguson, K. A. (1964). *Metab. Clin. Exp.*, **13**, 985.
16. Guttman, A.and Cooke, N. (1991). *J. Chromatogr.*, **559**, 285.
17. Guttman, A., Shieh, P., Hoang, D., Horvath, J., and Cooke, N. (1994). *Electrophoresis*, **15**, 221.
18. Ogston, A. G. (1958). *Trans. Faraday Soc.*, **54**, 1754.
19. Grossman, P. D., Menchan, S., and Hershey, D. (1992). *GATA*, **9**, 9.
20. Lumpkin, I. J., Dejardin, P., and Zimm, B. H. (1985). *Biopolymers*, **24**, 1573.
21. Guo, X. H. and Chen, S. H. (1990). *Phys. Rev. Lett.*, **21**, 2579.
22. De Gennes, P. G. (1979). *Scaling concept in polymer physics.* Cornell University Press, Ithaca, NY.
23. Lerman, L. S. and Frisch, H. L. (1982). *Biopolymers*, **21**, 995.
24. Viovy, J. L. and Duke, T. (1993). *Electrophoresis*, **14**, 322.
25. Slater, G. W. and Noolandi, J. (1989). *Biopolymers*, **28**, 1781.
26. Chiari, M., Nesi, M., and Righetti, P. G. (1994). *Electrophoresis*, **15**, 616.
27. Guttman, A. (1995). *Electrophoresis*, **16**, 611.
28. Tsuji, K. (1994). *J. Chromatogr. A*, **661**, 257.
29. Werner, W., Demorest, D., Stevens, J., and Wictorowicz, J. E. (1993). *AnL. Biochem.*, **212**, 253.
30. Guttman, A., Horváth, J., and Cooke, N. (1993). *Anal. Chem.*, **65**, 199.
31. Shief, P., Hoang, D., Guttman, A., and Cooke, N. (1994). *J. Chromatogr. A*, **676**, 219.
32. Guttman, A. (1983). *U.S. Patent*: 5 332 481.
33. Shieh, P. (1995). *U.S. Patent*: 5 462 646.
34. Nakatani, M., Shibukawa, A., and Nakagawa, T. (1994). *J. Chromatogr. A*, **672**, 213.
35. Tsuji, K. (1991). *J. Chromatogr.*, **550**, 823.
36. Neuhoff, V. (1984). *Electrophoresis*, **5**, 251.
37. Tanaka, T. (1981). *Sci. Am.*, **244**, 124.
38. Widhalm, A., Schwer, C., Blass, D., and Kenndler, E. (1991). *J. Chromatogr.*, **546**, 446.

39. Wu, D. and Regnier, F. (1992). *J. Chromatogr.*, **608**, 349.
40. Werner, W., Demorest, D., and Wictorowicz, J. E. (1993). *Electrophoresis*, **14**, 759.
41. Hebenbrock, K., Schugerl, K., and Freitag, R. (1993). *Electrophoresis*, **14**, 753.
42. Guttman, A. and Nolan, J. (1994). *Anal. Biochem.*, **221**, 285.
43. Simo-Alfonso, E. F., Conti, M., Gelfi, C., and Righetti, P. G. (1995). *J. Chromatogr A*, **689**, 85.
44. Guttman, A., Shieh, P., Lindahl, J. and Cooke, N. (1994). *J. Chromatogr. A*, **676**, 227.
45. Gump, E. L. and Monning, C. A. (1995). *J. Chromatogr. A*, **715**, 167.
46. Liu, J., Abid, S., and Lee, M. S. (1995). *Anal. Biochem.*, **229**, 221.
47. Benedek, K. and Guttman, A. (1994). *J. Chromatogr. A*, **680**, 375.
48. Laush, R., Scheper, T., Reif, O. W., Schlosser, J., Fleischer, J., and Freitag, R. (1993). *J. Chromatogr.*, **654**, 190.
49. Noolandi, J. (1992). *Annu. Rev. Phys. Chem.*, **43**, 237.
50. Guttman, A., Shieh, P., Lindahl, J., and Cooke, N. (1994). *J. Chromatogr. A*, **676**, 227.
51. Pinto, D. M., Arriaga, E. A., Craig, D., Angelova, J., Sharma, N., Ahmadzadeh, H. *et al.* (1997). *Anal. Chem.*, **69**, 3015.

5

Conventional isoelectric focusing in gel slabs, in capillaries, and immobilized pH gradients

PIER GIORGIO RIGHETTI, ALESSANDRA BOSSI, and
CECILIA GELFI

1. Introduction

This chapter both updates our chapter in the previous edition of this book (1) and covers the most recent aspects of the technique, including capillary iso-electric focusing (cIEF), the rising star in the field of IEF methodologies. We also draw the reader's attention to two textbooks on conventional IEF (2) and on IPGs (3). For capillary IEF (cIEF), we also recommend two recent reviews (4, 5).

Fractionations which rely on differential rates of migration of sample molecules, for example along the axis of a chromatographic column or along the electric field lines in electrophoresis, generally lead to concentration bands or zones which are essentially always out of equilibrium. The narrower the band or zone the steeper the concentration gradients and the greater the tendency of these gradients to dissipate spontaneously. This dissipative transport is thermodynamically driven; it relates to the tendency of entropy to break down all gradients, to maximize dilution and, during this process, to thoroughly mix all components (6). Most frequently, entropy exerts its effects via diffusion which causes molecules to move down concentration gradients and so produces band broadening and component intermixing.

The process of isoelectric focusing (IEF) in carrier ampholytes (CA) (7) and in immobilized pH gradients (IPG) (8) provides an additional force which counteracts diffusion of CAs and so maximizes the ratio of separative to dissi-pative transports. This substantially increases the resolution of the fraction-ation method. The sample focuses towards its isoelectric point (pI) driven by the voltage gradient and by the shape of the pH gradient along the separation axis (*Figure 1*). The separation can be optimized by using thin or ultrathin matrices (0.5 mm or less in thickness) and by applying very low sample loads (as permitted by high sensitivity detection techniques, such as silver and gold

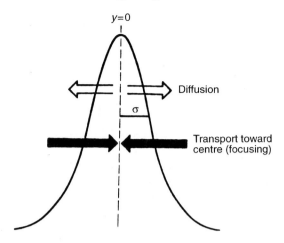

Figure 1. Illustration of the forces acting on a condensed zone in isoelectric focusing (IEF). The focused zone is represented as a symmetric Gaussian peak about its focusing point (pI; y = 0). Migration of sample towards the pI position is driven by the voltage gradient and by the slope of the pH gradient. σ is the standard deviation of the peak. (Courtesy of Dr O. Vesterberg.)

staining, radioactive labelling, immunoprecipitation followed by amplification with peroxidase- or alkaline phosphatase-linked secondary antibodies).

2. Conventional IEF in amphoteric buffers

2.1 General considerations

2.1.1 The basic method

IEF is an electrophoretic technique by which amphoteric compounds are fractionated according to their pIs along a continuous pH gradient (9). Contrary to zone electrophoresis, where the constant (buffered) pH of the separation medium establishes a constant charge density at the surface of the molecule and causes it to migrate with constant mobility (in the absence of molecular sieving), the surface charge of an amphoteric compound in IEF keeps changing, and decreasing, according to its titration curve, as it moves along a pH gradient until it reaches its equilibrium position, i.e. the region where the pH matches its pI. There, its mobility equals zero and the molecule comes to a stop.

The gradient is created, and maintained, by the passage of an electric current through a solution of amphoteric compounds which have closely spaced pIs, encompassing a given pH range. The electrophoretic transport causes these carrier ampholytes (CA) to stack according to their pIs, and a pH gradient, increasing from anode to cathode, is established. At the beginning of the run, the medium has a uniform pH, which equals the average pI of the CAs. Thus most ampholytes have a net charge and a net mobility. The most acidic CA moves toward the anode, where it concentrates in a zone whose pH equals its pI, while the more basic CAs are driven toward the cathode. A less acidic ampholyte migrates adjacent and just cathodal to the previous one and so on, until all the components of the system reach a steady state. After this stacking process is completed, some CAs still enter zones of higher, or lower, pH by diffusion where they are not any longer in isoelectric equilibrium. But as soon as they enter these zones, the CAs become charged and the applied voltage forces them back to their equilibrium position. This pendulum movement, diffusion versus electrophoresis, is the primary cause of the residual current observed under isoelectric steady state conditions. Finally, as time progresses, the sample protein molecules also reach their isoelectric point. If a macro ion is applied simultaneously above and below its pI, its negatively and positively charged species (whose net charge is defined by the macromolecule titration curve) migrate towards each other till they fuse, or merge (or focus) at the pI zone, having zero net charge.

2.1.2 Applications and limitations

The technique applies only to amphoteric compounds and more precisely to good ampholytes with a steep titration curve around their pI, *conditio sine qua non* for any compound to focus in a narrow band. This is very seldom a problem with proteins but it may be so for short peptides that need to contain at least one acidic or basic amino acid residue, in addition to the $-NH_2$ and $-COOH$ termini. Peptides which have only these terminal charges are isoelectric over the entire range of approximately pH 4 and pH 8 and so do not focus. Another limitation with short peptides is encountered at the level of the detection methods: CAs are reactive to most peptide stains. This problem may be circumvented by using specific stains, when appropriate (10, 11), or by resorting to immobilized pH gradients (IPG) which do not give background reactivity to ninhydrin and other common stains for primary amino groups (e.g. dansyl chloride, fluorescamine) (12).

In practice, notwithstanding the availability of CAs covering the pH 2.5–11 range, the practical limit of CA-IEF is in the pH 3.5–10 interval. Since most protein pIs cluster between pH 4 and 6 (13), this may pose a major problem only for specific applications.

When a restrictive support like polyacrylamide is used, a size limit is also imposed for sample proteins. This can be defined as the size of the largest molecules which retain an acceptable mobility through the gel. A conservative evaluation sets an upper molecular mass limit of about 750000 when using standard techniques. The molecular form in which the proteins are separated strongly depends upon the presence of additives, such as urea and/or detergents. Moreover, supramolecular aggregates or complexes with charged ligands can be focused only if their K_d is lower than 1 μM and if the complex is stable at pH = pI (14). An aggregate with a higher K_d is easily split by the pulling force of the current.

2.1.3 Specific advantages

(a) IEF is an equilibrium technique; therefore the results do not depend (within reasonable limits) upon the mode of sample application, the total protein load, or the time of operation.

(b) An intrinsic physicochemical parameter of the protein (its pI) may be measured.

(c) IEF requires only a limited number of chemicals, is completed within a few hours, and is less sensitive than most other techniques to the skill (or lack of it) of the operator.

(d) IEF allows excellent resolution of proteins whose pIs differ by only 0.01 pH units (with immobilized pH gradients, up to about 0.001 pH units); the protein bands are very sharp due to the focusing effect.

2.1.4 Carrier ampholytes

Table 1 lists the general properties of carrier ampholytes, i.e. of the amphoteric buffers used to generate and stabilize the pH gradient in IEF. The fundamental and performance properties listed in this table are usually required for a well-behaved IEF system, whereas the 'phenomena' properties are in fact the drawbacks or failures inherent to the technique. For instance, the 'plateau effect' or 'cathodic drift' is a slow decay of the pH gradient with time, whereby, upon prolonged focusing at high voltages, the pH gradient with the focused proteins drifts towards the cathode and is eventually lost in the cathodic compartment. There seems to be no remedy to this problem (except from abandoning CA-IEF in favour of the IPG technique), since there are

---

Table 1. Properties of carrier ampholytes

Fundamental 'classical' properties

1. Buffering ion has mobility of zero at pI
2. Good conductance
3. Good buffering capacity

Performance properties

1. Good solubility
2. No influence on detection systems
3. No influence on sample
4. Separable from sample

'Phenomena' properties

1. 'Plateau' effect (i.e. drift of the pH gradient)
2. Chemical change in sample
3. Complex formation

---

complex physicochemical causes underlying it, including a strong electro-osmotic flow generated by the covalently bound negative charges of the matrix (carboxyls and sulfate in both polyacrylamide and agarose) (as reviewed in ref. 15). In addition, it appears that basic CAs may bind to hydrophobic proteins, such as membrane proteins, by hydrophobic interaction. This cannot be prevented during electrophoresis, whereas ionic CA:protein complexes are easily split by the voltage gradient (16).

In chemical terms, CAs are oligoamino oligocarboxylic acids. They are available from different suppliers under different trade names (Ampholine or Pharmalyte from Pharmacia–Upjohn, Biolyte from Bio-Rad, Servalyte from Serva GmbH, Resolyte from BDH). There are two basic synthetic approaches: Vesterberg's approach, which involves reacting different oligoamines (tetra-, penta-, and hexa-amines) with acrylic acid (17); and the Pharmacia synthetic process, which involves the co-polymerization of amines, amino acids, and dipeptides with epichlorohydrin (2). The wide range synthetic mixtures (pH 3–10) contain hundreds, possibly thousands, of different amphoteric chemicals having pIs evenly distributed along the pH scale. Since they are obtained by different synthetic approaches, CAs from different manufacturers are bound to have somewhat different pIs. Thus, if higher resolution is needed, particularly for two-dimensional maps of complex samples, we suggest using blends of the different commercially available CAs. A useful blend is 50% Pharmalyte, 30% Ampholine and 20% Biolyte, by volume.

CAs from any source should have an average molecular mass of about 750 (size interval 600–900, the higher M_r referring to the more acidic CA species) (18). Thus CAs should be readily separable (unless they are hydrophobically complexed to proteins) from macromolecules by gel filtration. Dialysis is not recommended due to the tendency of CAs to aggregate. Salting out of proteins with ammonium sulfate seems to completely eliminate any contaminating CAs.

A further complication arises from the chelating effect of acidic CAs, especially towards Cu^{2+} ions, which may inactivate some metalloenzymes (19). In addition, focused CAs represent a medium of very low ionic strength (less than 1 mEq/litre at the steady state) (20). Since the isoelectric state involves a minimum of solvation, and thus of solubility, for the protein macro ion, there is a tendency for some proteins (e.g. globulins) to precipitate during the IEF run near their pI position. This is a severe problem in preparative runs. In analytical procedures it can be minimized by reducing the total amount of sample applied.

The hallmark of a 'carrier ampholyte' is the absolute value of pI – pK_{prox} (or ½ΔpK): the smaller this value is, the higher the conductivity and buffering capacity (at pH = pI) of the amphotere. A ΔpK = log 4 (i.e. pI – pK = 0.3) would provide an incredible molar buffering power (β) at the pI: 2.0 (unfortunately, such compounds do not exist in nature). A ΔpK = log 16 (i.e. pI – pK = 0.6) offers a β value of 1.35 at pH = pI. Let us take a practical example: Lys and His, two amino acids which can be considered good carrier ampholytes for IEF. For Lys, the pI value (9.74) is nested on a high saddle between two neighbouring protolytic groups (the α- and ϵ-amino; $pI - pK = 0.79$), thus providing an excellent β power (about 1.0). Conversely, the situation is not so brilliant with His: the pI value (7.47) is located down in a valley with a β value of only 0.24 (due to a $pI - pK$ of 1.5). When plotting the molar β power of a weak protolyte along the pH axis, one reaches a maximum of $\beta = 1.0$ at $pH = pK$. If one accepts, as a still reasonable β, a value of 1/3 of this maximum, this is located at $pH - pK = 0.996$. It is thus seen that even His, generally considered as a good 'carrier ampholyte', is in fact barely acceptable and falls just below this 1/3 limit of acceptance (21).

2.2 Equipment

2.2.1 Electrophoretic equipment

Three major items of apparatus are required: an electrophoresis chamber, a power supply, and a thermostatic unit.

i. Electrophoresis chamber

The optimal configuration of the electrophoresis chamber is for the lid to contain movable platinum wires (e.g. in the Multiphor 2, in the Pharmacia FBE3000, or in the Bio-Rad chambers models 1045 and 1415). This allows the use of gels of various sizes and the application of high field strengths across just a portion of the separation path. A typical chamber is shown in *Figure 2*.

ii. Power supply

The most suitable power supplies for IEF are those with automatic constant power operation and with voltage maxima at 2000–2500 V. The minimal requirements for good resolution are a limiting voltage of 1000 V and a reliable amperometer with a full scale not exceeding 50 mA. Lower field strengths cause the protein bands to spread (resolution is proportional to $\sqrt{E}$). The amperometer monitors the conductivity and so allows periodic manual adjustment of the electrophoretic conditions to keep the delivered power as close as possible to a constant value.

iii. Thermostatic unit

Efficient cooling is important for IEF because it allows high field strengths to be applied without overheating. Tap-water circulation is adequate for 8 M urea gels but not acceptable for gels lacking urea. Placing the electrophoresis apparatus in a cold room may be beneficial to prevent water condensation around the unit in very humid climates, but it is inadequate as a substitute for coolant circulation.

2.2.2 Polymerization cassette

The polymerization cassette is the chamber that is used to form the gel for IEF. It is assembled from the following elements: a gel supporting plate, a spacer, a cover (moulding) plate, and some clamps.

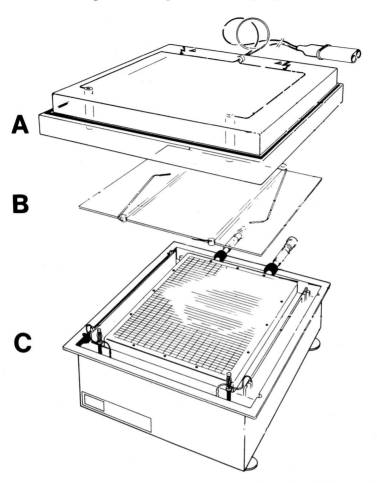

Figure 2. Drawing of the LKB Multiphor II chamber. (A) Cover lid. (B) Cover plate with movable platinum electrodes. (C) Base chamber with ceramic cooling block for supporting the gel slab. (Courtesy of LKB Produkter AB.)

i. Gel supporting plate

A plain glass plate is sufficient to support the gel when detection of the separated proteins does not require processing through several solutions (e.g. when the sandwich technique for zymograms or immunoblotting are to be applied) or when the polyacrylamide matrix is sturdy (gels > 1 mm thick, > 5%T). However, for thin soft gels, a permanent support is required. Glass coated with γ-methacryl-oxypropyl-trimethoxy-silane is the most reliable reactive substratum and is the most suitable for autoradiographic procedures (22) (see *Protocol 1* for the siliconization procedure). It is also the cheapest of such supports: dried-out gels can be removed with a blade and then by scrubbing with a brush. Unreacted silane may be hydrolysed by keeping the plates in Clorox for a few days. This step is unnecessary, however, if they have to go through successive cycles of siliconization. The glass plates used as a support should not be thicker than 1.0–1.2 mm. On the other hand, thin plastic sheets designed to bind polyacrylamide gel firmly (e.g. Gel-Bond PAG by Marine Colloids, PAG foils by Pharmacia–Upjohn, Gel Fix by Serva) are more practicable if the records of a large number of experiments have to be filed, or when different parts of the gel need to be processed independently (e.g. the first step of a two-dimensional separation or a comparison between different stains). The plastic sheet is applied to a supporting glass plate and the gel is cast onto this. The binding of the polyacrylamide matrix to these substrata, however, is not always stable and so care should be taken in using them, especially for detergent-containing gels and when using aqueous staining solutions. For good adherence, the best procedure is to cast 'empty' gels (i.e. polyacrylamide gel lacking CAs), wash and dry them, and then reswell with the solvent of choice (see Section 3.2.2).

Protocol 1. Siliconizing glass plates

Equipment and reagents
- IEF gel supporting plates
- Binding silane (γ-methacryloxy-propyl-trimethoxy-silane) prepared by Union Carbide (available through Pharmacia, Serva, or LKB) or repel silane (dimethyl-dichloro-silane) (available from suppliers Merck, Serva, etc.)

A. *Siliconizing with binding silane*

Two alternative procedures are available.

1. (a) Add 4 ml of binding silane to 1 litre of distilled water adjusted to pH 3.5 with acetic acid.
   (b) Leave the plates in this solution for 30 min.
   (c) Rinse with distilled water and dry in air.

2. (a) Dip the plates for 30 sec in a 0.2% solution of binding silane in anhydrous acetone.
   (b) Thoroughly evaporate the solvent using a hair drier.
   (c) Rinse with ethanol if required.

3. In either case, store the siliconized plates away from untreated glass.

B. *Siliconizing with repel silane*

1. Swab the glass plates with a wad impregnated with a 2% (w/v) solution of repel silane in 1,1,1-trichloroethane.

2. Dry the plates in a stream of air and rinse with distilled water.

---

ii. Spacer

U-gaskets of any thickness, between 0.2–5 mm, can be cut from rubber sheets (para-, silicone-, or nitrile-rubber). For thin gels, a few layers of Parafilm (each about 120 μm thick) can be stacked and cut with a razor blade. The width of such U-gaskets should be about 4 mm. In addition, cover plates with a permanent frame are commercially available (from Pharmacia–Upjohn) as are plastic trays with two lateral ridges for horizontal polymerization (Bio-Rad; see later, *Figure 4A*). A similar device may be home-made using Dymo tape strips, which are 250 μm thick to form the permanent spacer frame. Mylar foil strips or self-adhesive tape may be used as spacers for 50–100 μm thick gels. Rubber- or tape-gaskets should never be left to soak in soap (which they absorb) but just rinsed and dried promptly.

iii. Cover plate

Clean glass, glass coated with dimethyl-dichloro-silane (repel silane, *Protocol 1*), or a thick Perspex sheet are all suitable materials for the cover plate. If you wish to mould sample application pockets into the gel slab during preparation, attach Dymo tape pieces to the plate, or glue small Perspex blocks to the plate with drops of chloroform. Perspex should never be exposed unevenly to high temperatures (for example by being rinsed in running hot water) because it bends even if cut in thick slabs.

iv. Clamps

Clamps of adequate size and strength should be chosen for any gel thickness. Insufficient pressure may result in leakage of the polymerizing solution. The pressure of the clamps must be applied on the gasket, never inside it.

2.3 The polyacrylamide gel matrix

2.3.1 Reagents

Stocks of dry chemicals (acrylamide, bisacrylamide, ammonium persulfate) may be kept at room temperature provided they are protected from moisture by being stored in air-tight containers. Very large stocks are better sealed into plastic bags, together with Drierite (Merck), and stored in a freezer. TEMED stocks should also ideally be kept in a freezer, in an air-tight bottle or better, under nitrogen. Avoid contaminating acrylamide solutions with heavy metals, which can initiate its polymerization.

Acrylamide and bisacrylamide for IEF must be of the highest purity to avoid poor polymerization and strong electro-osmosis resulting from acrylic acid. Bisacrylamide is more hydrophobic and more difficult to dissolve than acrylamide, so start by stirring it in a little amount of luke-warm distilled water (the solution process is endothermic), then add acrylamide and water as required. Recently, novel monomers, endowed with extreme resistance to alkaline hydrolysis and with higher hydrophilicity, have been reported: they are *N*-acryloylamino-ethoxy ethanol (23) and *N*-acryloyl-amino propanol (24).

Table 2. Stock solutions for polyacrylamide gel preparation

Monomer solution[a,b]

30%T, 2.5%C	Mix 29.25 g of acrylamide and 0.75 g of bisacrylamide. Add water to 100 ml.
30%T, 3%C	Mix 29.1 g of acrylamide and 0.9 g of bisacrylamide. Add water to 100 ml.
30%T, 4%C	Mix 28.8 g of acrylamide and 1.2 g of bisacrylamide. Add water to 100 ml.

Initiator solution

40% (w/v) ammonium persulfate This reagent is stable at 4 °C for no more than one week.

Catalyst

TEMED This reagent is used undiluted. It is stable at 4 °C for several months.

[a] Monomer solutions can be stored at 4 °C for one month. Gels of $\geq$ 10%C (*Table 5*) require two monomer stock solutions, 30% acrylamide and 2% bisacrylamide.
[b] %T = g monomers per 100 ml; %C = g cross-linker per 100 g monomers.

Table 2 gives the composition and general storage conditions for monomers and catalyst solutions, while *Table 3* lists the most commonly used additives in IEF.

2.3.2 Gel formulations

In order to allow all the sample components to reach their equilibrium position at essentially the same rate, and the experiment to be terminated before the pH gradient decay process adversely affects the quality of the separation, it is best to choose a non-restrictive anti-convective support. There are virtually no theoretical but only practical lower limits for the gel concentration (the minimum being about 2.2%T, 2%C). Large pore sizes can be obtained both by decreasing %T and by either decreasing or increasing %C from the critical value of 5%. Although the pore size of polyacrylamide can be enormously enlarged by increasing the percentage of cross-linker, two undesirable effects also occur in parallel, namely increasing gel turbidity and proneness to syneresis (21, 25, 26). In this respect, *N,N'*-(1,2-dihydroxyethylene)bisacrylamide (DHEBA), with its superior hydrophilic properties, appears superior than bisacrylamide. In contrast, *N,N'*-diallyltartardiamide (DATD) inhibits the polymerization process and so gives porous gels just by reducing the actual %T of the matrix. Because unpolymerized acryloyl monomers may react with –NH$_2$ and –SH groups on proteins (27) and, once absorbed through the skin, act as neurotoxins, the use of DATD should be avoided altogether. *Table 4* gives the upper size of proteins which will focus easily in different %T gels. Because of these limitations, agarose is more suitable than polyacrylamide for the separation of very large proteins (M_r approx. 1 000 000), or supramolecular complexes. *Table 5* gives details of preparation for 4–7.5% polyacrylamide gels containing 2–8 M urea for total gel volumes ranging from 4–30 ml, together with the recommended amounts of catalysts.

Table 3. Common additives for IEF

Additive	Purpose	Concentration	Limitations
Sucrose, glycerol	To improve the mechanical properties of low %T gels and to reduce water transport and drift.	5–20%	The increased viscosity slightly slows the focusing process.
Glycine, taurine	To increase the dielectric constant of the medium. This increases the solubility of some proteins (e.g. globulins) and reduces ionic interactions.	0.1–0.5 M	Glycine is zwitterionic between pH 4 and 8, taurine between pH 3 and 7. Their presence somewhat slows the focusing process and shifts the resulting gradient.
Urea	Disaggregation of supramolecular complexes. Solubilization of water insoluble proteins, denaturation of hydrophilic proteins.	2–4 M 6–8 M	Unstable in solution especially at alkaline pH. Urea is soluble at ≥ 10°C; it accelerates polyacrylamide polymerization, so one should reduce the amount of TEMED added.
Non-ionic and zwitterionic detergents	Solubilization of amphiphilic proteins.	0.1–1%	To be added to the polymerizing solutions just before the catalysts to avoid foaming; they interfere with polyacrylamide binding to reactive substrata; they are precipitated by TCA, and require a specific staining protocol.

Table 4. Choice of polyacrylamide gel concentration

Protein M_r (upper limit for a given %T)	Gel composition
40000	%T = 7, %C = 5
75000	%T = 6, %C = 4
150000	%T = 5, %C = 3
800000	%T = 4, %C = 2.5

2.3.3 Choice of carrier ampholytes

A simple way to extend and stabilize the extremes of a wide (pH 3–10) gradient is to add acidic and basic (natural) amino acids. Thus lysine, arginine, aspartic acid, and glutamic acid are prepared as individual stock solutions containing 0.004% sodium azide and stored at 0–4°C. They are added in volumes sufficient to give 2–5 mM final concentration. To cover ranges spanning between 3–5 pH units, several narrow cuts of CAs need to be blended, with the proviso that the resulting slope of the gradient will be (over each segment of the pH interval) inversely proportional to the amount of ampholytes isoelectric in that region.

Table 5. Recipes for IEF gels (4–7.5%T and 2–8 M urea)

Gel vol. (ml)	30%T acrylamide:bisacrylamide monomer solution (ml)								2% carrier ampholytes[a]		Urea (g)				TEMED (µl)	40% APS[b] (µl)
	4%T	4.5%T	5%T	5.5%T	6%T	6.5%T	7%T	7.5%T	A	B	2 M	4 M	6 M	8 M		
30	4.0	4.5	5.0	5.5	6.0	6.5	7.0	7.5	1.5	1.88	3.6	7.2	10.4	14.4	9.0	30
25	2.34	3.75	4.17	4.58	5.0	5.42	5.83	6.25	1.25	1.56	3.0	6.0	9.0	12.0	7.5	25
20	2.66	3.0	3.34	3.66	4.0	4.33	4.66	5.0	1.0	1.25	2.4	4.8	7.2	9.6	6.0	20
15	2.0	2.27	2.5	2.75	3.0	3.25	3.5	3.75	0.75	0.94	1.8	3.6	5.4	7.2	4.5	15
10	1.33	1.5	1.66	1.83	2.0	2.16	2.33	2.5	0.5	0.63	1.2	2.4	3.6	4.8	3.0	10
8	1.06	1.21	1.33	1.46	1.6	1.73	1.86	2.0	0.4	0.5	0.96	1.92	2.88	3.84	2.4	8
7	0.93	1.05	1.17	1.28	1.4	1.51	1.63	1.75	0.35	0.44	0.84	1.68	2.52	3.36	2.1	7
6	0.8	0.9	1.0	1.1	1.2	1.3	1.4	1.5	0.3	0.38	0.72	1.44	2.16	2.88	1.8	6
5	0.66	0.75	0.83	0.91	1.0	1.08	1.16	1.25	0.25	0.31	0.6	1.2	1.8	2.4	1.5	5
4	0.53	0.6	0.66	0.73	0.8	0.86	0.93	1.0	0.2	0.25	0.48	0.96	1.44	1.92	1.2	4

[a] Use A for 40% solution (Ampholine, Servalve, Resolyte); B for Pharmalyte.
[b] APS: ammonium persulfate. To be added after degassing the solution and just before pouring it into the mould.

138

Shallow pH gradients are often used to increase the resolution of sample components. However, longer focusing times and more diffuse bands will result, unless the gels are electrophoresed at higher field strengths. Shallow pH gradients (shallower than the commercial 2 pH unit cuts) can be obtained in different ways:

(a) By subfractionating the relevant commercial carrier ampholyte blend. This can be done by focusing the CAs at high concentration in a multi-compartment electrolyser (28).

(b) By allowing trace amounts of acrylic acid to induce a controlled cathodic drift during prolonged runs. This is effective in the acidic pH region, but is rather difficult to obtain reproducible results from run-to-run.

(c) By preparing gels containing different concentrations of carrier ampholytes in adjacent strips (29) or with different thickness along the separation path (30).

(d) By adding specific amphoteric compounds (spacers) at high concentration (31).

In the last case, two kinds of ampholytes may be used for locally flattening the pH gradient: 'good' and 'poor'. Good CAs, those with a small $pI–pK_1$ (i.e. possessing good conductivity and buffering capacity at the pI), are able to focus in narrow zones. Low concentrations (5–50 mM) are sufficient to induce a pronounced flattening of the pH curve around their pIs. A list of these CAs is given in *Table 6*. Poor carrier ampholytes, on the other hand, form broad plateaus in the region of their pI, and should be used at high concentrations (0.2–1.0 M). Their presence usually slows down the focusing process. Some of them are listed in *Table 7*.

A note of caution: in an IEF system, the distribution of acids and bases is

Table 6. Good carrier ampholytes acting as spacers

Carrier ampholyte	pI	Carrier ampholyte	pI	Carrier ampholyte	pI
Aspartic acid	2.77	*p*-Aminobenzoic acid	3.62	Lysyl-glutamic acid	6.1
Glutathione	2.82	Glycyl-aspartic acid	3.63	Histidyl-glycine	6.81
Aspartyl-tyrosine	2.85	*m*-Aminobenzoic acid	3.93	Histidyl-histidine	7.3
o-Aminophenylarsonic acid	3.0	Diiodotyrosine	4.29	Histidine	7.47
Aspartyl-aspartic acid	3.04	Cystinyl-diglycine	4.74	L-Methylhistidine	7.67
p-Aminophenylarsonic acid	3.15	α-Hydroxyasparagine	4.74	Carnosine	8.17
Picolinic acid	3.16	α-Aspartyl-histidine	4.92	α,β-Diaminopropionic acid	8.2
Glutamic acid	3.22	β-Aspartyl-histidine	4.94	Anserine	8.27
β-Hydroxyglutamic acid	3.29	Cysteinyl-cysteine	4.96	Tyrosyl-arginine	8.38–8.68
Aspartyl-glycine	3.31	Tetraglycine	5.32	L-Ornithine	9.7
Isonicotinic acid	3.44	Pentaglycine	5.4	Lysine	9.74
Nicotinic acid	3.44	Triglycine	5.59	Lysyl-lysine	10.04
Anthranilic acid	3.51	Tyrosyl-tyrosine	5.6	Arginine	0.76
		Isoglutamine	5.85		

Table 7. Poor carrier ampholytes acting as spacers

Carrier ampholyte	pK_1	pK_2	pI
Carrier ampholytes with pIs 7–8			
β-Alanine	3.55	10.24	6.9
γ-Aminobutyric acid	4.03	10.56	7.3
δ-Aminovaleric acid	4.26	10.77	7.52
ε-Aminocaproic acid	4.42	11.66	8.04
'Good' buffers with acidic pIs			
Mes	1.3	6.1	3.7
Pipes	1.3	6.8	4.05
Aces	1.3	6.8	4.05
Bes	1.3	7.1	4.2
Mops	1.3	7.2	4.25
Tes	1.3	7.5	4.4
Hepes	1.3	7.5	4.4
Epps	1.3	8.0	4.65
Taps	1.3	8.4	4.85

according to their dissociation curve, in a pattern that may be defined as protonation (or deprotonation) stacking. If large amounts of these compounds originate from the samples (in the form of buffers), the limits of the pH gradient shift from the expected values. For example, 2-mercaptoethanol, as added to denatured samples for two-dimensional PAGE analysis, lowers the alkaline end from pH 10 to approx. pH 7.5 (32). This effect, however, may sometimes be usefully exploited. For example, high levels of TEMED in the gel mixture appear to stabilize alkaline pH gradients (33). In addition, TEMED is utilized in capillary IEF (cIEF) for blocking the capillary region after the detection point, so that basic proteins would focus in the region prior to the detector (34).

2.4 Gel preparation and electrophoresis

Protocol 2 outlines the series of steps required for an IEF run. The key steps are described in more detail below.

Protocol 2. CA-IEF flow sheet

1. Assemble the gel mould.
2. Mix all components of the polymerizing mixture, except ammonium persulfate.
3. Degas the mixture for a few minutes, and re-equilibrate (if possible) with nitrogen.

---

4. Add the required amount of ammonium persulfate stock solution and mix.

5. Transfer the mixture to the gel mould and overlay it with water.

6. Leave the mixture to polymerize (at least 1 h at room temperature or 30 min at 37 °C).

7. Open the gel mould and blot any moisture from the gel edges and surface.

8. Lay the gel on the cooling block of the electrophoretic chamber.

9. Apply the electrodic strips.

10. Pre-run the gel, if appropriate.

11. Apply the samples.

12. Run the gel.

13. Measure the pH gradient.

14. Reveal the protein bands using a suitable detection procedure.

2.4.1 Assembling the gel mould

Assembling the gel mould is visualized in *Figure 3A–C*. *Figure 3A* shows the preparation of the slot former which will give 20 sample application slots in the final gel. In the method shown, prepare the slot former by gluing a strip of embossing tape onto the cover plate and cutting rectangular tabs, with the dimensions shown, using a scalpel. Also glue a rubber U-gasket covering three edges of the cover plate to the plate. As shown in *Figure 3B*, apply a sheet of Gel-Bond PAG film to the supporting glass plate in a thin film of water. When using reactive polyester foils as the gel backing, use glycerol rather than water. Avoid leaving air pockets behind the gel backing sheet since this will produce gels of uneven thickness which will create distortions in the pH gradient. Also ensure that the backing sheet is cut flush with the glass support since any overhang easily bends. Finally, assemble the gel mould using clamps (*Figure 3C*). Assembly is usually made easier by wetting and blotting the rubber gasket just before use. Note that the cover plate has three V-shaped indentations on one edge which allow insertion of a pipette or syringe tip into the narrow gap between the two plates of the mould to facilitate opening the gel (see *Figure 3D*).

2.4.2 Preparation of the gel mixture

Polyacrylamide gels for IEF are cast on Gel-Bond film. *Protocol 2* is a flow sheet which gives the sequence of operations to be performed and *Protocol 3* describes precisely the polymerization procedure. A complete list of recipes for IEF gels is listed in *Table 5*.

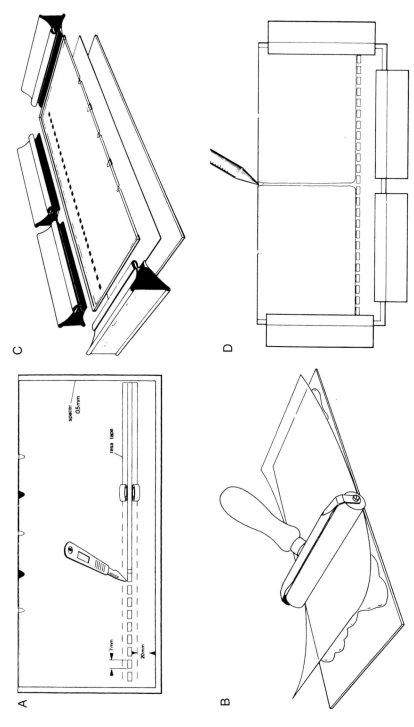

Figure 3. Preparation of the gel cassette. (A) Preparation of the slot former: onto the cover plate (bearing the rubber gasket U-frame) is glued a strip of tesa tape out of which rectangular tabs are cut with a scalpel. (B) Application of the Gel-Bond PAG film to the supporting glass plate. (C) Assembling the gel cassette. (D) Pouring the gelling solution in the vertically standing cassette using a pipette. (Courtesy of LKB Produkter AB.)

142

Protocol 3. Polymerization of the gel for IEF

Equipment and reagents

- Graduated glass measuring cylinder sufficient to hold the gel mixture during preparation
- Vacuum flask
- Mechanical vacuum pump
- Gel mould
- Urea

- 30%T acrylamide:bisacrylamide solution (see *Table 5*)
- 2% carrier ampholytes (Ampholine, Servalyte, Resolyte, or Pharmalyte)
- TEMED
- 40% ammonium persulfate (prepared fresh)
- Detergents (as desired, for the gel mixture)

Method

1. Mix all the components of the gel formulation (from *Table 5*, except TEMED, ammonium persulfate, and detergents, when used) in the glass measuring cylinder. Add distilled water to the required volume and transfer the mixture to a vacuum flask.

2. Degas the solution using the suction from a water pump; the operation should be continued as long as gas bubbles form. Manually swirl the mixture or use a magnetic stirrer during degassing. The use of a mechanical vacuum pump is desirable for the preparation of very soft gels but is unnecessarily cumbersome for urea gels (urea would crystallize). At the beginning of the degassing step the solution should be at, or above, room temperature, to decrease oxygen solubility. Its cooling during degassing is then useful in slowing down the onset of polymerization.

3. If possible, it is beneficial to re-equilibrate the degassed solution against nitrogen instead of air (even better with argon, which is denser than air). From this step on, the processing of the gel mixture should be as prompt as possible.

4. Add detergents, if required. If they are viscous liquids, prepare a stock solution beforehand (e.g. 30%) but do not allow detergent to take up more than 5% of the total volume. Mix briefly with a magnetic stirrer.

5. Add the volumes of TEMED and then ammonium persulfate as specified in *Table 5*. Immediately mix by swirling, and then transfer the mixture to the gel mould. Carefully overlay with water or butanol.

6. Leave to polymerize for at least 1 h at room temperature or 30 min at 37°C. Never tilt the mould to check whether the gel has polymerized; if it has not polymerized when you tilt it, then the top never will polymerize, because of the mixing caused by tilting. Instead, the differential refractive index between liquid (at the top and usually around the gasket) and gel phase is an effective, and safe, index of polymerization and is shown by the appearance of a distinct line after polymerization.

143

Protocol 3. *Continued*

7. The gels may be stored in their moulds for a couple of days in a refrigerator. For longer storage (up to two weeks for neutral and acidic pH ranges), it is better to disassemble the moulds and (after covering the gels with Parafilm and wrapping with Saran foil) store them in a moist box.

8. Before opening the mould, allow the gel to cool at room temperature if polymerized at a higher temperature. Laying one face of the mould onto the cooling block of the electrophoresis unit may facilitate opening the cassette. Carefully remove the overlay by blotting and also remove the clamps.

9. (a) If the gel is cast against a plastic foil, remove its glass support, then carefully peel the gel from the cover plate.

   (b) If the gel is polymerized on silanized glass, simply force the two plates apart, with a spatula or a blade.

   (c) If the gel is not bound to its support, lay the mould on the bench, with the plate to be removed uppermost and one side protruding a few centimetres from the edge of the table. Insert a spatula at one corner and force against the upper plate. Turn the spatula gently until a few air bubbles form between the gel and the plate then use the spatula as a lever to open the mould. Wipe any liquid from the gel surface with a moistened Kleenex tissue but be careful; keep moving the swab to avoid it sticking to the surface.

10. If bubbles form on both gel sides of the gel, try at the next corner of the mould.

11. If the gel separates from both glass plates and folds up, you may still be able to salvage it, provided the gel thickness is at least 400 μm and the gel concentration is at least 5%T. Using a microsyringe, force a small volume of water below the gel, then carefully make the gel lay flat again by manoeuvring it with a gloved hand. Remove any remaining air bubbles using the needle of the microsyringe. Cover the gel with Parafilm, and gently roll it flat with a rubber roller (take care not to damage it with excess pressure). Carefully remove any residual liquid by blotting.

As a variant to the above procedures, one can polymerize 'empty' gels (i.e. devoid of CAs), wash and dry them, and reswell them in the appropriate CA solution. This is a direct application of IPG technology (3). After polymerization (as in *Protocol 3* but in the absence of CAs), wash the gel three times in 300 ml distilled water each time to remove catalysts and unreacted monomers. Equilibrate the washed gel (20 min with shaking) in 1.5% glycerol and finally dry it onto Gel-Bond PAG foil. It is essential that the gel does not bend so, before drying, the foil should be made to adhere to a supporting glass plate (taking care to remove all air bubbles in between and fastening it in position with clean, rust-proof clamps). Drying must be at room temperature, in front of a fan. Finally, mount the dried gel back in the polymerization cassette and allow it to reswell in the appropriate CA solution (refer to *Figure 9* for more details).

In ultrathin gels, pH gradients are sensitive to the presence of salts, including TEMED and persulfate. Moreover, unreacted monomers are harmful to proteins. The preparation, washing, and re-equilibration of 'empty' gels removes these components from the gel and so avoids these problems.

2.4.3 Filling the mould

One of the three methods may be chosen: by gravity, capillarity, or the flap technique (35).

i. By gravity

This procedure uses a vertical cassette with a rubber gasket U-frame glued to the cover plate (*Figure 3A*). If the cover plate has V-indentations along its free edge as shown in *Figure 3A*, the gel mixture can be transferred simply by using a pipette or a syringe with its tip resting on one of these indentations (*Figure 3D*).

Avoid filling the mould too fast, which will create turbulence and trap air bubbles. If an air bubble appears, stop pouring the solution and try to remove the bubble by tilting and knocking the mould. If this manoeuvre is unsuccessful, displace the bubble with a 1 cm wide strip of polyester foil.

ii. By capillarity

This method is mainly used for casting reasonably thin gels (0.5–1 mm). It requires a horizontal sandwich with two lateral spacers (see *Figure 4A*). Feed the solution either from a pipette or from a syringe fitted with a short piece of fine-bore tubing. It is essential that the solution flows evenly across the whole width of the mould during casting. If an air bubble appears, do not stop pumping in the solution or this will produce more bubbles. Remove all the air bubbles at the end, using a strip of polyester foil. A level table is not mandatory but the mould should be left laying flat until the gel is completely polymerized.

iii. The flap technique (35)

This procedure is again mainly used for preparing thin gels. A 20–50% excess of gel mixture is poured along one edge of the cover (with spacers on both sides) (*Figure 4B*) and the support plate is slowly lowered on it (*Figure 4C*). Air bubbles can be avoided by using of clean plates. If bubbles do get trapped, remove them by lifting and lowering the cover plate once more. Since this method may lead to spilled unpolymerized acrylamide, take precautions for its containment (wear gloves and use absorbent towels to mop up excess).

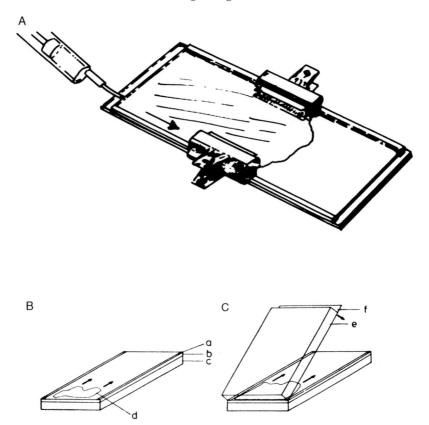

Figure 4. Casting of thin gel plates. (A) Capillary filling of an horizontally placed cassette. (B) and (C) The 'flap' technique: (a) spacer strips, (b) silanized glass plate or polyester film, (c) glass base plate, (d) polymerization mixture, (e) glass cover plate, (f) cover film. (A) Courtesy of LKB; (B) and (C) with permission from Radola (57).

2.4.4 Sample loading and electrophoresis

The electrophoretic procedure is described in *Protocol 4*. The gel is placed on the cooling plate of the electrophoresis chamber. It is necessary to perform the electrophoresis at a constant temperature and with well-defined conditions, since the temperature influences the pH gradient and consequently the separation positions of the proteins. Also the presence of additives (e.g. urea) strongly affects the separation positions by changing the physico-chemical parameters of the solution trapped between the gel matrix; some compensation factors are listed in *Table 8*. Electrodic strips filled with electrodic solutions (see *Table 9*) are placed on the surface of the gel (at anodic and cathodic extremes).

The electrophoretic procedure is devided into two steps. The first is a pre-

Table 8. Compensation factors

Effect of temperature		Effect of urea	
pH	$\Delta 25\text{–}4\,°C$	pH	Δ^a
–	–	3.0	+0.09
3.5	+0.0	3.5	+0.08
4.0	+0.06	4.0	+0.08
4.5	+0.1	4.5	+0.07
5.0	+0.14	5.0	+0.06
5.5	+0.2	5.5	+0.06
6.0	+0.23	6.0	+0.05
6.5	+0.28	6.5	+0.05
7.0	+0.36	7.0	+0.05
7.5	+0.39	7.5	+0.06
8.0	+0.45	8.0	+0.06
8.5	+0.52	8.5	+0.06
9.0	+0.53	9.0	+0.06
9.5	+0.54	9.5	+0.05
10.0	+0.55	–	–

[a] For 1 M urea. The effect is proportional to the urea concentration.

Table 9. IEF electrode solutions

Solution	Application	Concentration
H_3PO_4	Anolyte for all pH ranges	0.1 M
H_2SO_4	Anolyte for very acidic pH ranges ($pH_a < 4$)[a]	0.1 M
CH_3COOH	Anolyte for alkaline pH ranges ($pH_a > 7$)[a]	0.5 M
NaOH	Catholyte for all pH ranges	0.1 M[b]
Histidine	Catholyte for acidic pH ranges ($pH_c < 5$)[a]	0.2 M
Tris	Catholyte for acidic and neutral pH ranges ($pH_c < 5$)[a]	0.5 M

[a] pH_a, pH_c represent the lower and higher extremes of the pH range, respectively.
[b] Store in air-tight plastic bottles.

electrophoresis (characterized by the application of low voltages; see *Table 10*) which allows pre-focusing of carrier ampholytes and the elimination of all contaminants or unreacted catalysts from the gel matrix. Then the protein sample is loaded. In IEF gels, the sample should be applied along the whole pH gradient to determine its optimum application point. The second electrophoretic step is then started (see 'Electrophoresis of samples' in *Table 10*). This is carried out at constant power; low voltages are used for the sample entrance, followed by higher voltages during the separation and band sharpening. The actual voltage used in each step depends on gel thickness (see *Table 10*).

Table 10. Electrophoresis conditions

Pre-electrophoresis	Voltage (V)	Time (min)
Thick gels (> 1.5 mm)	200	30
Thin gels (< 1.5 mm)	400	30

Electrophoresis of samples	Voltage (V)[a]	Time (h)
Thick gels (2 mm)	200	5
Thick gels (1 mm)	400	3
Thin gels (0.5 mm)	600	2
Thin gels (0.25 mm)	800	1.5

[a] Electrophoresis is carried out at constant power so as to give an initial voltage as indicated.

Protocol 4. Electrophoretic procedure

Equipment and reagents

- Electrophoresis apparatus (see Section 2.2.1)
- Narrow-bore combination pH electrode (e.g. from Radiometer)
- Gel polymerized in the gel mould (from *Protocol 3*)
- Whatman No. 17 filter paper
- Electrode solutions (see *Table 9*)
- Protein samples
- Filter paper (e.g. Whatman No. 1, Whatman 3MM, Paratex II) if the sample is to be applied via filter paper tabs (see step 7)
- 10 mM KCl

Method

1. Set the cooling unit at 2–4 °C for normal gels, at 8–10 °C for 6 M urea, and at 10–12 °C for 8 M urea gels.

2. To enable rapid heat transfer, pour a few millilitres of a non-conductive liquid (distilled water, 1% non-ionic detergent, or light kerosene), onto the cooling block of the electrophoretic chamber. Form a continuous liquid layer between the gel support and the apparatus and gently lower the plate into place, avoiding trapping air bubbles and splashing water onto or around the gel. Should this happen, remove all liquid by careful blotting. When the gel is narrower than the cooling block, apply the plate on its middle. If this is not possible, or if the electrode lid is too heavy to be supported by just a strip of gel, insert a wedge (e.g. several layers of Parafilm) between the electrodes and cooling plate.

3. Cut electrode strips (e.g. from Whatman No. 17 filter paper). Note that most paper exposed to alkaline solutions becomes swollen and fragile, so we use only Whatman No. 17 paper strips approx. 5 mm wide and about 3 mm shorter than the gel width. Saturate them with electrode

solutions (see *Table 9*). However, they should not be dripping; blot them on paper towels if required.

4. Wearing disposable gloves, transfer the electrode strips onto the gel. They must be parallel and aligned with the electrodes. The cathodic strip firmly adheres to the gel; do not try to change its position once applied. Avoid cross-contaminating the electrode strips (including with your fingers). If any electrode solution spills over the gel, blot it off immediately, rinse with a few drops of water, and blot again. Check that the wet electrode strips do not exceed the size of the gel and cut away any excess pieces. Be sure to apply the most alkaline solution at the cathode and the most acidic at the anode (if you fear you have misplaced them, note that the colour of the NaOH soaked paper is yellowish but, of course, you may also check this with litmus paper). If you discover a mistake at this point, simply turn the plate around or change the electrode polarity. However, there is no remedy after the current has been on for a while.

5. The salt content of the samples should be kept as low as possible. If necessary, dialyse them against glycine (any suitable concentration) or dissolve in diluted CAs. When buffers are required, use low molarity buffers composed of weak acids and bases.

6. It is best to load the protein samples into pre-cast pockets. These should not be deeper than 50% of the gel thickness. If they are longer than a couple of millimetres and it is necessary to pre-electrophorese the gel (see step 8), the pockets should be filled with dilute CAs. After the pre-electrophoresis (see *Table 10* for conditions) remove this solution by blotting. Then apply the samples.

7. The amount of sample applied should fill the pockets. Try to equalize the volumes and the salt content among different samples. After about 30 min of electrophoresis at high voltage (see *Table 10* for conditions), the content of the pockets can be removed by blotting and new aliquots of the same samples loaded. The procedure can be repeated a third time. Alternatively, the samples can be applied to the gel surface, absorbed into pieces of filter paper. Different sizes and material of different absorbing power are used to accommodate various volumes of liquid (e.g. a 5 mm $\times$ 10 mm tab of Whatman No. 1 can retain about 5 μl, Whatman 3MM about 10 μl, Paratex II more than 15 μl). Up to three layers of paper may be stacked. If the exact amount of sample loaded has little importance, the simplest procedure is to dip the tabs into it, then blot them to remove excess sample. Otherwise, align dry tabs on the gel and apply measured volumes of each sample with a micropipette. For stacks of paper pieces, feed the solution slowly from one side rather than from the top. Do not allow a pool of sample to drag around its pad, but stop feeding liquid to add an extra tab when

Protocol 4. *Continued*

required. This method of application of samples is not suitable for samples containing alcohol.

8. Most samples may be applied to the gel near the cathode without pre-electrophoresis. However, pre-running is advisable if the proteins are sensitive to oxidation or unstable at the average pH of the gel before the run. Pre-running is not suitable for those proteins with a tendency to aggregate upon concentration, or whose solubility is increased by high ionic strength and dielectric constant, or which are very sensitive to pH extremes. Anodic application should be excluded for high salt samples and for proteins (e.g. a host of serum components) denatured at acidic pH. The optimal conditions for sample loading should be experimentally determined, together with the minimum focusing time, with a pilot run, in which the sample of interest is applied in different positions of the gel and at different times. Smears, or lack of confluence of the bands after long focusing time, denote improper sample handling and protein alteration.

9. After electrophoresis, the pH gradient in the focused gel may be read using a contact electrode. The most general approach, however, is to cut a strip along the focusing path (0.5–2 cm wide, with an inverse relation to the gel thickness), then cut segments between 3–10 mm long from this, and elute each for about 15 min in 0.3–0.5 ml of 10 mM KCl (or with the same urea concentration as present in the gel). For alkaline pH gradients, this processing of the gel should be carried out as quickly as possible, the elution medium should be thoroughly degassed and air-tight vials flushed with nitrogen should be used. Measure the pH of the eluted medium using a narrow-bore combination electrode. For most purposes, it is sufficient to note just the temperature of the coolant and the pH measurement in order to define an operational pH. For a proper physico-chemical characterization, the temperature differences should be corrected for as suggested in *Table 8* (which gives also corrections for the presence of urea).

2.5 General protein staining

Table 11 gives a list, with pertinent references, of some of the most common protein stains used in IEF. Detailed recipes are given below. An extensive review covering general staining methods in gel electrophoresis has recently appeared (41); see also Chapter 2. Additionally, although not specifically reported for IEF, some recent developments include: use of Eosin B dye (42); a mixed-dye technique comprising Coomassie blue R-250 and Bismark brown R (43); Stains-all for highly acidic molecules (44); fluorescent dyes for proteins, such as SYPRO orange and SYPRO red (45, 46); a two minute Nile red staining method (47).

Table 11. Protein staining methods

Stains	Application	Sensitivity	Reference
Coomassie blue G-250	General use	Low	36
Coomassie blue R-250/CuSO$_4$	General use	Medium	37
Coomassie blue R-250/sulfosalicylic acid	General use	High	38
Silver stain	General use	Very high	39
Coomassie blue R-250/CuSO$_4$	In presence of detergents	Medium	37
Coomassie blue R-250 at 60°C	In presence of detergents	High	40

2.5.1 Coomassie blue G-250 (36)

The advantages of this protocol (see *Protocol 5*) are that only one step is required (i.e. no protein fixation, no destain), peptides down to approx. 1500 molecular mass can be detected, there is little interference from CAs and, finally, the staining mixture has a long shelf-life. The small amount of dye that may precipitate with time can be removed by filtration or washed from the surface of the gels with liquid soap.

Protocol 5.   Coomassie blue G-250 staining procedure

Reagents

- Staining solution: mix 2 g Coomassie blue G-250 with 400 ml 2 M H$_2$SO$_4$, dilute the suspension with 400 ml distilled water, and stir for at least 3 h. Filter through Whatman No. 1 paper. Then add 89 ml 10 M KOH and 120 ml of 100% (w/v) TCA while stirring.

Method

1. Immerse the gel in staining solution until the required stain intensity is obtained.

2. To remove all salts and to increase the colour contrast, rinse the gel extensively with water.

2.5.2 Coomassie blue R-250/CuSO$_4$ (37)

Protocol 6 describes this stain procedure, which is highly recommended; it is easily carried out and has good sensitivity.

2.5.3 Coomassie blue R-250/sulfosalicylic acid (38)

Heat the gels for 15 min at 60°C in a solution of 1 g Coomassie R-250 in 280 ml methanol and 730 ml water, containing 110 g TCA and 35 g sulfo-salicylic acid (SSA). Destain at 60°C in 500 ml ethanol, 160 ml acetic acid, 1340 ml water. Precipitation of the dye at the gel surface is a common problem (remove it with alkaline liquid soap).

Protocol 6. Coomassie blue R-250/CuSO$_4$ staining procedure

Reagents

- Staining solution: dissolve 1.09 g CuSO$_4$ in 650 ml water and then add 190 ml of acetic acid. Mix this solution with 250 ml of ethanol containing 0.545 g of Coomassie R-250.
- Destaining solution: 600 ml ethanol, 140 ml acetic acid, 1260 ml water

Method

1. Stain the gel (without previous fixation) in staining solution for 30 min to a few hours, depending on its thickness.

2. During immersion in the staining solution, unsupported gels shrink and their surface becomes sticky. Therefore avoid any contact with dry surfaces.

3. Destain the gel in several changes of destaining solution.

2.5.4 Silver stain (39)

A typical method for silver staining is described in *Protocol 7*. A myriad of silvering procedures exist; others are described in Chapter 2 of this book. A good review with guide-lines can be found in ref. 48.

Protocol 7. Silver staining procedure

Reagents

The reagents required are indicated below

Method

The silver staining procedure involves exposing the gel to a series of reagents in a strict sequence, as described in the following steps.

1. 500 ml 12% TCA for 30 min.

2. 1 litre 50% methanol, 12% acetic acid for 30 min.

3. 250 ml 1% HIO$_4$ for 30 min.

4. 200 ml 10% ethanol, 5% acetic acid for 10 min (repeat three times).

5. 200 ml 3.4 mM potassium dichromate, 3.2 mM nitric acid for 5 min.

6. 200 ml water for 30 sec (repeat four times).

7. 200 ml 12 mM silver nitrate for 5 min in the light, then 25 min in the dark.

8. 300 ml 0.28 M sodium carbonate for 30 sec (repeat twice).

9. 300 ml 0.05% formaldehyde for several minutes.

---

10. 100 ml 10% acetic acid for 2 min.

11. 200 ml photographic fixative (e.g. Kodak Rapid Fix) for 10 min.

12. Several litres of water for extensive washes.

13. Finally, remove any silver precipitate on the gel surface using a swab.

2.5.5 Coomassie blue G-250/urea/perchloric acid (41)

This staining procedure is described in *Protocol 8.*

Protocol 8. Coomassie blue G-250/urea/perchloric acid staining procedure[a]

Reagents

- Fixative: dissolve 147 g TCA and 44 g sulfosalicylic acid in 910 ml of water
- Destaining solution: 100 ml acetic acid, 140 ml ethanol, 200 ml ethyl acetate, 1560 ml water

- Staining solution: dissolve 0.4 g Coomassie blue G-250 and 39 g urea in approx. 800 ml of water. Immediately before use, add 29 ml of 70% $HClO_4$ with vigorous stirring. Bring the volume to 1 litre with distilled water.

Method

1. Fix the protein bands for 30 min at 60 °C in fixative.

2. Stain the gel for 30 min at 60 °C in staining solution.

3. Destain for 4–22 h in destaining solution.

[a] This staining protocol may be used in the presence as well as in the absence of detergents.

2.6 Specific protein detection methods

Table 12 lists some of the most common specific detection techniques: it is not meant to be exhaustive.

Table 12. Specific protein detection techniques

Protein detected	Technique	Reference
Glycoproteins	PAS (periodic acid–Schiff) stain	49
Lipoproteins	Sudan black stain	50
Radioactive proteins	Autoradiography	51
	Fluorography	52
Enzymes	Zymograms[a]	53
Antigens	Immunoprecipitation *in situ*	54
	Print-immunofixation	55
	Blotting	56

[a] The concentration of the buffer in the assay medium usually needs to be increased in comparison with zone electrophoresis to counteract the buffering action by carrier ampholytes.

2.7 Quantitation of the focused bands

Some general rules about densitometry are:

(a) The scanning photometer used should have a spatial resolution of the same order as the fractionation technique.

(b) The stoichiometry of the protein:dye complex varies among different proteins.

(c) This stoichiometric relationship is linear only over a limited range of protein concentration.

(d) Photometric measurement is restricted at high absorbance by the limitations of the spectrophotometer.

2.8 Troubleshooting

2.8.1 Waviness of bands near the anode

This may be caused by:

(a) Carbonation of the catholyte: in this case, prepare fresh NaOH with degassed distilled water and store properly.

(b) Excess catalysts: reduce the amount of ammonium persulfate.

(c) Too long sample slots: fill them with dilute CAs.

(d) Too low a concentration of CAs: check the gel formulation.

To alleviate this problem, it is usually also beneficial to add low concentrations of sucrose (2–10%), glycine (e.g. 5–50 mM), or urea (up to 4 M) to the gel, and to apply the sample near the cathode. To salvage a gel during electrophoresis, as soon as the waves appear, apply a new anodic strip soaked with a weaker acid (e.g. acetic acid versus phosphoric acid) in front of the original one, and move the electrodes closer to one another.

2.8.2 Burning along the cathodic strip

This may be caused by:

(a) The formation of a zone of pure water at pH = 7: add to the acidic pH range a 10% solution of either the 3–10 or the 6–8 range ampholytes.

(b) The hydrolysis of the acrylamide matrix after prolonged exposure to alkaline pH: choose a weaker base, if adequate, and, unless a pre-run of the gel is strictly required, apply the electrodic strips after loading the samples.

2.8.3 pH gradients different from expected

It is a commonly experienced problem that the pH gradient experimentally obtained is different from the one expected (i.e. from the pH interval stated on the Ampholine bottle label). Points to note in this respect are:

(a) For acidic and alkaline pH ranges, the problem is alleviated by the choice of anolytes and catholytes whose pH is close to the extremes of the pH gradient.

(b) Alkaline pH ranges should be protected from carbon dioxide by flushing the electrophoretic chamber with moisture-saturated N_2 (or better with argon) and by surrounding the plate with pads soaked in NaOH. It is worth remembering that pH readings on unprotected alkaline solutions become meaningless within half an hour or so because of the absorption of CO_2.

(c) A large amount of a weak acid or base, supplied as sample buffer, may shift the pH range (2-mercaptoethanol is one of such bases). The typical effect of the addition of urea is to increase the apparent pIs of the CAs (see *Table 8*).

(d) The effect may be due to cathodic drift. To counteract this, try one or more of the following remedies:

   (i) Reduce the electrophoresis time to the required minimum (as experimentally determined for the protein of interest, or for a coloured marker of similar M_r).

   (ii) Increase the viscosity of the medium (with sucrose, glycerol, or urea).

   (iii) Reduce the amount of ammonium persulfate used to prepare the gel.

   (iv) Remove acrylic acid impurities by recrystallizing acrylamide and bisacrylamide, and by treating the monomer solutions with mixed-bed ion exchange resin.

   (v) For a final cure, incorporate into the gel matrix a reactive base, such as 2-dimethylamino-propyl-methacrylamide (Polyscience) (10); its optimal concentration (of the order of 1 μM) should be experimentally determined for the system being used.

2.8.4 Sample precipitation at the application point

If large amounts of material precipitate at the application point, even when the M_r of the sample proteins is well below the limits recommended in *Table 4*, the trouble is usually caused by protein aggregation.

(a) Try applying the sample in different positions on the gel, with and without pre-running; some proteins might aggregate only at a given pH.

(b) If there is evidence that the sample contains high M_r components, reduce the value of %T of the polyacrylamide gel.

(c) If it seems possible that the protein aggregation is brought about by the high concentration of the sample (for example, if the problem is also seen during zone electrophoresis) do not pre-run and set a low voltage (100–200 V) for several hours to avoid the concentrating effect of an established pH gradient at the beginning of the run. Also consider decreasing the protein load and switching to a more sensitive detection technique. The addition of detergent and/or urea is usually also beneficial.

(d) If the proteins are sensitive only to the ionic strength and/or the dielectric constant of the medium (in this case they perform well in zone electro-phoresis but are precipitated if dialysed against distilled water), increasing the CA concentration, adding glycine or taurine, and sample application without pre-running may overcome the problem.

(e) The correct choice of denaturing conditions (8 M urea, detergents, 2-mercaptoethanol) very often minimizes these solubility problems, disso-ciating the proteins (and macromolecular aggregates) to polypeptide chains.

3. Immobilized pH gradients

3.1 General considerations

3.1.1 The problems of conventional IEF

Table 13 lists some of the major problems associated with conventional IEF using amphoteric buffers. Some of them are quite severe; for example low ionic strength (*Table 13*, point 1) often induces near-isoelectric precipitation and smearing of proteins, even in analytical runs at low protein loads. The prob-lem of uneven conductivity is magnified in poor ampholyte mixtures, such as Poly Sep 47 (a mixture of 47 amphoteric and non-amphoteric buffers, claimed to be superior to CAs) (58). Due to their poor composition, huge conductivity gaps form along the migration path, against which proteins of different pIs are stacked. The results are simply disastrous (59). Cathodic drift (*Table 13*, point 6) is also a major unsolved problem of CA-IEF, resulting in extensive loss of proteins at the gel cathodic extremity upon prolonged runs. For all these reasons, in 1982 Bjellqvist *et al.* (8) launched the technique of immobilized pH gradients (IPGs).

3.1.2 The Immobiline matrix

IPGs are based on the principle that the pH gradient, which exists prior to the IEF run itself, is co-polymerized, and thus insolubilized, within the fibres of a polyacrylamide matrix. This is achieved by using, as buffers, a set of seven

---

Table 13. Problems with carrier ampholyte focusing

1. Medium of very low and unknown ionic strength
2. Uneven buffering capacity
3. Uneven conductivity
4. Unknown chemical environment
5. Not amenable to pH gradient engineering
6. Cathodic drift (pH gradient instability)

---

non-amphoteric, weak acids and bases, having the following general chemical composition: $CH_2=CH-CO-NH-R$, where R denotes either two different weak carboxyl groups, with pKs 3.6 and 4.6, or four tertiary amino groups, with pKs 6.2, 7.0, 8.5, and 9.3, available under the trade name Immobiline (from Pharmacia–Upjohn). A more extensive set, comprising ten chemicals (a pK 3.1 acidic buffer, a pK 10.3 basic buffer, and two strong titrants, a pK 1 acid and a pK > 12 quaternary base) is available as 'pI select' from Fluka AG, Buchs, Switzerland (see *Tables 14* and *15* for their formulas) (60). During gel polymerization, these buffering species are efficiently incorporated into the gel (84–86% conversion efficiency at 50°C for 1 h). Immobiline-based pH gradients can be cast in the same way as conventional polyacrylamide gradient gels, using a density gradient to stabilize the Immobiline concentration gradient, with the aid of a standard, two-vessel gradient mixer (see *Figure 8*). As shown in their formulas, these buffers are no longer amphoteric, as in conventional IEF, but are bifunctional. At one end of the molecule is located the buffering (or titrant) group, and at the other end is an acrylic double bond,

Table 14. Acidic acrylamido buffers

pK	Formula	Name	M_r
1.0	$CH_2=CH-CO-NH-\overset{\displaystyle CH_3}{\underset{\displaystyle CH_2-SO_3H}{C}}-CH_3$	2-Acrylamido-2-methylpropanesulfonic acid	207
3.1	$CH_2=CH-CO-NH-\underset{\displaystyle OH}{CH}-COOH$	2-Acrylamidoglycolic acid	145
3.6	$CH_2=CH-CO-NH-CH_2-COOH$	*N*-Acryloylglycine	129
4.6	$CH_2=CH-CO-NH-(CH_2)_3-COOH$	4-Acrylamidobutyric acid	157

Table 15. Basic acrylamido buffers

pK	Formula	Name	M_r
6.2	$CH_2=CH-CO-NH-(CH_2)_2-N\overset{\frown}{\underset{\smile}{}}O$	2-Morpholinoethylacrylamide	184
7.0	$CH_2=CH-CO-NH-(CH_2)_3-N\overset{\frown}{\underset{\smile}{}}O$	3-Morpholinopropylacrylamide	198
8.5	$CH_2=CH-CO-NH-(CH_2)_2-N(CH_3)_2$	*N,N*-Dimethylaminoethylacrylamide	142
9.3	$CH_2=CH-CO-NH-(CH_2)_3-N(CH_3)_2$	*N,N*-Dimethylaminopropylacrylamide	156
10.3	$CH_2=CH-CO-NH-(CH_2)_3-N(CH_2H_5)_2$	*N,N*-Diethylaminopropylacrylamide	184
>12	$CH_2=CH-CO-NH-(CH_2)_2-N(CH_2H_5)_3$	*N,N,N*-Triethylaminoethylacrylamide	198

which disappears during immobilization of the buffer on the gel matrix. The three carboxyl Immobilines have rather small temperature coefficients $(\mathrm{d}pK/\mathrm{d}T)$ in the 10–25 °C range, due to their small standard heats of ionization (≈ 1 kcal/mol) and thus exhibit negligible pK variations in this temperature interval. On the other hand, the five basic Immobilines exhibit rather large ΔpKs (as much as $\Delta pK = 0.44$ for the pK 8.5 species) due to their larger heats of ionization (6–12 kcal/mol). Therefore, for reproducible runs and pH gradient calculations, all the experimental parameters have been fixed at 10 °C.

Temperature is not the only variable that affects Immobiline pKs (and therefore the actual pH gradient generated). Additives in the gel that change the water structure (chaotropic agents, e.g. urea) or lower its dielectric constant, and the ionic strength of the solution, alter their pK values. The largest changes, in fact, are due to the presence of urea: acidic Immobilines increase their pK in 8 M urea by as much as 0.9 pH units, while the basic Immobilines increase their pK by only 0.45 pH unit (61). Detergents in the gel (2%) do not alter the Immobiline pK, suggesting that they are not incorporated into the surfactant micelle. For generating extended pH gradients, we use two additional chemicals which are strong titrants having pKs well outside the desired pH range. One is QAE (quaternary amino ethyl) acrylamide ($pK > 12$) and the other is AMPS (2-acrylamido-2-methyl-propane-sulfonic acid, $pK \approx 1.0$) (see *Tables 14* and *15*).

As shown in *Figure 5*, in IPG the proteins are placed on a gel with a pre-formed, immobilized pH gradient (represented by carboxyl and tertiary amino groups grafted to the polyacrylamide chains). When the field is applied, only the sample molecules (and any ungrafted ions) migrate in the electric field. Upon termination of electrophoresis, the proteins are separated into stationary, isoelectric zones. Due to the possibility of designing stable pH gradients at will, separations have been reported in only 0.1 pH unit-wide gradients over the entire separation axis leading to an extremely high resolving power ($\Delta pI = 0.001$ pH unit, see also Section 2.1.3).

3.1.3 Narrow and ultra narrow pH gradients

We define the gradients from 0.1–1 pH unit as narrow (towards the 1 pH unit limit) and those close to the 0.1 pH unit limit, as ultra narrow gradients. Within these limits we work on a tandem principle—that is, we choose a buffering Immobiline, either a base or an acid, with its pK within the pH interval we want to generate, and a non-buffering Immobiline, then an acid or a base, respectively, with its pK at least 2 pH units removed from either the minimum or maximum of our pH range. The titrant will provide equivalents of acid or base to titrate the buffering group but will not itself buffer in the desired pH interval. For these calculations, we used to resort to modified Henderson–Hasselbalch equations and to rather complex nomograms found in the LKB application note No. 321. In a later note (No. 324) are listed 58

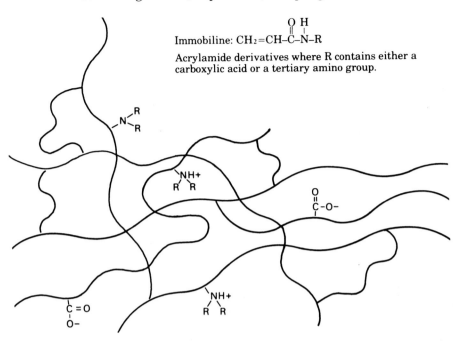

Figure 5. Isoelectric focusing in immobilized pH gradients (IPG). A hypothetical gel structure is depicted, where the lines represent the neutral acrylamide residues, the cross-over points the bisacrylamide cross-linking, and the positive and negative charges the grafted Immobiline molecules. (Courtesy of LKB Produkter AB.)

gradients each 1 pH unit wide, starting with the pH 3.8–4.8 interval and ending with the pH 9.5–10.5 range, separated by 0.1 pH unit increments. In *Table 16* are recipes giving the Immobiline volume which must be added to give 15 ml of mixture in the acidic (mixing) chamber to obtain pH_{min} and the corresponding volume for the basic (reservoir) chamber of the gradient mixer needed to generate pH_{max} of the desired pH interval. For 1 pH unit gradients between the limits pH 4.6–6.1 and pH 7.2–8.4 there are wide gaps in the pKs of neighbouring Immobilines and so three Immobilines need to be used to generate the desired pH_{min} and pH_{max} values (*Table 16*). As an example, consider the pH 4.6–5.6 interval. There are no available Immobilines with pKs within this pH region, so the nearest species, pKs 4.6 and 6.2, will act as both partial buffers and partial titrants. A third Immobiline is needed in each vessel, a true titrant that will bring the pH to the desired value. As titrant for the acidic solution (pH_{min}) we use pK 3.6 Immobiline and for pH_{max} we use pK 9.3 Immobiline (*Table 16*).

   If a narrower pH gradient is needed, it can be derived from any of the 58 1 pH intervals given in *Table 16* by a simple linear interpolation of intermediate Immobiline molarities. Suppose that from a pH 6.8–7.8 range, which is excellent for most haemoglobin (Hb) analyses, we want to obtain a pH gradient of

Table 16. 1 pH unit gradients: volumes of Immobiline for 15 ml of each starting solution[a,b]

Control pH at 20°C	Volume (μl) 0.2 M Immobiline pK acidic dense solution							pH range	Mid point	Control pH at 20°C	Volume (μl) 0.2 M Immobiline pK basic light solution						
	3.6	4.4	4.6	6.2	7.0	8.5	9.3				3.6	4.4	4.6	6.2	7.0	8.5	9.3
3.84±0.03	—	750	—	—	—	—	159	3.8–4.8	4.3	4.95±0.06	—	750	—	—	—	—	591
3.94±0.03	—	710	—	—	—	—	180	3.9–4.9	4.4	5.04±0.07	—	810	—	—	—	—	667
4.03±0.03	—	—	755	—	—	—	157	4.0–5.0	4.5	5.14±0.06	—	—	745	—	—	—	584
4.13±0.03	—	—	713	—	—	—	177	4.1–5.1	4.6	5.23±0.07	—	—	803	—	—	—	659
4.22±0.03	—	—	689	—	—	—	203	4.2–5.2	4.7	5.33±0.08	—	—	884	—	—	—	753
4.32±0.03	—	—	682	—	—	—	235	4.3–5.3	4.8	5.42±0.10	—	—	992	—	—	—	871
4.42±0.03	—	—	691	—	—	—	275	4.4–5.4	4.9	5.52±0.12	—	—	1133	—	—	—	1021
4.51±0.04	—	—	716	—	—	—	325	4.5–5.5	5.0	5.61±0.14	—	—	1314	—	—	—	1208
4.64±0.05	562	—	600	863	—	—	—	4.6–5.6	5.1	5.69±0.04	—	—	863	863	—	—	105
4.75±0.05	458	—	675	863	—	—	—	4.7–5.7	5.2	5.79±0.04	—	—	863	863	—	—	150
4.86±0.04	352	—	750	863	—	—	—	4.8–5.8	5.3	5.90±0.04	—	—	863	863	—	—	202
4.96±0.03	218	—	863	863	—	—	—	4.9–5.9	5.4	5.99±0.03	—	—	863	863	—	—	248
5.07±0.03	158	—	863	863	—	—	—	5.0–6.0	5.5	6.09±0.04	—	—	863	863	—	—	338
5.17±0.04	113	—	863	863	—	—	—	5.1–6.1	5.6	6.20±0.04	—	—	863	713	—	—	443
5.24±0.18	1251	—	—	1355	—	—	—	5.2–6.2	5.7	6.34±0.04	337	—	—	724	—	—	—
5.33±0.12	1055	—	—	1165	—	—	—	5.3–6.3	5.8	6.43±0.03	284	—	—	694	—	—	—
5.43±0.12	899	—	—	1017	—	—	—	5.4–6.4	5.9	6.53±0.03	242	—	—	682	—	—	—
5.52±0.09	775	—	—	903	—	—	—	5.5–6.5	6.0	6.63±0.03	209	—	—	685	—	—	—
5.62±0.07	676	—	—	817	—	—	—	5.6–6.6	6.1	6.73±0.03	182	—	—	707	—	—	—
5.71±0.06	598	—	—	755	—	—	—	5.7–6.7	6.2	6.82±0.03	161	—	—	745	—	—	—
5.81±0.06	536	—	—	713	—	—	—	5.8–6.8	6.3	6.92±0.03	144	—	—	803	—	—	—
5.91±0.05	486	—	—	689	—	—	—	5.9–6.9	6.4	7.02±0.03	131	—	—	884	—	—	—
6.01±0.05	447	—	—	682	—	—	—	6.0–7.0	6.5	7.12±0.03	120	—	—	992	—	—	—
6.10±0.04	416	—	—	691	—	—	—	6.1–7.1	6.6	7.22±0.03	112	—	—	1133	—	—	—
6.11±0.11	972	—	—	—	1086	—	—	6.2–7.2	6.7	7.21±0.03	262	—	—	—	686	—	—
6.21±0.09	833	—	—	—	956	—	—	6.3–7.3	6.8	7.31±0.03	224	—	—	—	682	—	—
6.30±0.08	722	—	—	—	857	—	—	6.4–7.4	6.9	7.41±0.03	195	—	—	—	694	—	—
6.40±0.07	635	—	—	—	783	—	—	6.5–7.5	7.0	7.50±0.03	171	—	—	—	724	—	—

pH (±)	V1	V2	V3	V4	pH range	control pH	pH (±)	V5	V6	V7	V8
6.49±0.06	565	732	—	—	6.6–7.6	7.1	7.60±0.03	152	771	—	—
6.59±0.05	509	699	—	—	6.7–7.7	7.2	7.70±0.03	137	840	—	—
6.69±0.05	465	683	—	—	6.8–7.8	7.3	7.80±0.03	125	934	—	—
6.78±0.04	430	684	—	—	6.9–7.9	7.4	7.80±0.03	116	1058	—	—
6.88±0.04	403	701	—	—	7.0–8.0	7.5	8.00±0.03	108	1217	—	—
6.98±0.04	381	736	—	—	7.1–8.1	7.6	8.09±0.03	103	1422	—	—
7.21±0.06	1028	750	—	—	7.2–8.2	7.7	8.36±0.05	548	750	—	750
7.31±0.06	983	750	—	—	7.3–8.3	7.8	8.46±0.05	503	750	—	750
7.41±0.05	938	750	—	—	7.4–8.4	7.9	8.56±0.05	458	750	—	750
7.66±0.15	1230	—	1334	—	7.5–8.5	8.0	8.76±0.04	331	—	720	—
7.75±0.12	1037	—	1049	—	7.6–8.6	8.1	8.85±0.03	279	—	692	—
7.85±0.10	885	—	1004	—	7.7–8.7	8.2	8.95±0.03	238	—	682	—
7.94±0.08	764	—	893	—	7.8–8.8	8.3	9.05±0.06	206	—	687	—
8.04±0.07	667	—	810	—	7.9–8.9	8.4	9.14±0.06	180	—	710	—
8.13±0.06	591	—	750	—	8.0–9.0	8.5	9.24±0.06	159	—	750	—
8.23±0.06	530	—	710	—	8.1–9.1	8.6	9.34±0.06	143	—	810	—
8.33±0.05	482	—	687	—	8.2–9.2	8.7	9.44±0.06	130	—	893	—
8.43±0.04	443	—	682	—	8.3–9.3	8.8	9.54±0.06	119	—	1004	—
8.52±0.04	413	—	692	—	8.4–9.4	8.9	9.64±0.06	111	—	1149	—
8.62±0.04	389	—	720	—	8.5–9.5	9.0	9.74±0.06	105	—	1334	—
8.40±0.14	1208	—	—	1314	8.6–9.6	9.1	9.50±0.06	325	—	—	716
8.49±0.12	1021	—	—	1133	8.7–9.7	9.2	9.59±0.06	275	—	—	691
8.59±0.10	871	—	—	992	8.8–9.8	9.3	9.69±0.06	235	—	—	682
8.68±0.08	753	—	—	884	8.9–9.9	9.4	9.79±0.06	203	—	—	689
8.78±0.07	659	—	—	803	9.0–10.0	9.5	9.88±0.06	177	—	—	713
8.87±0.06	584	—	—	745	9.1–10.1	9.6	9.98±0.06	157	—	—	755
8.97±0.05	525	—	—	707	9.2–10.2	9.7	10.08±0.06	141	—	—	817
9.07±0.04	478	—	—	686	9.3–10.3	9.8	10.18±0.06	129	—	—	903
9.16±0.07	440	—	—	682	9.4–10.4	9.9	10.28±0.06	119	—	—	1017
9.26±0.07	410	—	—	694	9.5–10.5	10.0	10.38±0.06	111	—	—	1165

[a] From LKB Application Note 324 (1984). The pH range (given in the middle column) is the one existing in the gel during the run at 10°C. For controlling the pH of the starting solutions, the values (control pH) are given at 20°C.

[b] When using the standard gel cassette (*Figure 8*), the volumes given are sufficient to prepare two gels.

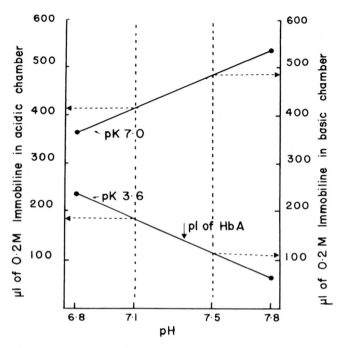

Figure 6. Graphic representation of the preparation of narrow (up to 1 pH unit) IPG gradients on the 'tandem' principle. The limiting molarities of p*K* 7.0 (buffering species) and p*K* 3.6 (titrant) Immobilines needed to generate a pH 6.8–7.8 interval, as obtained directly from *Table 19*, are plotted on a graph. These points are joined by straight lines, and the new molarities needed to generate any narrower pH gradient within the stated pH intervals are obtained by simple linear interpolation (broken vertical and horizontal lines). In this example a narrow pH 7.1–7.5 gradient is graphically derived.

7.1–7.5, which will resolve neutral mutants that co-focus with HbA. *Figure 6* shows the graphic method. The limiting molarities of the two Immobilines in the 1 pH unit interval are joined by a straight line (because the gradient is linear), and then the new pH interval is defined according to experimental needs (in our case pH 7.1–7.5). Two new lines are drawn from the two new limits of the pH interval, parallel to the ordinates (broken vertical lines). Where they intersect the two sloping lines defining the two Immobiline molarities, four new lines (dashed) are drawn parallel to the abscissa and four new molarities of the Immobilines defining the new pH interval are read directly on the ordinates. This process can be repeated for any desired pH interval down to ranges as narrow as 0.1 pH units.

3.1.4 Extended pH gradients: general rules for their generation and optimization

Linear pH gradients are obtained by arranging for an even buffering power throughout. The latter could be ensured only by ideal buffers spaced apart by

$\Delta pK = 1$. In practice, there are only eight buffering Immobilines with some wider gaps in ΔpKs, and so other approaches must be used to solve this problem. Two methods are possible. In one approach (*constant buffer concentration*), the concentration of each buffer is kept constant throughout the span of the pH gradient and 'holes' of buffering power are filled by increasing the amounts of the buffering species bordering the largest ΔpKs. In the other approach (*varying buffer concentration*), the variation in concentration of different buffers along the width of the desired pH gradient results in a shift in each buffer's apparent pK, together with the ΔpK values evening out. The second approach is preferred, since it gives much higher flexibility in the computational approach. In a series of papers (62–72), we have described a computer approach able to calculate and optimize any such pH interval, up to the most extended one (which can cover a span of pH 2.5–11. Tables for these recipes can be found in the book by Righetti (3) and in many of the above references (62–64, 71, 72). We prefer not to give such recipes here since anyone can easily calculate any desired pH interval with the user-friendly computer program (written on a Windows platform) of Giaffreda *et al.* (73) (available from Fluka Chemie). However, we will give here general guide-lines for the use of this program and the optimization of various recipes:

(a) When calculating recipes up to 4 pH units, in the pH 4–9 interval, there is no need to use strong titrants. As most acidic and basic titrants the pKs 3.1 and 10.3 Immobilines can be used, respectively.

(b) When optimizing recipes > 4 pH units (or close to the pH 3 or pH 11 extremes) strong titrants have to be used, otherwise linear pH gradients will never be obtained (since weak titrants will act as buffering ions as well).

(c) When calculating recipes of 4 pH units, it is best to insert in the recipe all the eight weak buffering Immobilines. The computer program will automatically exclude the ones not needed for optimization.

(d) The program of Giaffreda *et al.* (73) can calculate not only linear, but also concave or convex exponential gradients (including sigmoidal ones). In order to limit consumption of Immobilines (at high concentration in the gel they could give rise to reswelling and also interact with the macromolecule via ion exchange mechanisms), limit the total Immobiline molarity (e.g. to only 20 mM) and the average buffering power (β). In fact, these two items of information are specifically asked for by the program when preparing any recipe. In particular, note that recipes with an average β value of only 1–2 mequiv/litre/pH are quite adequate in IPGs. The protein macro ions, even at concentration > 10 mg/ml, rarely have β values > 1 microequiv/litre/pH.

(e) When working at acidic and alkaline pH extremes, however, please note that the average β power of the recipe should be progressively higher, so as to counteract the β value of bulk water. Additionally, at such pH extremes, the matrix acquires a net positive or net negative charge and this gives rise to strong electro-osmotic flow (EOF). In order to quench EOF, the washed and dried matrix should be reswollen against a gradient of viscous polymers (e.g. liquid linear polyacrylamide, hydroxyethyl cellulose) (74).

3.1.5 Non-linear, extended pH gradients

Although originally most IPG formulations for extended pH intervals had been described only in terms of rigorously linear pH gradients, this might not be the optimal solution in some cases. For example, the pH slope might need to be altered in pH regions that are overcrowded with proteins. This is particularly important in the separation of proteins in a complex mixture, such as cell lysates and so is imperative when performing two-dimensional (2D) maps. We have computed the statistical distribution of the pIs of water soluble proteins and plotted them in the histogram of *Figure 7*. From the histogram, given the relative abundance of different species, it is clear that an optimally resolving pH gradient should have a gentler slope in the acidic portion and a steeper profile in the alkaline region. Such a pH profile has been calculated by

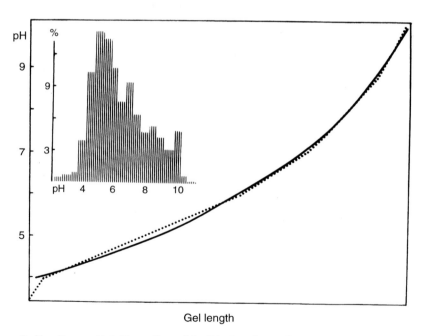

Gel length

Figure 7. Non-linear pH 4–10 gradient: 'ideal' (....) and actual (——) courses. The shape of the 'ideal' profile was computed from data on statistical distribution of protein pIs. The inset histogram shows the statistical distribution of the pIs of water soluble proteins and is redrawn with permission from Gianazza *et al.* (64).

assigning to each 0.5 pH unit interval in the pH 3.5–10 region a slope inversely proportional to the relative abundance of proteins in that interval. This generated the ideal curve (dotted line) in *Figure 7*. What is also important here is the establishment of a new principle in IPG technology, namely that the pH gradient and the density gradient stabilizing it need not be colinear, because the pH can be adjusted by localized flattening for increased resolution while leaving the density gradient unaltered. Although we have considered only the example of an extended pH gradient, narrower pH intervals can be treated in the same fashion.

3.1.6 Extremely alkaline pH gradients

We have recently optimized a recipe for producing an extremely alkaline immobilized pH gradient, covering non-linearly the pH 10–12 interval, for separation of very alkaline proteins, such as subtilisins and histones (75, 76). Successful separations were obtained in 6%T, 4%C polyacrylamide matrices, reswollen in 8 M urea, 1.5% Tween 20, 1.5% Nonidet P-40, and 0.5% Ampholine pH 9–11. Additionally, in order to quench the very high conductivity of the gel region on the cathodic side, the reswelling solution contained a 0–10% (anode to cathode) sorbitol gradient (or an equivalent 0–1% HEC gradient). Best focusing was obtained by running the gel at 17°C, instead of the customary 10°C temperature. In the case of histones, all of the major components had pI values between pH 11–12 and only minor components (possibly acetylated and phosphorylated forms) focused below pH 11. By summing up all bands observed in Arg- and Lys-rich fractions, eight to ten major components and at least 12 minor zones were clearly resolved. This same recipe could be used as first dimension run for a 2D separation of histones (77).

3.1.7 Storage of the Immobiline chemicals

There are two major problems with the Immobiline chemicals, especially with the alkaline ones; hydrolysis and spontaneous autopolymerization. Hydrolysis is quite a nuisance because then only acrylic acid is incorporated into the IPG matrix, with a strong acidification of the calculated pH gradient. Hydrolysis is an autocatalysed process for the basic Immobilines, since it is pH-dependent. For the pK 8.5 and 9.3 species, such a cleavage reaction on the amido bond can occur even in the frozen state, at a rate of about 20% per year. Autopolymerization, is also quite deleterious for the IPG technique. Again, this reaction occurs particularly with alkaline Immobilines, and is purely autocatalytic since it is greatly accelerated by deprotonated amino groups. Oligomers and n-mers are formed which stay in solution and can even be incorporated into the IPG gel. These products of autopolymerization, when added to proteins in solution, are able to bridge them via two unlike binding surfaces. A lattice is formed and the proteins (especially larger ones, such as ferritin, α_2-macroglobulin, and thyroglobulin) are precipitated out of solution.

This precipitation effect is quite marked and is noticeable even with short oligomers (> decamer).

These problems with basic Immobilines could potentially remove one of the major advantages of the IPG technique, namely its high reproducibility from run-to-run. As a remedy to these drawbacks, it has been shown that, when dissolved in anhydrous *n*-propanol (containing a maximum of 60 p.p.m. water), these species are stabilized against both hydrolysis and autopolymerization for a virtually unlimited period of time (less than 1% degradation per year even when stored at +4°C). Thus, present day alkaline Immobiline bottles are now supplied as 0.2 M solutions in *n*-propanol. The acidic Immobilines, being much more stable, are available as water solutions containing 10 p.p.m. of an inhibitor (79).

3.2 IPG methodology

The overall procedure is outlined in *Protocol 9*. Note that the basic equipment required is the same as for conventional CA-IEF gels. Thus the reader should consult Sections 2.2.1 and 2.2.2. In addition, as we use essentially the same polyacrylamide matrix, the reader is referred to Section 2.3 for a description of its general properties.

Protocol 9.   IPG flow sheet

1. Assemble the gel mould (*Protocol 10*) and mark the polarity of the pH gradient on the back of the supporting plate.
2. Mix the required amounts of Immobilines. Fill to one-half of the final volume with distilled water.
3. Check the pH of the solution and adjust as required.
4. Add the correct volume of 30%T acrylamide:bisacrylamide monomer (*Table 5*), glycerol (0.2–0.3 ml/ml of the 'dense' solution only), and TEMED (*Table 17*), and bring to final volume with distilled water.
5. For pH ranges removed from neutrality, titrate to about pH 7.5 using Tris base for acidic solutions and acetic acid for alkaline solutions.
6. Transfer the denser solution to the mixing chamber and the lighter solution to the other reservoir of the gradient mixer. Centre the mixer on a magnetic stirrer and check for the absence of air bubbles in the connecting duct.
7. Add ammonium persulfate to the solutions as specified in *Table 17*.
8. Allow the gradient to pour into the mould from the gradient mixer.
9. After pouring, allow the gel to polymerize for 1 h at 50°C.
10. Disassemble the mould and weigh the gel.
11. Wash the gel for 1 h for three times (20 min each) with 200 ml of distilled water with gentle shaking.

---

12. Reduce the gel back to its original weight using a non-heating fan.

13. Transfer the gel to the electrophoresis chamber (at 10°C) and apply the electrodic strips (as described in *Protocol 4*).

14. Load the samples and start the run.

15. After electrophoresis, stain the gel to detect the separated proteins.

Table 17. Working concentrations for the catalysts

Gel type	TEMED (μl/ml)		40% (w/v) ammonium persulfate (μl/ml)
	Acidic pH	Basic pH	
Lower limit	0.5	0.3	0.6
Standard, %T = 5[a]	0.5	0.3	0.8
Standard, %T = 3	0.7	0.5	1.0
For 5–10% alcohols	0.7	0.5	1.0
Higher limit[b]	0.9	0.6	1.4

[a] From ref. 8.
[b] From LKB Application Note No. 321.

3.2.1 Casting an Immobiline gel

When preparing for an IPG experiment, two pieces of information are required: the total liquid volume needed to fill the gel cassette, and the required pH interval. Once the first is known, this volume is divided into two halves: one-half is titrated to one extreme of the pH interval, the other to the opposite extreme. As the analytical cassette usually has a thickness of 0.5 mm and, for the standard 12 × 25 cm size (see *Figure 3*), contains 15 ml of liquid to be gelled, in principle two solutions, each of 7.5 ml, should be prepared. However, because the volume of some Immobilines to be added to 7.5 ml might sometimes be rather small (i.e. < 50 μl), we prefer to prepare a double volume, which will be enough for casting two gel slabs. The Immobiline solutions (mostly the basic ones) tend to leave droplets on the plastic disposable tips of micropipettes. For accurate dispensing, therefore, we suggest rinsing the tips once or twice with distilled water after each measurement. The polymerization cassette is filled with the aid of a two-vessel gradient mixer and thus the liquid elements which fill the vertically standing cassette have to be stabilized against remixing by a density gradient. In *Table 16* the two solutions are called 'acidic dense' and 'basic light' solutions. This choice is, however, a purely conventional one, and can be reversed, provided one marks the bottom of the mould as the cathodic side.

i. Assembling the gel mould

IPG gels are 0.5 mm thick. Therefore the polymerization is usually performed on a Gel-Bond PAG film. The cassette is assembled as described in *Protocol 10* and allowed to cool to 4 °C. The gradient mixer is carefully washed and cooled in the refrigerator to 4 °C. The light basic Immobiline solution and the dense acidic Immobiline solution are prepared according to the volumes listed in *Table 16* and also allowed to cool to 4 °C. Cooling the cassette and the solutions is an effective way to delay the onset of polymerization.

Protocol 10. Assembling the mould

Equipment
- Gel mould (Pharmacia)
- Dymo tape
- Repel silane (Sigma, Merck, Serva, etc.)
- Gel-Bond PAG film (Pharmacia)

Method

1. Wash the glass plate bearing the U-frame with detergent and rinse with distilled water.
2. Dry with paper tissue.
3. To mould sample application slots in the gel, apply suitably sized pieces of Dymo tape to the glass plate with the U-frame. A 5 × 3 mm slot can be used for sample volumes between 5–20 μl (this step is necessary only when preparing a new mould or rearranging an old one; see *Figure 3A*). To prevent the gel from sticking to the glass plates bearing the U-frame and slot former, coat them with repel silane according to *Protocol 1*. Make sure that no dust or fragments of gel from previous experiments remain on the surface of the gasket, since this can cause the mould to leak.
4. Use a drop of water on the Gel-Bond PAG film to determine the hydrophilic side. Apply a few drops of water to the plain glass plate and carefully lay the sheet of Gel-Bond PAG film on top with the hydrophobic side down (see *Figure 3B*). Avoid touching the surface of the film with fingers. Allow the film to extend 1 mm over one of the long sides of the plate, as a support for the tubing from the gradient mixer when filling the cassette with gel solution (but only if using a cover plate without V-indentations). Roll the film flat to remove air bubbles and to ensure good contact with the glass plate.
5. Clamp the glass plates together with the Gel-Bond PAG film and slot former on the inside, using clamps placed all along the U-frame, opposite to the protruding film. To avoid leakage, the clamps must be positioned so that the maximum possible pressure is applied (see *Figure 3C*).

Figure 8 gives the final assembly for cassette and gradient mixer. Note that inserting the capillary tubing conveying the solution from the mixer into the cassette is greatly facilitated when using a cover plate bearing three V-shaped indentations. As for the gradient mixer, it should be noted that one chamber contains a magnetic stirrer, while in the reservoir is inserted a plastic cylinder having the same volume, held by a trapezoidal rod. The latter, in reality, is a 'compensating cone' needed to raise the liquid level to such an extent that the two solutions (in the mixing chamber and in the reservoir) will be hydro-statically equilibrated. In addition, this plastic rod can also be utilized for manually stirring the reservoir after addition of TEMED and persulfate.

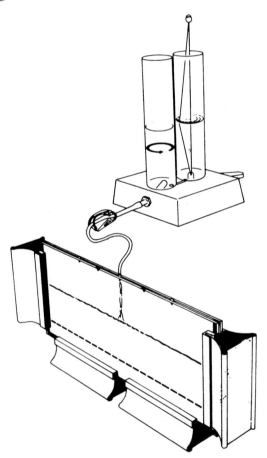

Figure 8. Set-up for casting an IPG gel. A linear pH gradient is generated by mixing equal volumes of a dense and light solution, titrated to the extremes of the desired pH interval. Note the 'compensating' rod in the reservoir, used as a stirrer after addition of catalysts and for hydrostatically equilibrating the two solutions. Insertion of the capillary convey-ing the solution from the mixer to the cassette is greatly facilitated by using modern cover plates, bearing three V-shaped indentations. (Courtesy of LKB Produkter AB.)

ii. *Polymerization of a linear pH gradient*

It is preferable to use 'soft' gels, i.e. with a low %T. Originally, all recipes were given for 5%T matrices, but today we prefer 3.5–4%T gels. These 'soft' gels can be easily dried without cracking and allow better entry of larger proteins. In addition, the local ionic strength along the polymer coil is increased, and this permits sharper protein bands due to increased solubility at the pI. A linear pH gradient is generated by mixing equal volumes of the two starting solutions in a gradient mixer. It is a must, for any gel formulation removed from neutrality (pH 6.5–7.5), to titrate the two solutions to neutral pH, so as to ensure reproducible polymerization conditions and avoid hydrolysis of the five alkaline buffering Immobilines. If the pH interval used is acidic, add Tris, if it is basic, add acetic acid. We recommend that a minimum of 15 ml of each solution (enough for two gels) is prepared and that the volumes of Immobiline needed are measured with a well-calibrated microsyringe to ensure high accuracy. Prepare the acidic dense solution and the basic light solution for the pH gradient as described in *Protocol 9* (stock acrylamide solutions are given in *Table 2*). Polymerize the gel by addition of the appropriate amounts of catalysts (see *Table 17*). The amounts of catalysts change as a function of the final concentration of monomers and in the presence of alcohols (which are responsible for a slight inhibitory effect on the polymerization reaction). If the same gradient is to be prepared repeatedly, the buffering and non-buffering Immobiline and water mixtures can be prepared as stock solutions and stored according to the recommendations for Immobiline. Prepared gel solutions must not be stored. However, gels with a pH less than 8 can be stored in a humidity chamber for up to one week after polymerization. An example of preparation of a linear pH gradient is given in *Protocol 11*.

Protocol 11. Polymerization of a linear pH gradient gel

Equipment and reagents

- Gradient maker (e.g. from Pharmacia–Hoefer): a model with vessels of 15 or 30 ml volume
- Magnetic stirrer and stirring bar
- Oven at 50°C

- Gel mould with V-indentations (see text and *Figure 8*)
- Non-heating fan
- Basic light and acidic dense gel mixtures (see Section 3.2.1): 7.5 ml of each

Method

1. Check that the valve in the gradient mixer and the clamp on the outlet tubing are both closed.

2. Transfer 7.5 ml of the basic, light solution to the reservoir chamber.

3. Slowly open the gradient maker valve just enough to fill the connecting channel with the solution and then quickly close it again. Transfer 7.5 ml of the acidic, dense solution to the mixing chamber.

4. Place the prepared mould upright on a levelled surface. The optimum

---

flow rate is obtained when the outlet of the gradient mixer is 5 cm above the top of the mould. Open the clamp of the outlet tubing, fill the tubing half-way with the dense solution, and close the clamp again.

5. Switch on the stirrer and set it to a speed of about 500 r.p.m.

6. Add the catalysts to each chamber as specified in *Table 17*.

7. Insert the free end of the tubing between the glass plates of the mould at the central V-indentation (*Figure 8*).

8. Open the clamp on the outlet tubing, then immediately open the valve between the dense and light solutions so that the gradient solution starts to flow down into the mould by gravity. Make sure that the levels of liquid in the two chambers fall at the same rate. The mould will be filled within 5 min. To assist the mould to fill uniformly across its width, the tubing from the mixer may be substituted with a two- or three-way outlet assembled from small glass or plastic connectors (e.g. spare parts of chromatographic equipment) and butterfly needles.

9. When the gradient mixer is empty, carefully remove the tubing from the mould. After leaving the cassette to rest for 5 min, place it on a levelled surface in an oven at 50 °C. Allow polymerization to continue for 1 h. Meanwhile, wash and dry the mixer and tubing.

10. When polymerization is complete, remove the clamps and carefully take the mould apart. Start by removing the glass plate from the supporting foil. Then hold the remaining part so that the glass surface is on top and the supporting foil underneath. Gently peel the gel away from the slot former, taking special care not to tear the gel around the slots.

11. Weigh the gel and then place it in 300 ml of distilled water for 1 h to wash out any remaining ammonium persulfate, TEMED, and un-reacted monomers and Immobilines. Change the water three times (changes are every 20 min).

12. After washing the gel, carefully remove any excess water from the surface with a moist paper tissue. To remove the water absorbed by the gel during the washing step, leave it at room temperature until the weight has returned to within 5% of the original weight. To shorten the drying time, use a non-heating fan placed at about 50 cm from the gel to increase the rate of evaporation. Check the weight of the gel after 5 min, and from this, estimate the total drying time. The drying step is essential since a gel containing too much water will 'sweat' during the electrofocusing run and droplets of water will form on the surface. However, if the gel dries too much, the value of %T will increase, resulting in longer focusing times and a greater sieving effect.

3.2.2 Reswelling dry Immobiline gels

Pre-cast, dried Immobiline gels, encompassing a few acidic ranges, are now available from Pharmacia–Hoefer Biotech. They all contain 4%T and they span the following pH ranges: pH 4–7; pH 4.2–4.8 (e.g. for α_1-antitrypsin analysis); pH 4.5–5.4 (e.g. for Gc screening); pH 5.0–6.0 (e.g. for transferrin analysis); and pH 5.6–6.6 (e.g. for phosphoglucomutase screening). Pre-cast, dried IPG gels in the alkaline region have not been introduced as yet, possibly because at high pHs the hydrolysis of both the gel matrix and the Immobiline chemicals bound to it is much more pronounced. However, a non-linear, pH 3–10 range for 2D maps is available.

It has been found that the diffusion of water through Immobiline gels does not follow a simple Fick's law of passive transport from high (the water phase) to zero (the dried gel phase) concentration regions, but it is an active phenomenon: even under isoionic conditions, acidic ranges cause swelling four to five times faster than alkaline ones. Given these findings, it is preferable to reswell dried Immobiline gels in a cassette similar to the one for casting the IPG gel. *Figure 9* shows the reswelling system: the dried gel is inserted in the cassette, which is clamped and allowed to stand on the short side. The reswelling solution is gently injected into the chamber via a small hole in the lower right side using a cannula and a syringe, until the cassette is completely filled. As the system is volume controlled, it can be left to reswell overnight, if needed. Gel drying and reswelling is the preferred procedure when an IPG gel containing additives is needed. In this case it is always best to cast an 'empty' gel (i.e. lacking the additives), wash it, dry it, and then reconstitute it in presence of the desired additive (e.g. urea, alkyl ureas, detergents, carrier ampholytes, and mixtures thereof).

3.2.3 Electrophoresis

A list of the electrode solutions in common use can be found in *Table 18*. A common electrophoresis protocol consists of an initial voltage setting of 500 V, for 1–2 h, followed by an overnight run at 2–2500 V. Ultra narrow gradients are further subjected to a couple of hours at 5000 V, or better at about 1000 V/cm across the region containing the bands of interest.

3.2.4 Staining and pH measurements

IPGs tend to bind strongly to dyes, so the gels are better stained for a relatively short time (30–60 min) with a stain of medium intensity, e.g. the second method listed in *Table 11*. For silver staining, a novel procedure optimized for IPG gels has been published (78) and is given in *Protocol 12*.

Accurate pH measurements are virtually impossible by equilibration between a gel slice and excess water, and not very reliable with a contact

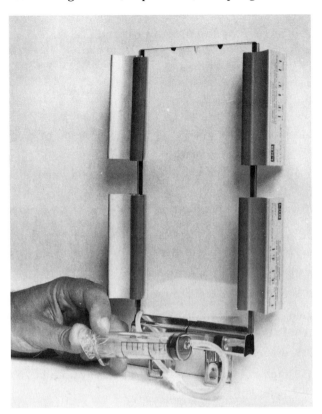

Figure 9. Reswelling cassette for dry IPG gels. The dried IPG gel (on its plastic backing) is inserted in the cassette, which is then gently filled with any desired reswelling solution via a bottom hole with the help of tubing and a syringe.

Table 18. IPG electrode solutions

Substance	Application	Concentration
Glutamic acid	Anolyte	10 mM
Lysine	Catholyte	10 mM
Carrier ampholytes[a,b]	Both electrolytes	0.3–1%
Distilled water	Both electrolytes	–

[a] Of the same or of a narrower range than the IPG.
[b] For mixed-bed gels or for samples with high salt concentration.

electrode. One can preferably either refer to the banding pattern of a set of marker proteins, or elute CAs from a mixed-bed gel and measure their pH (see Section 3.2.5; for the correction factors to be applied for electrophoresis temperatures other than 10 °C and for the effect of urea, see Section 3.1.2).

Protocol 12. Silver staining procedure for IPG gels

Reagents
The reagents required are indicated below

Method
The procedure involves exposing the gels to a series of reagents in a strict sequence, as described in the following steps.

1. Fixation: 40% ethanol, 10% acetic acid for 5 min (repeat four times in total).

2. Rinse: 20% ethanol for 5 min.

3. Rinse: water for 5 min.

4. Sensitization: 12.5% glutaraldehyde for 15 min.

5. Rinse: water for 10 min (repeat twice).

6. Rinse: 20% ethanol for 10 min (repeat twice).

7. Silver stain: 2 g/litre $AgNO_3$, 10 ml 5% NH_4OH, 50 ml/litre NaOH (in 20% ethanol), for 15 min.

8. Rinse: 20% ethanol for 5 min (repeat twice).

9. Rinse: 150 mg/litre sodium thiosulfate pentahydrate in 20% ethanol for 30 sec.

10. Development: 1 ml/litre formaldehyde, 0.1 g/litre citric acid in 20% ethanol, for 2–5 min.

11. Stop: 5% acetic acid, 20% ethanol for 30 min.

12. Storage before drying: 20% ethanol for > 1 h.

3.2.5 Mixed-bed, CA-IPG gels

In CA-IPG gels, the primary immobilized pH gradient is mixed with a secondary soluble carrier ampholyte-driven pH gradient. It sounds strange that, given the problems connected with the CA buffers (discontinuities along the electrophoretic path, pH gradient decay, etc.), which the IPG technique was supposed to solve, one should resurrect this past methodology. In fact, when working with membrane proteins (80) and with microvillar hydrolases, partly embedded in biological membranes (16), we found that the addition of CAs to the sample and IPG gel increases protein solubility, possibly by forming mixed-micelles with the detergent used for membrane solubilization (80) or by directly complexing with the protein itself (16). It is a fact that, in the absence of CAs, these same proteins essentially fail to enter the gel and mostly precipitate or give elongated smears around the application site (in general cathodic sample loading). More recently, it has been found that, on a relative hydrophobicity scale, the five basic Immobilines (pKs 6.2, 7.0, 8.5, 9.3, and 10.3) are decidedly more hydrophobic than their acidic counterparts (pKs 3.1, 3.6, and 4.6). Upon incorporation in the gel matrix, the phenomenon becomes co-operative and could lead to the formation of hydrophobic patches on the surface of such a hydrophilic gel as polyacrylamide. Since the strength of a hydrophobic interaction is directly proportional to the product of the cavity area times its surface tension, it is clear that experimental conditions which lead to a decrement of molecular contact area axiomatically weaken such interactions. Thus CAs might quench the direct hydrophobic protein: IPG matrix interaction, effectively detaching the protein from the surrounding polymer coils and allowing good focusing into sharp bands. For this to happen, the CA shielding species should already be impregnated in the Immobiline gel and present in the sample solution as well. In other words, CAs can only prevent the phenomenon and cannot cure it a posteriori. It has been additionally found that addition of CAs to the sample protects it from strongly acidic and alkaline boundaries originating from the presence of salts in the sample zone (especially 'strong' salts, such as NaCl, phosphates, etc.) (81).

However, a note of caution should be mentioned concerning the indiscriminate use of the CA-IPG technique: at high CA levels (> 1%) and high voltages (> 100 V/cm) these gels start exuding water with dissolved carrier ampholytes, with severe risks of short circuits, sparks, and burning on the gel surface. The phenomenon is minimized by incorporating chaotropes (e.g. 8 M urea) or polyols (e.g. 30% sucrose) and by lowering the CA molarity in the gel (82). As an answer to the basic question of when and how much CAs to add, we suggest the following guide-lines:

(a) If the sample focuses well, ignore the mixed-bed technique (which presumably will be mostly needed with hydrophobic proteins and in alkaline pH ranges).

(b) Add only the minimum amount of CAs (in general about 1%) needed to avoid sample precipitation and for producing sharply focused bands.

3.3 Troubleshooting

One could cover pages with a description of all the troubles and possible remedies in any methodology. These are summarized in *Table 19*. We highlight the following points:

(a) When the gel is 'gluey' and there is poor incorporation of Immobilines, the biggest offenders are generally the catalysts (e.g. too old persulfate, crystals wet due to adsorbed humidity, wrong amounts of catalysts added to the gel mix). Check in addition the polymerization temperature and the pH of the gelling solutions.

(b) Bear in mind the last point in *Table 19*: if you have done everything right, and still you do not see any focused protein, you might have simply positioned the platinum wires on the gel with the wrong polarity. Unlike conventional IEF gel, in IPGs the anode has to be positioned at the acidic

Table 19. Troubleshooting guide for IPGs

Symptom	Cause	Remedy
Drifting of pH during measurement of basic starting solution	Inaccuracy of glass pH electrodes (alkaline error)	Consult information supplied by electrode manufacturer
Leaking mould	Dust or gel fragment on the gasket	Carefully clean the gel plate and gasket
The gel consistency is not firm, gel does not hold its shape after removal from the mould	Inefficient polymerization	Prepare fresh ammonium persulfate and check that the recommended polymerization conditions are being used
Plateau visible in the anodic and/or cathodic section of the gel during electrofocusing, no focusing proteins seen in that part of the gel	High concentration of salts in the system	Check that the correct amounts of ammonium persulfate and TEMED are used
Overheating of gel near sample application when beginning electrofocusing	High salt content in the sample	Reduce salt concentration by dialysis or gel filtration
Non-linear pH gradient	Back-flow in the gradient mixer	Find and mark the optimal position for the gradient mixer on the stirrer
Refractive line at pH 6.2 in the gel after focusing	Unincorporated polymers	Wash the gel in 2 litres of distilled water; change the water once and wash overnight
Curved protein zones in that portion of the gel which was at the top of the mould during polymerization	Too rapid polymerization	Decrease the rate of polymerization by putting the mould at $-20\,°C$ for 10 min before filling it with the gel solution, or place the solutions at $4\,°C$ for 15 min before casting the gel
Uneven protein distribution across a zone	Slot or sample application not perpendicular to running direction	Place the slot or sample application pieces perfectly perpendicular to the running direction
Diffuse zones with unstained spots, or drops of water on the gel surface during the electrofocusing	Incomplete drying of the gel after the washing step	Dry the gel until it is within the 5% of its original weight
No zones detected	Gel is focused with the wrong polarity	Mark the polarity on the gel when removing it from the mould

(or less alkaline) gel extremity, while the cathode has to be placed at the alkaline (or less acidic) gel end.

3.4 Analytical results with IPGs

We will limit this section to some examples of separations in ultra narrow pH intervals, where the tremendous resolving power (ΔpI) of IPGs can be fully appreciated. The ΔpI is the difference, in surface charge, in pI units, between two barely-resolved protein species. Rilbe (21) has defined ΔpI as:

$$\Delta(\text{pI}) = \sqrt[3]{\frac{D[d(\text{pH})/dx]}{E[-du/d(\text{pH})]}}$$

where D and $du/d(\text{pH})$ are the diffusion coefficient and titration curve of proteins, E is the voltage gradient applied, and $d(\text{pH})/dx$ is the slope of the pH gradient over the separation distance. Experimental conditions that minimize ΔpI will maximize the resolving power. Ideally, this can be achieved by simultaneously increasing E and decreasing $d(\text{pH})/dx$, an operation for which IPGs seem well suited. As stated previously (see Section 2.1.3), with conventional IEF it is very difficult to engineer pH gradients that are narrower than 1 pH unit. One can push the ΔpI, in IPGs, to the limit of 0.001 pH unit. The corresponding limit in CA-IEF is only 0.01 pH unit. We began to investigate the possibility of resolving neutral mutants, which carry a point mutation involving amino acids with non-ionizable side chains and are, in fact, described as 'electrophoretically silent' because they cannot be distinguished by conventional electrophoretic techniques. The results were quite exciting. As shown in *Figure 10*, HbF Sardinia (which carries an Ile → Thr substitution in γ-75) is not quite resolved from the wild-type HbF in a 1 pH unit span CA-IEF (top panel). In a shallow IPG range spanning only 0.25 pH units, HbF Sardinia and the wild-type are now well resolved (central panel). There is, however, a more subtle mutation that could not be resolved in the present case. The lower band is actually an envelope of two components, called Aγ and Gγ, carrying a Gly → Ala mutation in γ-136. These two tetramers, normal components during fetal life, are found in approximately an 80:20 ratio. If the pH gradient is further decreased to 0.1 pH unit (over a standard 10 cm migration length), even these two tetramers can be separated (bottom panel) with a resolution close to the practical limit of ΔpI = 0.001 (83).

4. Capillary isoelectric focusing (cIEF)

4.1 General considerations

In addition to the reviews suggested at the beginning of this chapter (4, 5), we recommend the following reviews: by Mazzeo and Krull (84), Righetti and Chiari (85), Hjertèn (86), Pritchett (87), Kilàr (88), and Wehr *et al.* (89).

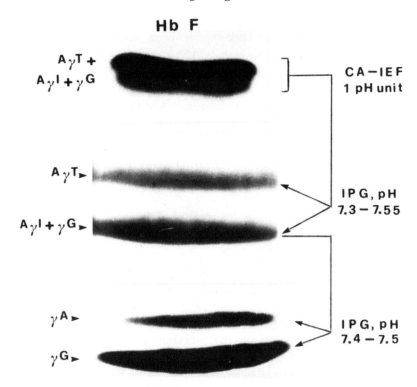

Figure 10. Focusing of umbilical cord lysates from an individual heterozygous for fetal haemoglobin (HbF) Sardinia (for simplicity, only the HbF bands are shown, and not the two other major components of cord blood, i.e. HbA and HbF_{ac}). Upper panel: focusing performed in a 1 pH unit span in CA-IEF. Note that broadening of the HbF occurs, but not the splitting into well-defined zones. Central panel: same sample as above, but focused over an IPG range spanning 0.25 units. Bottom panel: same as above, but in an IPG gel spanning 0.1 pH unit. The resolved Aγ/Gγ bands are in a 20:80 ratio, as theoretically predicted from gene expression. Their identity was established by eluting the two zones and fingerprinting the polypeptides. With permission from Cossu and Righetti (83).

Capillary electrophoresis offers some unique advantages over conventional gel slab techniques: the amount of sample required is truly minute (a few microlitres at the injection port, but only a few nanolitres in the moving zone); the analysis time is in general very short (often just a few minutes) due to the very high voltages applicable; analyte detection is on-line and is coupled to automatic storage of electropherograms on a magnetic support.

A principal difference between IEF in a gel and in a capillary is that, in the latter, the focused proteins have to be mobilized past the detector unless an on-line imaging detection system is being used. Three techniques are mainly used: chemical and hydrodynamic flow mobilization (in coated capillaries) and mobilization utilizing electro-osmotic flow (EOF, in uncoated or partially coated capillaries). We do not encourage the last approach (90) since the transit times of the focused zones change severely from run-to-run and so we will describe only the cIEF approach in coated capillaries, where EOF is completely suppressed.

4.2 cIEF methodology

4.2.1 General guide-lines for cIEF

Table 20 gives some guide-lines for cIEF in coated capillaries. Since a large number of procedures for silanol deactivation has been reported (91) and a good coating of the standard required is very difficult to achieve in a general biochemical laboratory, we recommend buying pre-coated capillaries, e.g. from Beckman or from Bio-Rad.

The following general guide-lines are additionally suggested:

(a) All solutions should be degassed.

(b) The ionic strength of the sample may influence dramatically the length of the focusing step and completely ruin the separation. Therefore, desalting of the sample prior to focusing or use of a low buffer concentration (ideally made of a weak buffering ion and a weak counterion) is preferable. Easy sample desalting can be achieved via centrifugation through Centricon membranes (Amicon).

(c) The hydrolytic stability of such coatings is poor at alkaline pH; therefore mobilization with NaOH may destroy the coating after a few runs. Zwitterions can be used as mobilizers instead.

(d) Ideally, non-buffering ions should be excluded in all compartments for cIEF. This means that in the electrodic reservoirs one should use weak acids (at the anode) and weak bases (at the cathode) instead of phosphoric acid and NaOH, as used today by most cIEF users. This includes the use of zwitterions (e.g. Asp pI = 2.77, or Glu pI = 3.25 at the anode, and Lys pI = 9.74, or Arg pI = 10.76 at the cathode).

(e) When eluting the focused bands past the detector, we have found that resolution is maintained better by a combination of salt elution (e.g. adding 20 mM NaCl or sodium phosphate to the appropriate compartment) and a syphoning effect, obtained by having a higher liquid level in one compartment and a lower level in the other. The volumes to use will depend on the apparatus. For the BioFocus 2000 apparatus from Bio-Rad, the volumes to use are 650 μl and 450 μl, respectively.

4.2.2 Increasing the resolution by altering the slope of the pH gradient

Methods have not yet been devised for casting IPGs in a capillary format and so it is difficult in cIEF to achieve the resolution typical of IPGs, namely

---

Table 20. Guidelines for cIEF separation of proteins in coated capillaries

1. Ampholytes: commercially available ampholyte solutions (e.g. Pharmalyte, Biolyte, Servalyte, etc.) in the desired pH range.
2. Sample: 1–2 mg/ml protein solution, mixed with 3–4% carrier ampholytes (final concentration) in the desired pH range. The sample solution should be desalted or equilibrated in a weak buffer–counterion system.
3. Capillary: 25–30 cm long, 50–75 μm i.d., coated, filled with the sample–CA solution.
4. Anolyte: 10 mM phosphoric acid (or any other suitable weak acid, such as acetic, formic acids) or low pI zwitterions.
5. Catholyte: 20 mM NaOH (or any other suitable weak base, such as Tris, ethanolamine) or high pI zwitterions.
6. Focusing: 8–10 kV, constant voltage, for 5–10 min.
7. Mobilizer: 50–80 mM NaCl, or 20 mM NaOH (cathodic) or 20 mM NaOH (anodic).
8. Mobilization: 5–6 kV, constant voltage.
9. Detection: 280 nm, near to the mobilizer (or any appropriate visible wavelength for coloured proteins).
10. Washing: after each run, with 1% neutral detergent (such as Nonidet P-40, Triton X-100) followed by distilled water.

$\Delta pI = 0.001$. In the best cases, one can achieve only what CA-IEF can do, i.e. $\Delta pI = 0.01$. Nevertheless, some spectacular results can be obtained by adding spacers (or pH gradient modifiers) to the commercial CA buffer mixture, as suggested in Section 2.3.3 and in *Tables 6* and *7*. Some examples are given below.

In screening for thalassaemia syndromes, one approach is to separate and quantify the three main Hb components of umbilical cord blood (fetal, acetylated fetal, and adult Hbs; HbF, F_{ac}, and A, respectively). In a standard pH 6–8 carrier ampholyte mixture, only poor separations are obtained (*Figure 11*, upper panel). When the same separation is repeated in the same Ampholine pH range, but with added 50 mM β-alanine (which acts by locally flattening the pH gradient in the region where F and A focus), excellent separation is now obtained (*Figure 11*, lower panel) (92). Another relevant case in clinical chemistry is the proper separation and quantitation between adult Hb and its glycated form (A_{1c}), high levels of the latter being indicative of diabetic conditions. Again this separation is very poor in plain pH 6–8 Ampholines (*Figure 12*, upper panel), but it is excellent when the latter are added with an equimolar mixture of 0.33 M β-alanine and 0.33 M 6-aminocaproic acid (*Figure 12*, lower panel) (93).

4.2.3 On the problem of protein solubility at their pI

In all focusing techniques, the biggest problem is protein precipitation at and near the pI value. Proteins have a minimum of charges and solvation at $pH = pI$ and so in the low ionic strength environment typical of focusing conditions, they tend to precipitate and often produce smears which can encom-

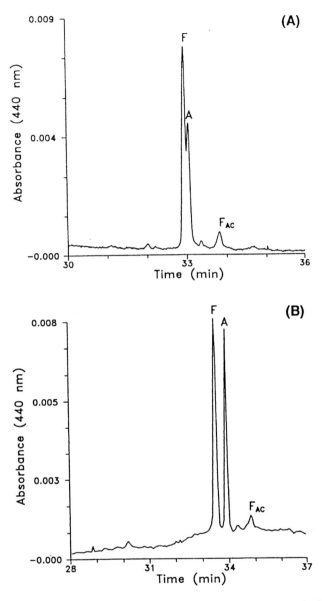

Figure 11. Separation of HbF, A, and F_{ac} by capillary IEF. Background electrolyte: 5% Ampholine, pH 6–8, added with 0.5% TEMED (A) and additionally with 3% short chain polyacrylamide and 50 mM β-Ala (B). Anolyte: 20 mM H_3PO_4; catholyte: 40 mM NaOH. Sample loading: by pressure, for 60 sec. Focusing run: 20 kV constant at 7 μA (initial) to 1 μA (final current), 20 °C. Capillary: coated with poly(*N*-acryloylamino propanol), 25 μm i.d., 23.6/19.1 total/effective length. Mobilization conditions: with 200 mM NaCl added to anolyte, 22 kV. Detection at 415 nm. From Conti *et al.* (92), with permission.

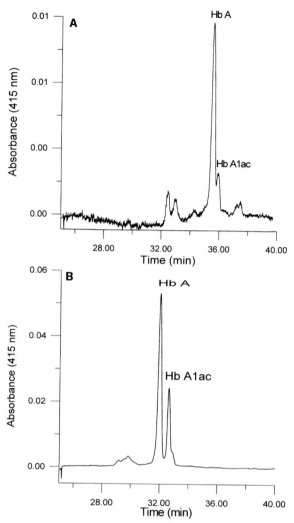

Figure 12. Separation of HbA from A_{1c} by capillary IEF in the absence (A) and in presence (B) of 3% short chain polyacrylamide and an equimolar mixture of separators, 0.33 M β-Ala and 0.33 M 6-aminocaproic acid. Background electrolyte: 5% Ampholine, pH 6–8, added with 0.5% TEMED. Anolyte: 20 mM H_3PO_4; catholyte: 40 mM NaOH. Sample loading: by pressure, for 60 sec. Focusing run: 20 kV constant at 7 μA (initial) to 1 μA (final current), 20°C. Capillary: coated with poly(*N*-acryloylamino propanol), 25 μm i.d., 23.6/19.1 total/effective length. Mobilization conditions: with 200 mM NaCl added to anolyte, 22 kV. Detection at 415 nm. From Conti *et al.* (93), with permission.

pass as much as 1 pH unit. When, this occurs, the experiment is completely ruined. We have studied this problem extensively in the last few years, and finally we have obtained some remarkable results with a number of very difficult proteins (94). In addition to mild solubilizers proposed in the past, such as glycols (glycerol, ethylene, and propylene glycols) (95) non-detergent sulfo-betaines, in concentrations up to 1 M, have been found to be effective in reducing this problem in a number of cases (96, 97). Other common zwitterions, such as taurine and a few of the Good's buffers (e.g. Bicine, CAPS) are also quite useful in acidic pH gradients and up to pH 8. Addition of sugars, notably saccharose, sorbitol and, to a lesser extent, sorbose (20% in cIEF) also greatly improve protein solubility in the proximity of the pI. The improvement was dramatic if these sugars were admixed with 0.2 M taurine (98). Thus, a whole array of solubilizers, compatible with native structure and maintenance of enzyme activity, is now available. These additives, unlike non-ionic or zwitterionic surfactants, have the advantage of remaining monomeric, i.e. of being unable to form micelles, even at concentrations up to 1 M. Thus, subsequent to fractionation, they can be easily eliminated from the protein by gel filtration or centrifugation through dialysis membranes.

5. Conclusions

Conventional IEF in amphoteric buffers is now a well established technique and further major improvements are unlikely. In addition, IPG methodology, now in the 15th year after its invention, has reached a plateau in terms of innovation but is still gaining momentum in use in most laboratories around the world, especially in its unique application as a first dimension in 2D maps. The new rising star is cIEF, which is still in its infancy but has a lot to offer for future users. Particularly appealing is the fact that cIEF offers a fully instrumental approach to electrophoresis, thus lessening dramatically the experimental burden and the labour-intensive approach of gel slab operations. While capillary electrophoresis equipment is currently available mainly as single channel units, the new generation of equipment will offer multichannel capabilities, in batteries from 20 to 100 capillary arrays. Thus, rapid growth is expected in this field.

Acknowledgements

Our research reported here has been supported by Agenzia Spaziale Italiana (ASI, Roma, contract No. ARS-96–214), by Progetto Strategico No. 96.05076.ST74, and by Comitato tecnologico No. 96.01895.CT11 from Consiglio Nazionale delle Ricerche (CNR, Roma). We thank colleagues who have collaborated with us over the years (e.g. Drs A. Görg, B. Bjellqvist, E. Gianazza, C. Gelfi, and M. Chiari, to name just a few) whose work has greatly helped in establishing the modern techniques.

Pier Giorgio Righetti et al.

References

1. Righetti, P.G., Gianazza, E., Gelfi, C., and Chiari, M. (1990). In *Gel electrophoresis of proteins. a practical approach* (ed. B.D. Hames and D. Rickwood), pp. 149–216. IRL Press, Oxford.
2. Righetti, P.G. (1983). *Isoelectric focusing: theory, methodology and applications*. Elsevier, Amsterdam.
3. Righetti, P.G. (1990). *Immobilized pH gradients: theory and methodology*. Elsevier, Amsterdam.
4. Righetti, P.G. and Gelfi, C. (1994). *J. Cap. Electrophoresis*, **1**, 27.
5. Righetti, P.G., Gelfi, C., and Chiari, M. (1996). In *Capillary electrophoresis in analytical biotechnology* (ed. P.G. Righetti), pp. 509–38. CRC Press, Boca Raton.
6. Giddings, J.C. (1991). *Unified separation science*, pp. 86–109. John Wiley & Sons, New York.
7. Svensson, H. (1961). *Acta Chem. Scand.*, **15**, 325; (1962). *Ibid.*, **16**, 456.
8. Bjellqvist, B., Ek, K., Righetti, P.G., Gianazza, E., Gorg, A., Postel, W., *et al.* (1982). *J. Biochem. Biophys. Methods*, **6**, 317.
9. Rilbe, H. (1973). *Ann. N. Y. Acad. Sci.*, **209**, 11.
10. Gianazza, E., Chillemi, F., Gelfi, C., and Righetti, P.G. (1979). *J. Biochem. Biophys. Methods*, **1**, 237.
11. Gianazza, E., Chillemi, F., and Righetti, P.G. (1980). *J. Biochem. Biophys. Methods*, **3**, 135.
12. Gianazza, E., Chillemi, F., Duranti, M., and Righetti, P.G. (1984). *J. Biochem. Biophys. Methods*, **8**, 339.
13. Gianazza, E. and Righetti, P.G. (1980). *J. Chromatogr.*, **193**, 1.
14. Krishnamoorthy, R., Bianchi-Bosisio, A., Labie, D., and Righetti, P.G. (1978). *FEBS Lett.*, **94**, 319.
15. Righetti, P.G. and Tonani, C. (1992). In *Theoretical advancement in chromatography and related separation techniques* (ed. F. Dondi and G. Guiochon), Vol. 383, pp. 581–605. NATO ASI Series C: Mathematical and Physical Sciences, Kluwer Academic Publisher, Dordrecht.
16. Sinha, P.K. and Righetti, P.G. (1986). *J. Biochem. Biophys. Methods*, **12**, 289.
17. Vesterberg, O. (1973). *Ann. N. Y. Acad. Sci.*, **209**, 23.
18. Bianchi-Bosisio, A., Snyder, R.S., and Righetti, P.G. (1981). *J. Chromatogr.*, **209**, 265.
19. Galante, E., Caravaggio, T., and Righetti, P.G. (1975). In *Progress in isoelectric focusing and isotachophoresis* (ed. P.G. Righetti), pp. 3–12. Elsevier, Amsterdam.
20. Righetti, P.G. (1980). *J. Chromatogr.*, **190**, 275.
21. Rilbe, H. (1996). *pH and buffer theory. A new approach*, pp. 1–192. J. Wiley & Sons, Chichester.
22. Bianchi-Bosisio, A., Loehrlein, C., Snyder, R.S., and Righetti, P.G. (1980). *J. Chromatogr.*, **189**, 317.
23. Chiari, M., Micheletti, C., Nesi, M., Fazio, M., and Righetti, P.G. (1994). *Electrophoresis*, **15**, 177.
24. Simò-Alfonso, E., Gelfi, C., Sebastiano, R., Citterio, A., and Righetti, P.G. (1996). *Electrophoresis*, **17**, 723; (1996). *Ibid.*, **17**, 732.
25. Righetti, P.G., Brost, B.C., and Snyder, R.S. (1981). *J. Biochem. Biophys. Methods*, **4**, 347.

26. Gelfi, C. and Righetti, P.G. (1981). *Electrophoresis*, **2**, 213.
27. Chiari, M., Righetti, P.G., Negri, A., Ceciliani, F., and Ronchi, S. (1992). *Electrophoresis*, **13**, 882.
28. Bossi, A. and Righetti, P.G. (1995). *Electrophoresis*, **16**, 1930.
29. Låås, T. and Olsson, I. (1981). *Anal. Biochem.*, **114**, 167.
30. Altland, K. and Kaempfer, M. (1980). *Electrophoresis*, **1**, 57.
31. Caspers, M.L., Posey, Y., and Brown, R.K. (1977). *Anal. Biochem.*, **79**, 166.
32. Righetti, P.G., Tudor, G., and Gianazza, E. (1982). *J. Biochem. Biophys. Methods*, **6**, 219.
33. Yao, J.G. and Bishop, R. (1982). *J. Chromatogr.*, **234**, 459.
34. Zhu, M.D., Rodriguez, R., and Wehr, T. (1991). *J. Chromatogr.*, **559**, 479.
35. Radola, B.J. (1980). *Electrophoresis*, **1**, 43.
36. Righetti, P.G. and Chillemi, F. (1978). *J. Chromatogr.*, **157**, 243.
37. Righetti, P.G. and Drysdale, J.W. (1974). *J. Chromatogr.*, **98**, 271.
38. Neuhoff, V., Stamm, R., and Eibl, H. (1985). *Electrophoresis*, **6**, 427.
39. Merril, C.R., Goldman, D., Sedman, S.A., and Ebert, M.H. (1981). *Science*, **211**, 1438.
40. Vesterberg, O. (1972). *Biochim. Biophys. Acta*, **257**, 11.
41. Wirth, P.J. and Romano, A. (1995). *J. Chromatogr. A*, **698**, 123.
42. Waheed, A.A. and Gupta, P.D. (1996). *Anal. Biochem.*, **233**, 249.
43. Choi, J.K., Yoon, S.H., Hong, H.Y., Choi, D.K., and Yoo, G.S. (1996). *Anal. Biochem.*, **236**, 82.
44. Myers, J.M., Veis, A., Sabsay, B., and Wheeler, A.P. (1996). *Anal. Biochem.*, **240**, 300.
45. Steinberg, T.H., Jones, L.J., Haugland, R.P., and Singer, V.L. (1996). *Anal. Biochem.*, **239**, 223.
46. Steinberg, T.H., Haugland, R.P., and Singer, V.L. (1996). *Anal. Biochem.*, **239**, 238.
47. Javier Alba, F., Bermudez, A., Bartolome, S., and Daban, J.R. (1996). *BioTechniques*, **21**, 625.
48. Rabilloud, T. (1990). *Electrophoresis*, **11**, 785.
49. Hebert, J.P. and Strobbel, B. (1974). *LKB Application Note* No. 151.
50. Godolphin, W.J. and Stinson, R.A. (1974). *Clin. Chim. Acta*, **56**, 97.
51. Laskey, R.A. and Mills, A.D. (1975). *Eur. J. Biochem.*, **56**, 335.
52. Laskey, R.A. (1980). In *Methods in enzymology*, (ed. L. Grossman and K. Moldave),Vol. 65, pp. 363–71. Academic Press, San Diego.
53. Harris, H. and Hopkinson, D.A. (1976). *Handbook of enzyme electrophoresis in human genetics*. Elsevier, Amsterdam.
54. Richtie, R.F. and Smith, R. (1976). *Clin. Chem.*, **22**, 497.
55. Arnaud, P., Wilson, G.B., Koistinen, J., and Fudenberg, H.H. (1977). *J. Immunol. Methods*, **16**, 221.
56. Towbin, H., Staehelin, T., and Gordon, J. (1979). *Proc. Natl. Acad. Sci. USA*, **76**, 4350.
57. Radola, B.J. (1980). *Electrophoresis*, **1**, 43.
58. Cuono, C.B. and Chapo, G.A. (1982). *Electrophoresis*, **3**, 65.
59. Righetti, P.G., Fazio, M., and Tonani, C. (1988). *J. Chromatogr.*, **440**, 367.
60. Righetti, P.G., Gelfi, C., and Chiari, M. (1996). In *Methods in enzymology* (ed. B.L. Karger and W.S. Hancock), Vol. 270, pp. 235–55. Academic Press, San Diego.

61. Gianazza, E., Artoni, G., and Righetti, P.G. (1983). *Electrophoresis*, **4**, 321.
62. Gianazza, E., Celentano, F., Dossi, G., Bjellqvist, B., and Righetti, P.G. (1984). *Electrophoresis*, **5**, 88.
63. Gianazza, E., Astrua-Testori, S., and Righetti, P.G. (1985). *Electrophoresis*, **6**, 113.
64. Gianazza, E., Giacon, P., Sahlin, B., and Righetti, P.G. (1985). *Electrophoresis*, **6**, 53.
65. Mosher, R.A., Bier, M., and Righetti, P.G. (1986). *Electrophoresis*, **7**, 59.
66. Righetti, P.G., Gianazza, E., and Celentano, F. (1986). *J. Chromatogr.*, **356**, 9.
67. Celentano, F., Gianazza, E., Dossi, G., and Righetti, P.G. (1987). *Chemometr. Intell. Lab. Syst.*, **1**, 349.
68. Celentano, F.C., Tonani, C., Fazio, M., Gianazza, E., and Righetti, P.G. (1988). *J. Biochem. Biophys. Methods*, **16**, 109.
69. Righetti, P.G., Fazio, M., Tonani, C., Gianazza, E., and Celentano, F.C. (1988). *J. Biochem. Biophys. Methods*, **16**, 129.
70. Gianazza, E., Celentano, F.C., Magenes, S., Ettori, C., and Righetti, P.G. (1989). *Electrophoresis*, **10**, 806.
71. Tonani, C. and Righetti, P.G. (1991). *Electrophoresis*, **12**, 1011.
72. Righetti, P.G. and Tonani, C. (1991). *Electrophoresis*, **12**, 1021.
73. Giaffreda, E., Tonani, C., and Righetti, P.G. (1993). *J. Chromatogr.*, **630**, 313.
74. Bianchi-Bosisio, A., Righetti, P.G., Egen, N.B., and Bier, M. (1996). *Electrophoresis*, **7**, 128.
75. Bossi, A., Righetti, P.G., Vecchio, G., and Severinsen, S. (1994). *Electrophoresis*, **15**, 1535.
76. Bossi, A., Gelfi, C., Orsi, A., and Righetti, P.G. (1994). *J. Chromatogr. A*, **686**, 121.
77. Righetti, P.G., Bossi, A., Görg, A., Obermaier, C., and Boguth, G. (1996). *J. Biochem. Biophys. Methods*, **31**, 81.
78. Rabilloud, T., Brodard, V., Peltre, G., Righetti, P.G., and Ettori, C. (1992). *Electrophoresis*, **13**, 264.
79. Gåveby, B.M., Pettersson, P., Andrasko, J., Ineva-Flygare, L., Johannesson, U., Görg, A., *et al.* (1988). *J. Biochem. Biophys. Methods*, **16**, 141.
80. Rimpilainen, M. and Righetti, P.G. (1985). *Electrophoresis*, **6**, 419.
81. Righetti, P.G., Chiari, M., and Gelfi, C. (1988). *Electrophoresis*, **9**, 65.
82. Astrua-Testori, S. and Righetti, P.G. (1987). *J. Chromatogr.*, **387**, 121.
83. Cossu, G. and Righetti, P.G. (1987). *J. Chromatogr.*, **398**, 211.
84. Mazzeo, J.R. and Krull, I.S. (1993). In *Capillary electrophoresis technology* (ed. A.N. Guzman), pp. 795–818. Dekker, New York.
85. Righetti, P.G. and Chiari, M. (1993). In *Capillary electrophoresis technology* (ed. A.N. Guzman), pp. 89–116. Dekker, New York.
86. Hjertèn, S. (1992). In *Capillary electrophoresis: theory and practice* (ed. P.D. Grossman and J.C. Colburn), pp. 191–214. Academic Press, San Diego.
87. Pritchett, T.J. (1996). *Electrophoresis*, **17**, 1195.
88. Kilàr, F. (1994). In *Handbook of capillary electrophoresis* (ed. J.P. Landers), pp. 95–109. CRC Press, Boca Raton.
89. Wehr, T., Zhu, M.D., and Rodriguez-Diaz, R. (1996). In *Methods in enzymology* (ed. B.L. Karger and W.S. Hancock), Vol. 270, pp. 358–74. Academic Press, San Diego.
90. Thormann, W., Caslavska, J., Molteni, S., and Chmelik, J. (1992). *J. Chromatogr.*, **589**, 321.

91. Chiari, M., Nesi, M., and Righetti, P.G. (1996). In *Capillary electrophoresis in analytical biotechnology* (ed. P.G. Righetti), pp. 1–36. CRC Press, Boca Raton.
92. Conti, M., Gelfi, C., and Righetti, P.G. (1995). *Electrophoresis*, **16**, 1485.
93. Conti, M., Gelfi, C., Bianchi-Bosisio, A., and Righetti, P.G. (1996). *Electrophoresis*, **17**, 1590.
94. Esteve-Romero, J.S., Bossi, A., and Righetti, P.G. (1996). *Electrophoresis*, **17**, 1242.
95. Ettori, C., Righetti, P.G., Chiesa, C., Frigerio, F., Galli, G., and Grandi, G. (1992). *J. Biotechnol.*, **25**, 307.
96. Vuillard, L., Marret, N., and Rabilloud, T. (1995). *Electrophoresis*, **16**, 295.
97. Vuillard, L., Braun-Breton, C., and Rabilloud, T. (1995). *Biochem. J.*, **305**, 337.
98. Conti, M., Galassi, M., Bossi, A., and Righetti, P.G. (1997). *J. Chromatogr. A*, **757**, 237.

6

Two-dimensional gel electrophoresis

SAMIR M. HANASH

1. Introduction

Two-dimensional gel electrophoresis (2DE) has been and remains a unique and powerful approach to resolve proteins and polypeptides in complex protein mixtures. However, in the past this approach has caused as much frustration as satisfaction to investigators because of its complexity, the difficulty of achieving good resolution and reproducibility, and the perceived existence of alternative molecular approaches, primarily at the nucleic acid level, to uncover gene products or gene expression patterns of interest. Thus, 2DE as a methodology has had a limited appeal to investigators. This situation has gradually been reversed in the past few years, due to a number of important developments, some technological and others related to a renewed emphasis on 'going back to the proteins' for a greater understanding of important biological processes, such as signal transduction, growth and differentiation, and cellular responses to a variety of agents and conditions. The renaissance in 2DE can also be linked to the greater availability of ready-made products for this approach, to an enhanced resolution over a wider pH range from the most acidic to the most basic, to the increased feasibility of identifying proteins separated by 2DE, and to the burgeoning availability of relevant resources on the Internet. A description of approaches and resources for the two-dimensional separation of proteins is provided in this chapter. The procedures described are all in use in the author's laboratory and have been developed or adapted from published monographs.

2. World Wide Web (WWW) resources

An amazing array of resources and information is available on the Internet related to 2DE, and the investigator contemplating the use of this procedure may benefit from a WWW search under the heading of 'two-dimensional-electrophoresis' or 'two-dimensional-gel electrophoresis'. Information that can be obtained falls under a number of categories and covers issues that include:

(a) Procedures for 2DE provided by a number of groups and laboratories as practiced by them, as well as references to the literature.
(b) An ever increasing number of specialized 2DE databases maintained on the web.
(c) Tools that can be downloaded for gel analysis.
(d) Descriptions of training courses and upcoming meetings.

A partial listing of 2D databases available on the web is provided in *Table 1* and a more current listing may be obtained at the following address: `http://www-lmmb.ncifcrf.gov/EP/EPmail.html`. It should be noted that the quality and utility of the databases on the web is highly variable. Some do not represent more than a set of 2D images of gels obtained by an investigator or a group, with rudimentary or no annotation. Others are annotated with respect to proteins identified by the group. A word of caution is in order, with respect to the latter point. A careful review of some patterns in databases on the WWW reveals discrepancies between databases in the identity of protein spots that appear to be a part of the same spot constellations, in different databases. Naturally there could be a number of explanations for such discrepancies. Some relate to the means utilized to determine identity (e.g. Western blot analysis versus microsequencing). A number of recent special issues of the journal *Electrophoresis* also contains information relevant to the practice and application of 2DE.

3. 2DE reagents

'Electrophoresis grade' acrylamide, bisacrylamide, TEMED, ammonium persulfate, and agarose are available from a number of suppliers. Acrylamide in particular should be of the highest quality available. The highest quality urea, 'ultrapure' is used. Ampholytes are supplied under various manufacturers' trade names and are available in several pH ranges. The use of mixtures of ampholytes from different suppliers may minimize lot-to-lot variability of any one manufacturer and also reduce 'gaps' in the pH gradient. Special blends of wide and narrow range ampholytes may be used to optimize particular separations. The non-ionic detergent Nonidet P-40 differs among suppliers, and results in different focusing patterns. Nonidet P-40 and Triton X-100 may be considered equivalent but not identical. Dithiothreitol (DTT) and dithioerythritol (DTE) may be considered equivalent in redox potential for disulfide bond reduction. Immobilines, repel silane, and Gel-Bond PAG film are all obtainable from Pharmacia. All other chemicals used are reagent grade. MilliQ water (Millipore) or double distilled deionized water is used for all solutions.

Table 1. List of two-dimensional electrophoretic protein gel databases on the World Wide Web (WWW)

1. A 2D PAGE protein database of *Drosophila melanogaster*
   `http://tyr.cmb.ki.se/`
2. EsPASy Server (2D liver, plasma, etc. SWISS-(PROT, 2DPAGE, 3DIMAGE), BIOSCI, Melanie software)
   `http://expasy.hcuge/ch/`
3. CSH QUEST Protein Database Center (2D REF52 rat, mouse embryo, yeast, Quest software)
   `http://siva.cshl.org/`
4. NCI/FCRDC LMMB Image Processing Section (GELLAB software)
   `http://www-ips.ncicrf.gov/lemkin/gellab.html`
5. 2-D Images Meta-Database
   `http://www-lmmb.ncifcrf.gov/2dwgDB/`
6. *E. coli* Gene–Protein Database Project—ECO2DBASE (in NCBI repository)
   `ftp://ncbi.nlm.nih.gov/repository/ECO2DBASE/`
7. Argonne Protein Mapping Group Server (mouse liver, human breast cell, etc.)
   `http://www.anl.gov/CMB/PMG/`
8. Cambridge 2D PAGE (a rat neuronal database)
   `http://sunspot.bioc.cam.uk/Neuron.html`
9. Heart Science Centre, Harefield Hospital (Human Heart 2D gel Protein DB)
   `http://www.harefield.nthames.nhs.uk/nhli/protein/`
10. Berlin Human Myocardial 2D Electrophoresis Protein Database
    `http://www.chemie.fu-berlin.de/user/pleiss/`
11. Max Delbruck Ctr. for Molecular Medicine—Myocardial 2D gel DB
    `http://www.mdc-berlin.de/~emu/heart/`
12. The World of Electrophoresis (EP literature, ElphoFit)
    `http://www.uni-giessen.de/~gh43/electrophoresis.hyml`
13. Human Colon Carcinoma Protein Database (Joint Protein Structure Lab)
    `http://www.ludwig.edu.au/www/jpsl/jpslhome.html`
14. Large Scale Biology Corp (2D maps: rat, mouse, and human liver)
    `http://www.lsbc.com/`
15. Yeart 2D PAGE
    `http://yeast-2dpage.gmm.gu.se/`
16. Proteome Inc. YPD Yeast Protein Database
    `http://www.proteome.com/YPDhome.html`
17. Keratiunocyte, cDNA database (Danish Centre for Human Genome Research)
    `http://biobase.dk/cgi-bin/celis/`
18. Institut de Biochemie et Genetique—Yeast 2D gel DB
    `http://www.ibgc.u-bordeaux2.fr/YPM/`
19. Swiss-Flash newsletter
    `http://expasy.hcuge.ch/www/swiss-flash.html`
20. Phosphoprotein Database
    `http://www-lmmb.ncifcrf.gov/phosphoDB/`

4. Equipment

4.1 First-dimension isoelectric focusing apparatus

A conventional tube gel electrophoresis apparatus is used for isoelectric focusing (IEF) with carrier ampholytes. The upper unit (cathode chamber) consists of a cylindrical or rectangular container with 200 ml capacity. 10–20

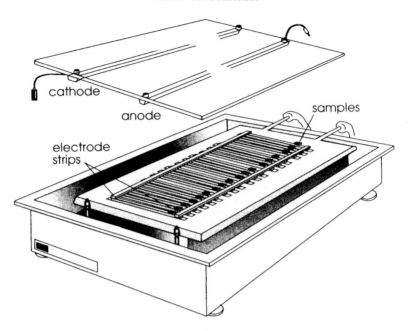

Figure 1. Apparatus for electrophoresis of IPG strips. (Reprinted with permission from ref. 5.)

holes are bored into the bottom of the chamber with tight-fitting rubber grommets to hold the gel tubes. The lower (anode) chamber may be simply a large beaker or Plexiglas box with two litres capacity. The platinum electrode in each chamber is positioned so as to be equidistant from the ends of each tube. Isoelectric focusing is done at room temperature. No special provision for cooling is necessary because the IEF gels in the tubes are completely immersed in two litres of anode solution that acts as an effective heat sink. Immobiline strips are run in Pharmacia Multiphor II apparatus using a dedicated kit (*Figure 1*).

4.2 Second-dimension slab gel apparatus

Any commercially available slab gel electrophoresis apparatus may be used in conjunction with first-dimension tube gels or strips. Preference is given to a unit in which the running buffer may be cooled with either cold tap-water or a circulating cooling bath. The primary consideration is the size format of the final 2D pattern, which is largely determined by experimental design considerations (as opposed to technical design factors). Considerations of technical design include whether the second-dimension slab gel is cast and run sandwiched between two glass plates, in which case the thickness of the second-dimension gel must be sufficient to accommodate the first-dimension gel. In this type of apparatus, the first-dimension IEF or IPG (immobilized pH gradient) gel is placed onto the top end of the SDS slab gel and held there with agarose. An SDS slab gel may also be cast on a glass plate or plastic backing (e.g. Gel-Bond PAG film) and run in a horizontal apparatus, such as that used for isoelectric focusing of IPG gels. In this instance, the electrophoresis chamber must have buffer chambers on each end of the cooling plate and the gel is connected to the buffer electrode chambers with several thicknesses of buffer-saturated filter paper. The IEF gel is then placed directly onto the surface of the SDS gel. For forming an acrylamide gradient for the second-dimension SDS gel electrophoresis, a two-chambered gradient mixer is used. An integrated system for casting and running 2D protein gels referred to as ISO-DALT, marketed by Pharmacia, has been fairly widely used, including in the author's laboratory, for carrier ampholyte-based separations. The second-dimension component of the system is also used for the second-dimension separation of proteins in IPG strips.

4.3 Apparatus for electroblotting

Electrophoretic transfer of protein from an acrylamide slab gel to a transfer membrane may be performed using either a tank system or a so-called semi-dry apparatus. A tank system consists of a rectangular buffer tank of about two litre capacity. The transfer 'sandwich', consisting of the slab gel juxtaposed to the transfer membrane, is firmly held between two perforated Plexiglas sheets, and is held in a vertical position between platinum wire electrodes. In the semi-dry apparatus (e.g. Pharmacia) the sandwich is placed directly between two flat carbon electrode plates.

4.4 Power supply

A power supply capable of delivering 1500 V (< 10 mA for 20 IEF gels) is sufficient for isoelectric focusing of tube gels with carrier ampholytes. For isoelectric focusing on IPG gels, a 3000–5000 V power supply is required. For electrophoresis in the second dimension, the current capacity becomes important when multiple gels are run from one power supply. Each second-dimension gel will draw about 120 mA (at 400 V) at the beginning of electrophoresis (~ 60 mA at the end of a run as resistance in the gel decreases). Thus if ten gels are run, the power supply current capacity must be 1.2 A. Otherwise lower initial voltages will have to be used, resulting in longer electrophoresis times.

4.5 Platform shaking table

Platform shaking tables should have a horizontal surface capable of accepting four to six glass trays during the silver staining procedure. Reciprocating or orbital shakers may be used and should have variable speed control between 60–120 cycles or revolutions per minute respectively. When gels are to be equilibrated with most fluorographic cocktails the shaking table should be installed inside a fume-hood.

4.6 Apparatus for gel drying

A gel dryer (e.g. Bio-Rad or Pharmacia) consists of a flat heating block with channels on the upper surface to transport vapours from the gel to a connecting vacuum system. The heating block should be regulated to 60°C (or lower) when drying gels for fluorography, since higher temperatures result in artefacts on the X-ray film. A porous polyethylene sheet lies on the heating block and supports the gel during drying. A sheet of silicone rubber covers the gel and forms a seal at the edges of the heating block. The vacuum pump must be protected from water and acid vapours by a cold trap (e.g. acetone/dry ice). Additionally it is prudent to include a chemical trap (e.g. NaOH or Na_2CO_3) between the cold trap and the pump. A water aspirator may be used in place of a vacuum pump (no traps are then required) but sudden changes in water pressure may result in loss of vacuum that produces undesirable cracking of the gel. Drying times are also one and a half to two times longer when using a water aspirator vacuum.

5. Sample preparation and solubilization

5.1 Sample preparation

For most 2D studies, it is desirable to separate and visualize as many of the protein constituents in the sample as possible. In that case, fractionation of the constituents, such as into subcellular fractions, may not be necessary. In the case of cell suspensions, cells are thoroughly washed with phosphate-buffered saline (PBS) and pelleted. Care is exercised to remove all traces of excess wash solution or protein-containing medium. It is desirable to prepare cell pellets in 15 ml microcentrifuge tubes with each pellet containing sufficient protein for one to a few 2D gels. Cell pellets are stored frozen at −80°C and are solubilized immediately before electrophoresis. Solid tissue samples are frozen immediately and kept at −70°C until solubilization. Because isoelectric focusing is sensitive to charge modification, it is important to minimize protein alterations (e.g. proteolysis, deamidation of glutamine and asparagine, oxidation of cysteine to cystic acid, and carbamylation) that can result from improper sample preparation. It is elementary that a sample in which proteins have been allowed to degrade will yield a poor quality 2D gel! The sample solubilization procedure is described in *Protocol 1*. Once solubilized, samples may be stored frozen at −80 °C for short periods (< one month) without significant protein modification.

Protocol 1.   Sample solubilization

Reagents

- Solution A: 9 M urea, 2% (v/v) Nonidet P-40, 0.8% (w/v) ampholine pH 3–10, 2% (v/v) 2-mercaptoethanol—store in 1 ml aliquots at −80 °C
- Solution B: 15.5 mg/ml phenylmethylsulfonyl fluoride (PMSF) in 95% ethanol—store at 4 °C for up to one month
- Solubilization solution: immediately prior to sample solubilization, add 100 μl solution B to 1 ml solution A

Method

1. For the total lysis of cells, add 10 μl solubilization solution for each 1 × 10^6 frozen cells pelleted in a microcentrifuge tube ($\sim$ 3 μg protein/μl).
2. Vortex five to six times for 4–5 sec each time over a 20 min period at room temperature to solubilize the proteins.[a] Avoid causing excessive foaming while vortexing.
3. Centrifuge for 3 min at 15 000 *g* in a microcentrifuge.

[a] Solubilization of solid tissue may require more vigorous effort or possibly homogenization.

For reproducible data, it is important to standardize protein loads. Cell equivalents of supernatant to load onto first-dimension gels are shown in *Table 2*. For solid tissue, add 10 μl of solubilization solution per mg wet weight of tissue. Incubate with intermittent vortexing for 20 min. If proteins are already in solution in water or buffer, add at least an equal volume of solubilization solution, perferably two volumes, and incubate for 20 min at room temperature. The ratio of solubilization solution to sample and the amounts to load onto first-dimension gels (*Table 2*) are for silver stained 2D gels. There is no need to perform a preliminary DNase and RNase digestion, as originally described by O'Farrell (1), because of the small number of cells. If gels are to be Coomassie blue stained, more protein must be applied to the gel and DNase/RNase digestion prior to addition of solubilization solution might be necessary to achieve acceptable resolution. With isotopically labelled samples, 1–2 × 10^5 c.p.m. of ^{125}I or 2.5–5 × 10^5 c.p.m. of ^{35}S is sufficient to give autoradiographic or fluorographic two-dimensional patterns after four to seven days of X-ray film exposure. Protein-specific radioactivity is usually sufficiently high (0.5–1 × 10^5 c.p.m./μg) that fewer cell equivalents of protein are loaded than for silver stained gels.

6. First-dimension electrophoresis

6.1 First-dimension IEF with carrier ampholytes

The investigator contemplating the use of 2DE has the choice of two first-dimension separation modes. One is the traditional carrier ampholyte-based

Table 2. Sample loading for various cell types and subcellular fractions

Cell type	Amount loaded[a,b]
Ultrasmall (e.g. platelets)	1×10^8 c.e.
Small (e.g. lymphocytes)	2.5×10^6 c.e.
Large (e.g. neutrophils/epithelial cells)	1.2×10^6 c.e.
Plasma	$3\ \mu l^c$
Subfractions from a small cell type	
Cytosol	3×10^6 c.e.[d]
Nuclei	20×10^6 c.e.
Soluble nuclear	20×10^6 c.e.[d]

[a] Amounts are determined empirically for silver stained gels.
[b] c.e. = cell equivalents, the amount of protein obtained from the indicated number of cells.
[c] The amount refers to the volume of plasma. Add two volumes of solubilization solution (see *Protocol 1*).
[d] For each volume of cytosol or soluble nuclear proteins, add a minimum of one volume of solubilization solution (See *Protocol 1*).

system and the second is based on the preparation of immobilized pH gradients (IPG) (2). A major advantage of IPG for novel users is the availability of pre-cast strips from Pharmacia, thus providing an easy-to-use 2D technique. Both separation modes continue to be utilized in the author's laboratory. The standard carrier ampholyte-based gels are more economical when cast in a set of 20 and have been somewhat more dependable in our hands than IPG-based gels. The latter are utilized primarily when large amounts of protein are loaded for preparative runs or when basic proteins are to be separated. IPG gels do not suffer from the cathodic drift observed with carrier ampholyte-based gels.

Procedures are described in *Protocol 2* for preparing and using first-dimension IEF tube gels with wide range ampholytes, pH 3–10; the recipe for these gels is given in *Table 3*. The focusing time, as measured in V-h, is increased in the author's laboratory from the 7000–10 000 V-h, reported by most investigators, to 22 000 V-h. Increasing V-h takes advantage of cathodic drift (the time-dependent loss of the basic portion of the pH gradient) and increases the resolution of acidic polypeptides, which are the most abundant of total cellular proteins. Using the described procedures results in an effective separation range of pH 4–7.5 (*Figure 1*). Non-equilibrium pH gradient electrophoresis has been described for the separation of basic polypeptides (3) but we use IPG gels and a pH 7–10 gradient.

Protocol 2. First-dimension IEF with carrier ampholytes

Equipment and reagents

- Glass tubes (1.5 mm i.d. × 200 mm in length) for casting gels (1.5 mm × 160 mm). If glass tubes are to be reused, they must be cleaned in chromic acid, rinsed with alcoholic KOH, and thoroughly rinsed with water. A convenient and relatively inexpensive alternative for the first-dimension tubes is to cut 0.2 ml disposable glass pipette to 200 mm in length. The cut tubes may be used directly without cleaning, and are not reused, thus eliminating considerable time and effort.
- Stock solution for casting IEF tube gels: see Table 3

- Electrophoresis apparatus (see Section 4)
- Syringe fitted with disposable 200 μl micropipette tip (e.g. Eppendorf) with top collar cut off
- 2 litres bottom anode electrolyte: 0.015 M H_3PO_4—store at room temperature as a 10 × stock (0.15 M)
- 200 ml top cathode electrolyte: 0.01 M NaOH—store at room temperature as a 100 × stock (1 M)
- 2% Ponceau S in water
- SDS equilibration buffer (see *Protocol 5*)

Method

1. Prepare the gel solution using the recipes in *Table 3*.

2. Cover the lower ends of the tubes with Parafilm and introduce the gel solution into the tubes with a long needled syringe (e.g. a spinal tap needle), taking care not to include air bubbles.[a]

3. Clean the ends of the gel tubes and place them in the electrophoresis apparatus.

4. Fill the lower chamber with 2 litres bottom anode electrolyte. This volume is sufficient to cover the tube up to the level of the top of the gel. (The focusing is carried out at room temperature and the anolyte solution acts as a heat sink to cool the gels during the focusing.)

5. Degas about 200 ml of top cathode electrolyte for about 5 min. Pour this into the top chamber of the electrophoresis apparatus.

6. Use a syringe filled with top cathode electrolyte to displace air bubbles in the top of the tubes.

7. Pre-electrophorese the gels at 200–300 V for 600 V-h.

8. Load the sample (∼ 70 μg protein or the equivalent of protein from 1.0–2.5 × 10^6 cells, see *Table 1*) in a volume of 10–40 μl.

9. Focus at 1200 V for 16 h. Increase to 1500 V for the final 2 h.

10. Extrude the gels from the tubes using a syringe fitted with a disposable 200 μl pipette tip. Carry this out as follows. Fill the syringe with equilibration buffer and partially extrude the gel from the tube. Dip the bottom of the gel into 2% Ponceau S to mark the anode end of the gel. Then gently extrude the gel directly into a vial containing 2 ml of SDS equilibration buffer.

Protocol 2. *Continued*

11. After equilibration for 20 min, either load the gels onto second-dimension gels or freeze them in the vials and store indefinitely at −80°C until run.

---

[a] Alternatively, methods have been described for casting a number of gels simultaneously (14, 15). Basically, place the gel solution in a small container (e.g. a 25 ml beaker). Place the tubes in the beaker supported 1–2 mm from the bottom and slowly overlay the gel solution with water. The gel solution will slowly and uniformly rise part-way in the gel tubes. Place the beaker carefully into a larger beaker containing a volume of water such that the gel solution will rise further in the tubes, due to hydrostatic pressure. The desired level of the gel solution in the tube is achieved by controlling the volume of water in the beaker. It is not necessary to overlay the gel with water prior to polymerization. Allow 1 h for complete polymerization at room temperature.

Table 3. Recipe for first-dimension IEF tube gels

Stock solutions

Ampholytes: use as supplied, usually as 40% (w.v) solutions. Keep sterile and store at 4°C.
30% (w.v) acrylamide, 1.8% (w.v) bisacrylamide. Prepare fresh weekly. Store at 4°C in an amber bottle.
10% (w/v) ammonium persulfate. Prepare fresh weekly. Store at 4°C.

Recipe for casting 20 gels

Urea	8.25 g
pH 3–10 ampholytes[a]	0.33 ml
pH 3.5–10 ampholytes[a]	0.33 ml
pH 5–7 ampholytes[a]	0.1 ml
30% acrylamide, 1.8% bisacrylamide	2.0 ml
Water	6.0 ml

Dissolve the urea with gentle warming, ~ 37 °C. Degas for 2–3 min. Then add:

Nonidet P-40	0.3 ml
TEMED	7.0 µl
10% (w/v) ammonium persulfate	40.0 µl
Final volume 15 ml	

[a] Using the electrophoretic conditions described in the text (22 000 V-h), the effective separating range is ph 4.0–7.5. Use of wide pH range ampholytes from different suppliers is recommended to minimize lot-to-lot variability of any one supplier. Volumes given are for 40% (w/v) solutions as supplied.

6.2 Immobilized pH gradient gels

Pre-cast IPG strips, currently available from Pharmacia in the pH ranges of 4–7, 7–10, and 4–10, and in sizes of 11 and 18 cm, are extremely convenient to many investigators. Their use is described in *Protocol 3*. However, in occasional situations a special gradient, such as a narrow pH gradient, is required and gel casting may then need to be undertaken by the investigator. In that case, IPG slab gels are cast with a separation distance corresponding to the width of the planned second-dimension SDS gel. An additional 2 cm pH plateau is cast at the end of the IPG gel where the sample is applied. In

---

Protocol 3. First-dimension IEF with pre-cast IPG strips

Equipment and reagents

- Pre-cast IPG strips (Pharmacia) or a pre-cast IPG gel
- Pharmacia Multiphor II apparatus
- Pharmacia kit for running IPG strips
- Rehydration solution: 8 M urea, 2% NP-40, and 65 mM dithioerythritol
- U-shaped mould gasket made of 1/32" neoprene giving an average gel thickness of 0.7 mm: a newly cut gasket must first be thoroughly washed with detergent, soaked overnight in 8 M urea and detergent, and then thoroughly rinsed with distilled water to remove impurities that interfere with gel polymerization.
- Sample applicators made either from 1/8" silicone rubber or collars cut from 200 μl disposable pipette tips (e.g. Eppendorf): sample capacity is 20–80 μl as desired (applicators can also be purchased from Pharmacia).

Method

1. Cut the dried pre-cast IPG slab gel into 4–5 mm wide strips with a paper cutter or use commercially available IPG strips.

2. Engrave gel numbers on the plastic backing.

3. Rehydrate the gel in a casting mould to which a U-gasket, cut from mylar or Gel-Bond PAG film, is added so as to achieve the original gel thickness.

4. Rehydrate with the rehydration solution for 6–16 h.

5. Lightly blot the rehydrated gel strips between water-saturated filter paper to remove excess urea solution.

6. Place the gel strips 2 mm apart on the flat bed cooling plate (15°C) of the electrophoresis apparatus.

7. Saturate thick paper electrode strips with 10 mM glutamic acid or 10 mM lysine (for the anode or cathode respectively) and then blot to near dryness.

8. Place sample applicators on the surface of the gel near the anode and within the pH plateau region.

9. Place the IPG strip containing box on the cooling plate and coat the surface with silicone oil.

10. Begin focusing at 300 V, with maximum power settings of 1000 V, $0.05n$ W, and $0.1n$ mA, where n = number of gel strips. After 1 h, increase the settings to 5000 V, 5 W and 2 mA and continue focusing for 16–18 h.

11. Either equilibrate the gels and run these in the second dimension separating, or store them frozen at −80°C.

---

general, the plateau is cast at the anode end for pH 4–7 and 7–10 IPG gels since anodal sample application for these broad range gradients gives improved sample penetration into the gel with resultant increases in spot intensities and numbers of spots. Procedures for casting IPG gels have been described in refs 2, 4, and 5.

7. Second-dimension SDS–polyacrylamide gel electrophoresis

7.1 Casting gradient SDS gels

Procedures are described below for casting individual gradient slab gels. Procedures for the simultaneous casting of large numbers of gradient gels (10–20) have been described elsewhere (6).

The second dimension separates proteins on the basis of molecular weight in an SDS gel. An 11.5–14%T (2.6% cross-linking) polyacrylamide gradient gel is used as the second dimension in the author's laboratory and provides effective separation of proteins of mass from 10000–100000 daltons. Proteins outside this range are less well resolved. Thus it may be necessary to increase

---

Protocol 4. Casting the second-dimension gradient SDS gel

Equipment and reagents

- Two-chambered gradient mixer with about 30 ml capacity per chamber
- Second-dimension slab gel apparatus (see Section 4.2)
- Reagents for second-dimension gel (see Table 4)
- Water-saturated *sec*-butanol

Method

1. Assemble the mould using two glass plates (17.5 × 17.5 cm) and a silicone or neoprene U-gasket 1.5 mm thick (1/16″) supplied by the manufacturer. Do not siliconize the glass plates.

2. Prepare the dense (14%T) and light 11.5%T) acrylamide solutions as described in *Table 4*.

3. Pour the 14%T acrylamide into the gradient market chamber that leads to the gel mould. Pour the 11%T solution into the other chamber. Cast the acrylamide gradient and overlay the gel with 1–2 ml of water-saturated *sec*-butanol.

4. Allow the gel to polymerize 1 h at room temperature. Remove the clamps from the gel sandwich and carefully remove the U-gasket.

5. Rinse the surface of the gel briefly with water. Place the gel upside down in a vertical position to drain the surface of excess water.

6. Support the slab gel in a vertical position to facilitate the application of the first-dimension gel.

---

or decrease the acrylamide gradient range, or even use gels of constant acryl-amide concentration, to achieve the desired resolution. Proteins with molec-ular weights less than 10000 electrophorese close to the dye front and are not resolved. The recipe for the second-dimension SDS gradient gel is given in *Table 4* and the procedure for casting is described in *Protocol 4*. It does not use a stacking gel. Use of a stacking gel yields a more compact spot shape in the molecular weight dimension, but often results in spot elongation in the focusing dimension, due to some lateral band spreading of the protein as it leaves the first-dimension focusing gel and forms a sharp zone in the stacking gel. Interpretation of gel patterns is less confusing with slight spot elongation in the molecular weight axis compared with elongation in the focusing dimension. Consequently the second-dimension SDS gel is run without a stacking gel.

7.2 Application of the first-dimension gel to the SDS slab gel and electrophoresis

The procedure depends on the type of isoelectric focusing gel used for the first dimension, IPG or IEF. Both procedures are described in *Protocol 5* together with the conditions for electrophoresis.

Following electrophoresis, the gel is stained to reveal the separated proteins. The various approaches for protein visualization are discussed in Chapter 2. For most routine applications, silver staining is the preferred

---

Table 4. Second-dimension gradient SDS slab gel mixture

Stock solutions

30% (w/v) acrylamide, 0.8% (w.v) bisacrylamide. Filter and store at 4°C for one to two weekls in an amber bottle.
Gel buffer: 1.5 M Tris–HCl pH 8.5. Store at 4°C.
87% (w/w) glycerol. Store at room temperature.
10% (w.v) SDS. Store at room temperature.
40% (v/v) ammonium persulfate. Store at 4°C for one week.

Recipe for casting a gel 16.5 cm × 16.5 cm × 1.5 mma

	Light solution (11.4%T)	Dense solution (14%T)
30% acrylamide, 0.8% bis	7.5 ml	9.1 ml
Gel buffer	5.0 ml	5.0 ml
87% glycerol	–	5.6 ml
10% SDS	0.2 ml	0.2 ml
TEMED	5 μl	3 μl
10% ammonium persulfate	0.15 ml	0.1 ml
Water	7.1 ml	–
Final volume	20 ml	20 ml

a The separating gel is an 11.4%T–14%T (2.6%C) acrylamide gradient, containing 0.375 M Tris–HCl pH 8.5 and 0.1% (w/v) SDS.

---

approach. *Figure 2* shows a silver-stained 2D gel that involved first-dimension IEF focusing whilst *Figure 3* shows a silver-stained 2D gel of the same sample except that the protein separation in the first dimension was carried out using IPG strips. Alternatively the proteins may be blotted onto a membrane (Western blotting) prior to visualization (see Section 9).

Protocol 5. Second-dimension electrophoresis

Equipment and reagents

- Nylon sieve or strainer
- Second-dimension gel (from *Protocol 4*)
- First-dimension IPG gel strip or IEF tube gel
- SDS equilibration buffer (for analysis of first-dimension IEF gel): 0.125 Tris–HCl pH 6.8, 2% (w/v) SDS, 10% (v/v) glycerol and 65 mM dithioerythritol. Add a small spatula tip (~ 50 mg) of bromophenol blue. Place the solution in an automatic dispenser (e.g. adjustable delivery up to ~ 2 ml) and store at room temperature. Storage at 4 °C results in SDS precipitation. Incomplete resolubilization of the SDS may interfere with the dispensing mechanism.

- Second-dimension slab gel apparatus (see Section 4.2)
- SDS–urea equilibration buffer (for analysis of first-dimension IPG gels): add 26 g solid urea per 50 ml SDS equilibration buffer (6 M urea final concentration)
- 0.5% (w.v) agarose in 0.025 M Tris, 0.192 M glycine, 0.1% (w.v) SDS: microwave or boil (> 90 min) to dissolve agarose, and store at −20°C
- Running buffer: 0.025 M Tris, 0.192 M glycine, 0.1% (w.v) SDS
- Urea
- Water-saturated *sec*-butanol

A. *Loading a first-dimension IEF tube gel*

1. Thaw the gel, if frozen in SDS equilibration buffer (see *Protocol 2*), and equilibrate for 20 min with gentle agitation (e.g. on an aliquot mixer).

2. Drain the SDS equilibration solution from the gel using a small nylon sieve or strainer.

3. Spread the gel on a piece of plastic or Parafilm. Note the orientation (the anode is marked with Ponceau S).

4. Transfer the gel, without stretching, onto the top surface of the second-dimension gel (from *Protocol 4*) and ensure complete contact of the IEG gel to the slab gel.

5. Heat the 0.5% agarose to 90°C. Embed the gel with 2 ml agarose solution. Allow the agarose to cool for 5 min before placing the slab gel in the electrophoresis apparatus.

B. *Loading a first-dimension IPG gel*

1. Place the gel strip (3–5 cm wide) in 6–8 ml SDS–urea equilibration buffer and equilibrate for 30 min with agitation.

2. Place the strip on its long edge on a lintless absorbent towel.

3. Handling the gel strip by the edge of the plastic backing, cut off the excess backing and about 5 mm of gel occupied by the electrode strips. An additional portion of gel from the plateau region may be cut off it necessary so the IPG strip will fit on the second-dimension gel. Note the orientation, anode, and cathode, after trimming the gel strip.

4. Place 2 ml of agarose at 90°C onto the top surface of the slab gel and immediately immerse the IPG gel strip in the agarose solution. Carefully press down the IPG strip (use a spatula and press against the plastic backing, not the gel) onto the surface of the slab gel ensuring complete contact.

5. Allow the agarose to solidify for 5 min, then place the slab gel in the electrophoresis apparatus.

C. *Electrophoresis conditions*

1. Immerse the slab gel completely in cold electrophoresis running buffer so that electrophoresis is performed at 10°C.

2. Initial voltage conditions depend on the type of first-dimension gel. For IEF tube gels, electrophorese at 400 V constant voltage until the tracking dye reaches the bottom of the slab gel (~ 3 h). For IPG gels, electrophorese initially at 100 V until the bromophenol blue tracking dye has entered the separating gel about 1 cm (~ 30 min), then increase the voltage to 400 V.[a]

[a] The current is initially 120 mA per slab gel and decreases to 60 mA at the end of electrophoresis. Thus if many gels are electrophoresed simultaneously, e.g. ten gels per electrophoresis unit using the DALT apparatus of Anderson and Anderson (6), the power supply must be capable of delivering 1.2 A at 400 V.

8. Preparative 2DE, special gradients, and gel formats

8.1 Preparative IPG

The amount of protein that can be loaded on a single carrier ampholyte-based tube gel generally does not exceed 100–200 μg. Optimal resolution is achieved with loads of 30 μg or less. The identification of protein spots by microsequencing for a protein of average size and at average abundance in a cell extract requires a greater amount of protein to be loaded on individual first-dimension gels. A major advantage of IPG is the ability to scale-up protein loads without changing gel formats.

Figure 4 demonstrates the large number of polypeptide spots that can be visualized by Coomassie staining of an electroblot of an IPG-based preparative 2D gel, with a lymphoid protein loading of 1 mg (7). A volume of up to 80 μl of solubilized sample was loaded per IPG strip. There was no detectable loss of resolution in gels at the highest protein loads. However, some variability is observed in the extent to which certain polypeptides, particularly actin, are focused in some gels. Actin streaking is observed between the application site and the stationary position of actin in the gels and does not appear to be related to protein load. At loadings of up to 1 mg of cellular proteins that varied in abundance, the amount of protein per spot did not appear to be a

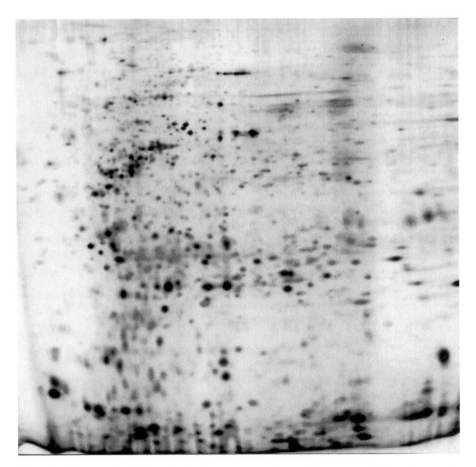

Figure 2. Silver stained 2D gel of a human oesophageal tumour protein using carrier ampholytes. The first-dimension isoelectric focusing was performed in a tube gel using wide range ampholytes (pH 3–10) supplemented with pH 5–7 ampholytes for 22 000 V-h. The effective separation range is pH 4 (*left*) to pH 7 (*right*). The second-dimension SDS gel was an 11.4% (*top*) to 14% (*bottom*) acrylamide gradient.

limiting factor for sharp focusing for any of the proteins. At such a protein load, substantial streaking occurs in carrier ampholyte-based 2D gels.

Adequate protein solubilization and entry into the IPG gel is essential for achieving good resolution. We have therefore examined various means of solubilizing proteins for preparative IPG-based separations. Equally satisfactory results have been obtained with the use of either mercaptoethanol or dithioerythritol as reducing agents. Inclusion of 1% SDS in the solubilization mixture, which required the samples to be loaded at the cathode, resulted in streaking of some proteins and had no obvious benefit.

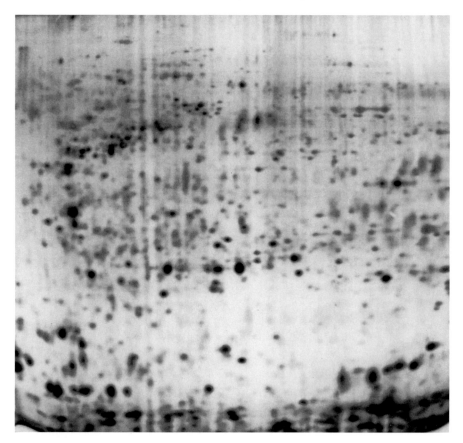

Figure 3. Silver-stained 2D gel of the same tumour as in *Figure 2* except that the proteins were separated in the first-dimension using IPG strips pH 4–7. An identical second-dimension gradient was used as for the gel shown in *Figure 2*.

8.2 Special gel formats

8.2.1 Large gel format

The technique developed by Klose in 1975 has been updated (8), resulting from the extensive experience in Dr Klose's laboratory, since introduction of the method. This particular version of 2DE requires special equipment able to handle the large (46 × 30 cm) gel format. The resolution achieved is quite impressive. Features of the technique include:

(a) Sample loading into the acidic side of the IEF gel allowing both acidic and basic proteins to be resolved in the same gel.

(b) Preparation of ready-made gel solutions that can be stored frozen.

The patterns obtained have included several thousand polypeptide spots from mouse tissues.

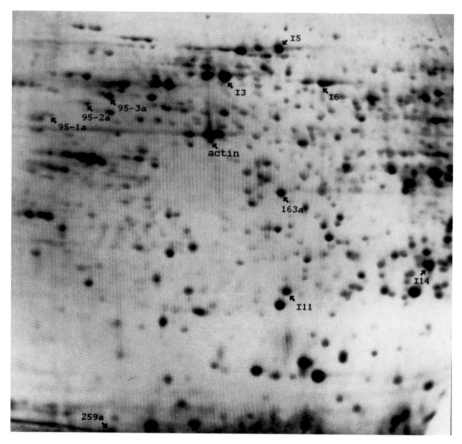

Figure 4. Electroblot of a pH 4–7 preparative IPG gel, loaded with 1 mg of lymphoid protein and stained with Coomassie blue. *Arrows* point to the location of some of the proteins with unmodified N-termini that were readily sequenced from the blots.

8.2.2 Micro 2DE

Dr Manabe's group has applied micro 2DE primarily to the analysis of plasma proteins under native or denaturing conditions (9). The gels are only a few centimetres long in both directions. Recently, improvements in the micro 2DE technique have been reported with extended applications to the analysis of cellular proteins (9).

8.2.3 Analysis of very basic proteins

Applications involving the analysis of very basic proteins, such as histones and ribosomal proteins, require first-dimension gel gradients that extend to pH 12. Görg *et al.* have recently reported the use of very alkaline IPG gradients for 2DE of ribosomal and nuclear proteins (10). The pH gradients were: IPGs 8–12, 9–12, and 10–12. These gradients have necessitated different optimization steps with respect to pH engineering, gel composition, and running conditions. The substitution of acrylamide by dimethylacrylamide and addition of 2-propanol or methyl cellulose to the IPG rehydration solution were necessary to achieve high resolution.

9. Electrophoretic transfer of proteins (Western blotting)

To understand a variety of biological processes it is necessary to identify and characterize specific proteins found in complex mixtures on 2D gels. The resolution of the 2D protein pattern is maintained when proteins are transferred from the second-dimension slab gel to a suitable membrane. Several types of transfer membranes may be used. Nitrocellulose membranes are available from a number of suppliers. Nitrocellulose with 0.45 μm pore size is most commonly used but 0.2 μm pore size membranes have a greater binding capacity. Immobilon membranes are available from Millipore. We use nitrocellulose for immunological detection and for *in situ* proteolytic digestion of the transferred protein for subsequent peptide mapping. Immobilon membranes are used when amino acid analysis of N-terminal microsequencing of isolated proteins is to be performed.

9.1 Electroblotting

Immediately following SDS slab gel electrophoresis, electrophoretically transfer the separated proteins to a nitrocellulose or Immobilon sheet (16.5 × 16.5 cm). Many types of blotting apparatus are available commercially and hence the exact blotting procedure will depend on the specific apparatus used. Follow the manufacturer's instructions.

9.2 Protein staining of electroblots

Protein may be detected on transfer membranes using several staining procedures depending on the subsequent use of the membrane and the sensitivity desired. Nitrocellulose may be stained with India ink and Ponceau S. India ink stained blots have a sensitivity greater than the equivalent sensitivity of a Coomassie blue stained gel but less than a silver stained gel. Ponceau S has less sensitivity and is used to stain blots of 'preparative' 2D gels in which individual polypeptides are to be isolated for further chemical characterization (e.g. peptide mapping and/or *in situ* digestion of the protein) and purification of peptides by high-performance liquid chromatography for microsequencing (11). Immobilon membrane blots are prepared when the chemical characterization involves amino acid analysis or direct N-terminal microsequencing of isolated proteins (12). On Immobilon membranes, proteins are visualized by Ponceau S or Coomassie blue staining. Gloves must be worn whenever handling transfer membranes.

(a) India ink staining. Wash the blot briefly with water and then place it in staining solution [50 μl of high quality India ink, e.g. Pelican, in 50 ml PBS containing 0.3% (v/v) Tween 20] for 30–60 min. The blot must be completely submerged at all times. Rinse in water, place on an absorbent paper, and air dry.

(b) Ponceau S staining. Place the blot in staining solution [0.1% (w/v) Ponceau S in 1% acetic acid] for 5 min. Destain with 1% acetic acid. If protein spots are to be cut from the blot, keep the membrane moist.

(c) Coomassie blue staining. Stain the blot for 2 min in 0.2% (w/v) Coomassie blue R-250 in 40% methanol, 5% acetic acid. Destain with several changes of 40% methanol, 5% acetic acid. The gel may be air dried or kept moist in water prior to cutting out protein spots.

10. Gel analysis

The most common use of 2D gels is to obtain quantitative assays for a large number of proteins, or for particular proteins in a complex mixture. However, two major limitations of 2D gels should also be considered. First, only relatively abundant proteins are detected. Secondly, although many simultaneous protein measurements can be obtained, these measurements may not be precise enough for some investigations. Thus, in trying to detect small quantitative differences in the amount of a protein, one may need large numbers of duplicate gels to achieve a statistically significant result. Since there are many steps in making a gel, each of which can cause various types of spot variability, many spot size differences may be caused by the technique and hence one must be wary of concluding that differences observed are biological in nature without careful checking.

10.1 Causes of variability in 2D patterns

Since there are many potential sources of experimentally caused variability in 2D gels, it is invariably necessary to replicate experiments by preparing duplicate gels or obtaining new samples. Although there are exceptions, such as in group experiments in which there are many individuals in each group, it is safe to state that a two-gel comparison never proves anything in a convincing manner.

Problems that may be experienced using 2D gels include:

(a) Spots near the acidic or basic ends of the gel, or near 'gaps' in the pattern caused by large spots, can vary considerably in their size compared with spots in more reproducible regions of the gel.

(b) Variability in the equilibration of first-dimension gels, as well as in some other steps, can cause varying degrees of protein loss, with more loss of low molecular weight proteins (or greater diffusion) being the typical result.

(c) Variability in the degree to which samples are solubilized may also affect spot sizes on the gel.

(d) Underlying spots (i.e. co-migration of two or more spots) should be considered as a possibility in case of unexpected quantitative results.

(e) Temperature has been shown to affect 2D spot patterns, rather profound differences being observed when the temperature is raised or lowered by 10°C (13).

(f) Old samples, even if meticulously stored, should cause suspicion, since new spots corresponding to degradation products or other post-translational modifications may be observed.

(g) Sample contamination is a serious problem. The potential sources of contaminants are dependent on sample preparation procedures that may be unique to an experiment. Therefore, only the most general advice can be offered: 'know your contaminants'. For example, if you suspect that serum proteins in cell culture medium may be making their way onto the gels, deliberately prepare gels of samples with varying levels of the medium. If studying human lymphocytes, as in our case, deliberately prepare samples spiked with platelets, red cells, or other suspected contaminant cells, so as to detect such contamination, and have a knowledge of which spots will be affected to any degree. If fractionating cells or other samples, it is prudent to run the 'uninteresting' fractions as well as the fraction of interest to avoid concluding that a spot is bigger for a treated sample in a case where this sample was simply not fractionated as cleanly.

The potential problems cited above have been of the type in which single actions lead to a difference in the spot sizes. More usual, however, are small amounts of variability in each of many steps, that together lead to variability in the spot sizes that is complex. It becomes difficult to single out any particular step as being the one causing the most trouble. Usually one observes a certain amount of variation from gel-to-gel when identical samples are loaded, slightly greater variation if the same sample is independently prepared twice, and more variability still if the two samples are electrophoresed on different days. This variability occurs even if all steps of the process are apparently standardized. When comparing two groups of samples for group-related differences, there will also be irrelevant biological differences from sample-to-sample within each group that constitutes an additional source of spot size variability.

Variation in the sample preparation procedure, the amount of protein loaded, the spot visualization technique, the spot measuring procedure, and batch-to-batch gel differences are some of the usual sources of annoying spot size variability that one must consider. Effort should clearly be directed at decreasing these sources of error by improved procedures. When analysing gels and designing experiments, one must therefore consider all these sources of variability to be present to some degree, and be convinced that such variation is not leading to wrong conclusions. In most experiments some variability may be reduced by judicious choice of experimental design or incorporated into a statistical model.

10.2 Computerized 2D gel scanning and analysis

The classical approach to 2D gel analysis, namely visual comparison, continues to have important merits stemming from the discerning power of the analyst's brain. However the visual approach does not yield quantitatively reliable data. Computerized scanning devices and densitometers have become widely accessible, and quantitative analysis has allowed an assessment of the significance of variation in spot intensity between samples to be made. If direct visual scoring or interactive quantification procedures are used, it is a good idea to perform such measurements without knowledge of which particular sample is being judged, since otherwise bias in measuring becomes possible. Having more than one person assign spot size scores is also recommended if visual methods are used.

Several computerized systems specifically designed for 2D gel analysis have been available for some time and have been quite diverse in the specific hardware and software used. More recently, there has been a trend to acquire software from vendors who specialize in the analysis of electrophoretic gels. A list of suppliers/developers of 2D analysis software and their web sites appears in *Table 5*. For all systems, first computer images must be obtained through a digitizing camera, or scanner. Algorithms to detect and quantify spots and also edit the results of such procedures form the most basic software needs. The next important piece of software on such systems allows the comparison of spots between two or more gels. These algorithms can vary from simple displays of two gels, which assist an interactive user in identifying gel differences, to complex automatic algorithms that match spots on large groups of gels.

A presentation and discussion of the plethora of software and computer systems available for 2D gel analysis is beyond the scope of this chapter. It should be pointed out however that the expanding number of 2D gel patterns that can be visualized on the Internet opens up the possibility of visually

---

Table 5. Web sites for imaging software developers

Bioimage 2D analysis software	http://www.bioimage.com
Flicker 2D gel image comparator	http://www-lmmb.ncifcrf.gov/flicker/
Kepler 2D analysis software	http://www.lsbc.com/kepler/kpl.htm
Melanie II 2D analysis software	http://www.expasy.ch/melanie
Protein Databases analysis software	http://www.pdiq.com
Phoretix 2D gel analysis software	http://www.phoretix.com
GELLAB-II 2D Gel Analysis	http://www-lmmb.ncifcrf.gov/lempkin/gellab.html
Itti, I.c. 2D analysis softaware	http://www.itti.com/2d.htm
Advanced American Biotechnology	http://members.aol.com/aabsoft/aab.html
Scanalytics	http://www.scanalytics.com/software.htm

---

comparing different 2D gels including the investigator's own. At the recent 2D meeting, P. Lemkin (lemkin@nic.fcrf.gov), proposed implementation of a distributed gel comparison method that runs on any WWW connected computer and is invoked from Netscape or Mosaic. For this approach to work, investigators will need to scan their 2D images onto a computer that is connected to the WWW. Thus, any one gel image from one of the Internet gel databases could be compared with the investigator's gel image. This concept has wide appeal and it is likely that in one manner or the other, gel comparisons could be made directly on the Internet in the near future.

References

1. O'Farrell, P. H. (1975). *J. Biol. Chem*, **250**, 4007.
2. Strahler, J. R., Hanash, S. M., Somerlot, L., Görg, A., Weser, J., and Postel, W. (1987). *Electrophoresis*, **8**, 165.
3. O'Farrell, P. Z., Goodman, H. M., and O'Farrell, P. H. (1977). *Cell*, **12**, 1133.
4. Hanash, S. M., Strahler, J. R., Somerlot, L., and Görg, A. (1987). *Electrophoresis*, **8**, 229.
5. Westermeier, R. (1997). *Electrophoresis in practice: a guide to methods and applications of DNA and protein separations*, xvii and 331. VCH, Weinheim.
6. Anderson, N. L. and Anderson, N. G. (1978). *Anal. Biochem.*, **85**, 341.
7. Hanash, S. M., Strahler, J. R., Neel, J. V., Hailat, N., Melhem, R., Keim, D., *et al.* (1991). *Proc. Natl. Acad. Sci. USA*, **88**, 5709.
8. Klose, J. and Kobalz, U. (1995). *Electrophoresis*, **16**, 1034.
9. Manabe, T., Yamamoto, H., and Kawai, M. (1995). *Electrophoresis*, **16**, 407.
10. Görg, A., Obermaier, C., Boguth, G., Csordas, A., Diaz, J.-J., and Madjar, J.-J. (1997). *Electrophoresis*, **18**, 328.
11. Aebersold, R. H., Leavitt, J., Saavedra, R. A., Hood, L. E., and Kent, S. B. (1987). *Proc. Natl. Acad. Sci. USA*, **84**, 6970.
12. Matsudaira, P. (1987). *J. Biol. Chem.*, **262**, 10035.
13. Görg, A., Postel, W., Friedrich, C., Kuick, R., Strahler, J., and Hanash, S. (1991). *Electrophoresis*, **12**, 653.
14. Anderson, N. G. and Anderson, N. L. (1978). *Anal. Biochem.*, **85**, 331.
15. Garrels, J. I. (1979). *J. Biol. Chem.*, **254**, 7961.

7

Peptide mapping

ANTHONY T. ANDREWS

1. Introduction

Current versions of one-dimensional techniques such as polyacrylamide gel electrophoresis (PAGE), polyacrylamide gel electrophoresis in the presence of detergents such as sodium dodecyl sulfate (SDS-PAGE), immuno-electrophoresis (IE), and isoelectric focusing in polyacrylamide gels (IEF) have very high resolution, are reproducible, reliable, rapid, and relatively inexpensive and simple to set-up and use. Although slower and more complex two-dimensional separations give unparalleled resolution that no other analytical method can match, the number of samples which can be examined in a given time is much reduced. In spite of these capabilities, however, there are situations where neither one- nor two-dimenstional methods can give an unequivocal answer. Most of these devolve into a question of 'relatedness' between two or more proteins and this is where peptide mapping can make a major contribution. When working with intact proteins, identifying an un-known or following changes in protein components typically involves compari-son with known standard proteins. In most one-dimensional electrophoretic systems it is usual to run a least one standard sample together with a series of unknowns on a single gel slab under the same conditions of time, voltage, current, temperature, gel composition, etc. Two unrelated proteins can be similar in size (and hence fail to separate by SDS–PAGE) or in molecular charge (and hence have similar pIs, so separate poorly by IEF), but perhaps less obviously, in non-denaturing PAGE a large highly charged protein may have a similar mobility to a small molecule with a low net molecular charge. Traditional physical, chemical and biochemical methods of analysis can be applied to solving such problems but they usually require substantial amounts of material (e.g. milligram scale) and there are many occasions where this is not available. A major advantage of electrophoretic methods of analysis is the very small amounts of material needed.

Of course, all the one dimensional methods provide some evidence on similarities and differences between proteins, so the more types of analysis that are applied to a sample the more rigorous this evidence becomes. For example, in PAGE, separations in concentrated gels (high values of %T) are determined largely by molecular sieving (size differences), while charge differences predominate in low %T gels. Thus running samples on a series of gel slabs of different %T can be used to demonstrate homogeneity in terms of both size and charge. This approach also enables Ferguson plots to be constructed (1–3) and molecular masses to be measured but it is rather time-consuming.

Unfortunately, even this is not sufficient because there are a few occasions when unrelated proteins are not distinguished. Much more importantly there are quite frequent examples of when related proteins may run very differently and give the misleading impression that there are many impurities. This is often the case with protein–precursor relationships such as enzymes and their zymogens, for example, or when proteins are subject to some form of post-synthetic modification (e.g. glycosylation, phosphorylation). Degradation during sample preparation is another potential source of heterogeneity because virtually all biological materials contain at least traces of proteinase activity.

The best approach to dealing with such problems will often be peptide mapping (3–6). This consists of breaking down the unknown protein into a number of peptide fragments in a very specific and controlled manner, separating the peptide mixture, and then comparing the pattern of separated zones with those of one or more standard proteins treated in the same way.

With such a general definition of peptide mapping, it is obvious that there are a great many possible ways of doing it. Procedures can be varied at the peptide production stage as well as by using different peptide separation methods. Initially, two decisions have to be made:

(a) What fragmentation method is to be used?

(b) Should it be performed before or after the sample protein has been applied to the electrophoresis gel?

If one is confident that the sample protein is relatively pure, it will usually be preferable first to generate the peptide mixture *in vitro* and then separate the components because this makes it easy to define and control the hydrolysis reaction. Either chemical (7, 8) or enzymic methods (9, 10) can be used to hydrolyse the protein. A high degree of specificity in peptide bond cleavage is perhaps less easy to achieve with chemical methods than with enzymic methods but, with care, some reasonably selective methods are available (*Table 1*). Almost any proteinase could, in theory, be used for enzymic cleavage of polypeptide chains, but the most popular ones are *Staphylococcus aureus* V8 proteinase, trypsin, chymotrypsin, papain, and pepsin (*Table 2*). Enzymic hydrolysis has the advantages that it is quick in most cases, more specific than chemical methods, and the buffers used are often compatible with the subsequent electrophoretic separation of peptides, so digests may be applied directly to the gels for analysis. Unfortunately, the enzymes themselves may contribute peptides to the mixture so a blank (enzyme only) digest

Table 1. Chemical methods for selective peptide bond cleavage in proteins

Reagent	Bonds cleaved	References
Cyanogen bromide	– Met – X –	7, 11–14
Hydroxylamine	– Asn – Gly –	7, 8, 11, 14, 15
BNPS – skatole	– Trp – X –	7, 14, 16
N-chlorosuccinimide (N-bromosuccinimide)	– Trp – X –	7, 17
Partial acid hydrolysis	– Asp – Pro –	7, 8, 11, 14, 18
Partial basic cleavage	– Ser – X –	7, 19
Heat (110 °C, pH 6.8, 1–2 h)	– Asp – Pro –	20
2-nitro-S-thiocyanobenzoate	– X – Cys –	8, 14

must always also be analysed. If chemical methods are used for peptide gener-ation, it may be necessary to remove excess chemical reagents from the samples and/or reduce the ionic strength and change the pH before analysis. Dialysis will not usually be practicable for this if small peptides are present, so other techniques, such as column desalting, precipitation, ion exchange treat-ment, etc will be needed. These have their own drawbacks and are at the very least, time-consuming. In such cases it may be best to use capillary electro-phoresis (29, 30) or non-electrophoretic methods of separating peptides such as reversed-phase HPLC (31–40) or hydrophobic interaction chromatography by HPLC or FPLC.

Frequently, however, sample proteins are not readily available in a purified form, so it is then often convenient to apply the samples directly to gel electrophoresis and, once the protein components have been separated into their respective zones, to perform the hydrolysis to peptides *in situ* and map the peptides in a second stage. Again, there are many possible permutations as to how this is done. For example, PAGE, SDS–PAGE, IEF, etc. can all be used for the initial separation of the protein components in the sample and a number of questions then arise. If a gel matrix is present, as is usually the case, are the protein zones then to be individually cut out and the protein eluted or not before hydrolysis; are the proteins to be blotted onto an immobilizing membrane; is a chemical or enzymic hydrolysis stage to be used; what method of mapping the resulting peptides is intended, etc.? In spite of such a large range of possibilities, many experiments employ SDS–PAGE in the final stage to separate the peptides into a one-dimensional map and usually the peptides have been generated from protein zones obtained by subjecting the sample to PAGE or SDS–PAGE as a first step. *In situ* hydrolysis of the protein zones in the gel without extraction is the most common pro-cedure and is performed enzymically with proteinases. Chemical cleavage methods are difficult to use in this context and are seldom employed.

Although there are many other peptide mapping procedures, the most quoted method is that described by Cleveland *et al.* (21). For microscale work, such as protein sequencing, the protein components are separated by either

Table 2. Proteinases used for selective peptide bond hydrolysis of proteins for peptide mapping

Proteinase	pH optimum	Bond specificity	References
Staphylococcus aureus V8 proteinase	4–8	Glu – X, Asp – X	9, 11, 23, 24
α-Chymotrypsin	7–9	Trp – X, Tyr – X, Phe – X, Leu – X	10, 11, 21–23
Trypsin	8–9	Arg – X, Lys – X	10, 11, 22, 23
Pepsin	2–3	N- and C-sides of Leu, aromatic residues, Asp, and Gluy	10
Thermolysin	~ 8	Ile – X, Leu – X, Val – X, Phe – X, Ala – X, Met – X, Tyr – X	10, 23
Subtilisin	7–8	Broad specificity	21, 23
Pronase (*Streptomyces griseus* proteinase)	7–8	Broad specificity	21
Ficin	7–8	Lys – X, Arg – X, Leu – X, Gly – X, Tyr – X	21, 23
Elastase	7–8	C-side of non-aromatic neutral amino acids	21, 23
Papain	7–8	Lys – X, Arg – X, (Leu – X), (Gly – X), (Phe – X)	23, 25
Clostripain	7–8	Arg – X	10, 23
Lys–C endoproteinase (*Lysobacter enzymogenes*)		Lys – X	26
Asp–N endoproteinase (*Pseudomonas fragi*)		X – Asp, X – Cys	27
Arg–C endoproteinase (mouse submaxillary gland)		Arg – X	28

216

one-dimensional methods, or very frequently by two-dimension electrophoresis (2DE), blotted onto an immobilizing membrane, hydrolysed to peptides, and the resulting peptides fractionated by HPLC (31–40) or capillary electrophoresis (29, 30), ideally with mass spectroscopic identification and sequencing (41).

2. Apparatus

The apparatus required for peptide mapping is the same as that required for PAGE, SDS–PAGE, IEF etc. (see Chapters 1, 5, and 6). Since the peptide map from a sample protein is compared to the pattern of peptides obtained from one or more standard proteins, the use of the slab gel format is virtually obligatory for both one-dimensional separations as well as two-dimensional maps. For micromapping by capillary electrophoresis, the standard apparatus is used, usually employing capillary zone electrophoresis (CZE) or micellar electrokinetic chromatography (MECC) but gel-filled capillaries have also been used to separate peptides by capillary SDS–PAGE (42).

3. The standard technique

3.1 Using pure protein samples

If the samples contain a number of component proteins which require separation from each other before protein mapping can be applied, this can easily be done by any of the one- or two-dimensional methods described in Chapters 1 and 6. The separated protein zones are detected (Chapter 2) and the protein(s) of interest are recovered as indicated in Chapter 3.

In the original protocol (21) which is described in *Protocol 1*, purified proteins are made up as 0.5 mg/ml samples in 0.125 M Tris–HCl buffer pH 6.8 containing 0.5% SDS, 10% (v/v) glycerol, and 0.001% bromophenol blue, and heated at 100°C for 2 min to inhibit extraneous proteinase activity and to unfold and denature the proteins, to ensure optimum SDS binding. Known amounts of proteinase are then added and the samples incubated at 37°C. Commercially available proteinases such as many of those in *Table 2* are usually employed in this type of work. Since proteinases vary greatly in activity depending upon their specificity and purity, the identity of the substrate proteins, temperature, pH, etc., it is difficult to give universally ideal conditions for generating peptide digests. However, it is worth emphasizing that complete digestion to the smallest peptides that can be achieved with the proteinase in question (a so-called limit digest) is not required or even desirable. What is needed is a partial hydrolysate with a maximal number of different peptides and preferably a considerable proportion of large peptides, since many small 'limit peptides' are likely to be poorly fixed in the gel and may be washed out and lost, so contributing nothing to the peptide map.

Typical amounts of proteinase required with incubation times of about 10–60 min may be of the order of 1–100 µg/ml.

Protocol 1.   Standard peptide mapping procedure using purified proteins[a]

Equipment and reagents

- Slab gel electrophoresis equipment
- Purified sample protein
- SDS polyacrylamide slab gel: either use a uniform concentration 15%T 2.6%C gel, or a concentration gradient 5–20%T (or 10–25%T) gel in the Laemmli buffer system (Chapter 1, Section 5)

- SDS sample buffer: 0.125 M Tris–HCl pH 6.8, 0.5% SDS, 10% (v/v) glycerol, 0.001% bromophenol blue
- Proteinases of choice (see *Table 2*)
- Protein standards of known molecular weight in SDS sample buffer

Method

1. Prepare the protein samples (0.5 mg/ml) in SDS sample buffer and heat them at 100°C for 2 min.

2. Add proteinase to each sample, typically to 1–100 µg/ml final concentration. Incubate at 37°C for a set period of time (typically 10–60 min).

3. Stop proteolysis by adding 2-mercaptoethanol to 10% final concentration, SDS to 2% final concentration, and boiling for 2 min.

4. Load 20–30 µl of each sample (10–15 µg of peptide) onto the SDS–PAGE slab gel. Also load standard proteins of known molecular weights in parallel tracks of the gel.

5. Carry out electrophoresis until the bromophenol blue nears the end of the gel.

6. Detect the separated peptides by Coomassie blue staining or silver staining (see Chapter 2). Radiolabelled peptides can be detected by autoradiography or fluorography.

[a] Adapted from Cleveland *et al.* (21).

In order to establish good hydrolysis conditions, it may be useful to do a preliminary experiment in which the substrate is incubated with proteinase and samples withdrawn at different time intervals, boiled briefly to end the hydrolysis, and examined by PAGE or SDS–PAGE. An example of this is shown in *Figure 1*. It can be seen that the qualitative appearance of the peptide band patterns is surprisingly not very sensitive to incubation time. Most of the peptides that can be seen after 10 min of hydrolysis are also observable after 90 min or even 3 h, although naturally there are marked quantitative differences as the initial large peptide products are subsequently broken down to smaller peptides. In this example, one might choose a 30 min hydrolysis time to give a good spread of peptides for mapping. A similar trial experiment could be conducted by choosing a fixed incubation time and

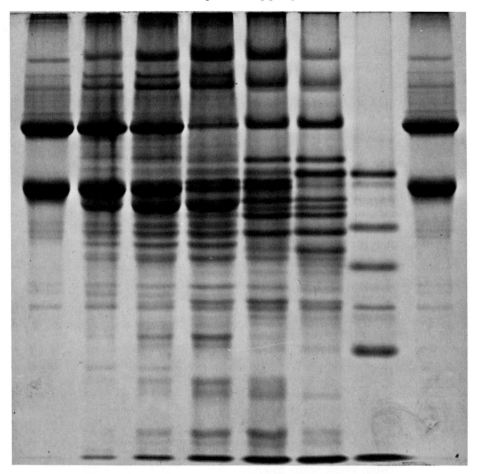

Figure 1. A typical preliminary experiment in peptide mapping in which the substrate protein and a proteinase are incubated for various times to establish conditions for optimum hydrolysis. In this example a 1% solution of total casein (Hammarsten casein, BDH Ltd.) in 0.1 M sodium phosphate buffer (pH 7.5) was incubated at 37°C with 3 μg/ml of trypsin (Sigma Type XII, TPCK treated). After the desired incubation time samples were withdrawn, heated at 100°C for 2 min to inactivate the trypsin and diluted five-fold with stacking gel buffer containing 5% sorbitol and 0.001% bromophenol blue. The digests were then applied to a $T = 12.5\%$, $C = 4\%$ PAGE gel for analysis. Samples *left* to *right* were: control (0 min incubation), 3 min, 10 min, 30 min, 90 min, 3 h, 24 h, and control (0 min).

varying the amount of proteinase added. Again it will be found that the quantitative pattern of peptide zones is not very sensitive to changes in proteinase level. This shows the relative insensitivity of both the protein digestion and the subsequent peptide mapping steps to the hydrolysis conditions. This is very useful when dealing with proteins of unknown susceptibility to the proteinase used or when examining samples containing variable amounts of contaminating proteinases, for example, because reproducible and characteristic peptide patterns are easily obtained.

After a suitable incubation time has elapsed, proteolysis is stopped by adding 2-mercaptoethanol and SDS to the samples and boiling them for about 2 min (*Protocol 1*). The samples are then loaded onto the SDS–PAGE slab. Cleveland *et al.* (21) used a conventional SDS–PAGE slab gel arrangement with a T = 15%, C = 2.6% gel made up in the Laemmli (43) buffer system which was then run, stained and destained, or subjected to autoradiography or fluorography in the usual way to reveal the zones of separated peptides. Sharper peptide bands and greater resolution is often obtained if a concentration gradient gel is used instead of one with a uniform concentration. *Figure 2* shows that this improvement in band sharpness can be very pronounced. The preparation of concentration gradient gels has been described earlier (Chapter 1). Depending upon the molecular mass of the original protein and size of the peptide fragments produced, gels with a linear T = 5–20% or 10–25% gradient should be chosen.

3.2 Standard *in situ* hydrolysis procedure

Sometimes it will not be possible to prepare solutions of pure sample proteins for analysis. In these cases, the sample proteins are best separated by a

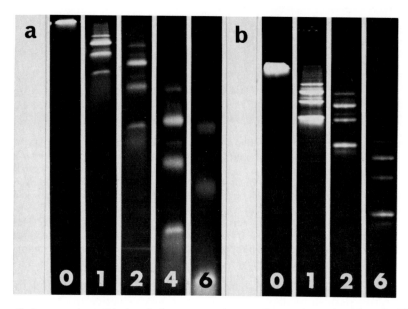

Figure 2. Increased peptide band sharpness using gradient polyacrylamide gels. Digestion products of p49 were separated on either (a) a uniform acrylamide gel (15% acrylamide, 2.66% bisacrylamide) or (b) a gradient gel (10–20% acrylamide, 5% bisacrylamide). The numbers indicate the electrophoresis time in hours after penetration of the p49 digest into the resolving gel. (Reproduced from ref. 8 with permission.)

preliminary gel electrophoresis step using virtually any of the usual methods. The gels are then briefly stained and destained and placed on a transparent plastic sheet over a light box. Individual protein zones are then cut out with a scalpel or razor blade and trimmed to a size small enough to fit easily into the sample wells of an SDS–PAGE gel to be used for the peptide mapping stage (see *Protocol 2*). Before doing this, however, the pieces of gel are soaked for 30 min in 0.125 M Tris–HCl pH 6.8 containing 0.1% SDS, and in the original method (21) also containing 1 mM EDTA. This can inhibit some proteinases, such as *S. aureus* V8 proteinase for example, so in these cases no EDTA should be added.

Protocol 2. Standard *in situ* peptide mapping

Equipment and reagents

- Light box and scalpel (or razor blade)
- Polyacrylamide slab gel containing the separated proteins
- Coomassie blue staining solution: 0.25% (w/v) Coomassie blue, 50% (v/v) methanol, 10% acetic acid
- Destaining solution: 5% acetic acid, 10% (v/v) methanol
- Proteinase solution: 1–100 μg/ml proteinase in 0.125 M Tris–HCl pH 6.8, 0.1% SDS, 10% glycerol, 0.005% bromophenol blue

- 0.125 M Tris–HCl pH 6.8, 0.1% SDS
- 0.125 M Tris–HCl pH 6.8, 0.1% SDS, 20% glycerol, 0.005% bromophenol blue
- Polyacrylamide peptide mapping slab gel: either a uniform concentration gel (e.g. 15%T, 2.6%C) or a concentration gradient gel (e.g. 5–20%T or 10–25%T) prepared in Laemmli buffer (Chapter 1, Section 5) with a 5 cm stacking gel
- Electrophoresis running buffer according to the Laemmli system (see Chapter 1, Section 5.2.4)

Method

1. Stain the polyacrylamide gel containing the separated proteins for 5–10 min in Coomassie blue staining solution and immediately destain for 10–15 min in destaining solution.[a]

2. Place the gel on the light box and cut out the protein zones of interest using a scalpel (or razor blade).

3. Keeping each protein sample separate, trim and/or cut up the recovered gel fragments such that they will (eventually) fit into the sample well of the second slab gel.

4. Add 10 ml of 0.125 M Tris–HCl pH 6.8, 0.1% SDS to each sample and leave at room temperature for 30 min.[b]

5. Discard this washing solution and place the gel fragments into separate wells of the second slab gel formed in the long (5 cm) stacking gel. Use a clean spatula to guide the fragments carefully to the bottom of the wells.

6. Fill the regions around the gel fragments with 0.125 M Tris–HCl pH 6.8, 0.1% SDS, 20% glycerol delivered from a syringe.

7. Overlay each sample with 10 μl of proteinase solution.

221

Protocol 2. *Continued*

8. Begin electrophoresis. When the bromophenol blue reaches near the bottom of the stacking gel, turn off the electrical current for 20–30 min to allow proteinase digestion of the sample proteins to occur.

9. Resume electrophoresis. Continue until the bromophenol blue nears the bottom the slab gel.

10. Detect the separated peptides by Coomassie blue or silver staining (see Chapter 2). Alternatively, detect radiolabelled peptides using autoradiography or fluorography.

[a] If the protein bands are too faint, recycle the gel through the staining and destaining procedure again or lengthen the period of staining.
[b] Reduce this washing time if the resulting peptide maps are too faint.

There have been reports that excessive losses of protein can occur during the 30 min incubation in the buffer. This is probably not important for most average proteins, but very small proteins, especially if they are highly glycosylated or have extreme isoelectric points, may be more susceptible to leaching out of the gel. If the final peptide maps are very faint, it may be worth reducing this washing time substantially (e.g. to 5–10 min) to see if it helps. The washing buffer is identical to the stacking gel buffer in the Laemmli system (43) so, once equilibrated by this washing stage, the gel pieces are simply pushed into place in the sample wells of the slab gel. The spaces around and immediately above and below the gel pieces are filled by adding a few microlitres of this same stacking gel buffer to which 20% glycerol has been added. Finally, the given amount of the proteinase to be used is dissolved in the same buffer but containing 10% glycerol and 10 µl is added to each sample well overlaying the sample protein gel piece. The SDS–PAGE mapping gel itself is the same as that used above with soluble protein digests (i.e. T = 15%, C = 2.6% or a gradient gel), the only difference being that a rather longer than normal stacking gel phase is preferred. Electrophoretic conditions are likewise the same as usual except that when the bromophenol blue tracking dye has reached a position close to the bottom of the stacking gel, the electrical current is switched off for 20–30 min. At this point both the sample proteins and the overlaid proteinase will have migrated down to the Kohlrausch boundary marked by the bromophenol blue and will be stacked in close proximity to each other. Switching off the current allows time for them to react and for the hydrolysis reaction to occur before separation of the resulting peptides is resumed. Having a longer than normal (e.g. 5 cm as opposed to 2.5 cm) stacking gel ensures that complete stacking together of the sample and proteinase has time to occur.

The extent of hydrolysis depends upon the length of time for which the electric current is switched off, the type and activity of the proteinase used, the ratio of proteinase to substrate, and the length of time the reactants are co-migrating in the stack itself even when the power is on. Indeed, some research workers do not switch the power off at all, but merely migrate the components slowly through the stacking gel at a low constant current and then increase it to full power once the bromophenol blue reaches the end of the stacking gel.

After the peptide separation, radiolabelled peptides can be detected by autoradiography of fluorography. Unlabelled peptides can be stained with Coomassie blue R-250 but nowadays very often by silver staining, because the former is often not sufficiently sensitive to reveal a good map. Depending upon the number of peptide zones formed, 10 μg of a starting protein per sample well is really about the minimum needed to give a peptide map of reasonable intensity, but silver staining (44) can reduce this requirement by a factor of about 100.

A typical result (*Figure 3*) shows that different proteins treated with the same proteinase give very different patterns of peptide bands. Likewise, of course, since different proteinases have different bond specificities, a series of enzymes applied to a single substrate protein also gives different peptide patterns.

4. Variations of the standard technique

Most dye or fluorescent labelling reagents for peptides react with terminal α-NH$_2$ groups and side chain ϵ-NH$_2$ groups of lysine residues and alter the net charge on the peptide molecules. When SDS–PAGE is used as the mapping gel, only size differences between the peptides determine their separation so that chemical modifications are without effect on the peptide band patterns even if they alter peptide molecular charge. This means that when *in vitro* chemical or enzymic hydrolysis is used to generate the peptide mixture a pre-labelling technique can be employed. However, this approach cannot be used if detergent-free PAGE or IEF is used in the subsequent mapping.

Most dye pre-labelling methods with dyes such as Remazol BBR (45) or Uniblue A (46) are less sensitive than conventional post-separation staining, but staining with dabsyl chloride is quite effective. A 10 mM solution in acetone of dabsyl chloride (4-dimethylamino-azobenzene-4'-sulfonyl chloride) is mixed with an equal volume of a 20 mg/ml peptide solution in borate buffer of pH 9 containing 5% SDS and heated at about 60 °C for 5 min (47). This labels the peptides with a strong orange colour. If this mixture is applied directly to the gel, excess reagent and by-products can act as a suitable tracking dye and no bromophenol blue need be added.

Fluorescent labelling is more sensitive than dye labelling. Reagents used are dansyl chloride (1-dimethylaminonaphthalene-5-sulfonyl chloride), fluorescamine (4-phenylspiro[furan-2(3H),1'-phthalan]-3,3'-dione), or *o*-phthaldialdehyde (48–50). These procedures are very simple. Fluorescamine

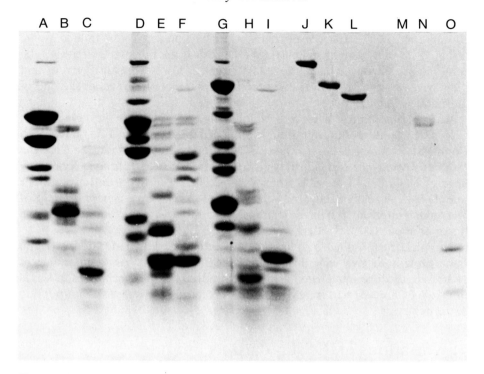

Figure 3. Peptide maps of albumin, tubulin, and alkaline phosphatase using several different proteases; proteolytic digestion allowed to occur in solution. Albumin, tubulin, and alkaline phosphatase, each at 0.67 mg/ml in sample buffer, were incubated at 37 °C for 30 min with the following proteases and then 30 μl of each sample loaded onto a 20% polyacrylamide slab gel. (A to C) Peptide maps of albumin, tubulin, and alkaline phosphatase after digestion with 33 μg/ml, 3.3 μg/ml, 33 μg/ml, respectively, of papain (final concentrations). (D to F) Peptide maps of albumin, tubulin, and alkaline phosphatase after digestion with 133 μg/ml, 67 μg/ml, 67 μg/ml, respectively, of *S. aureus* V8 protease. (G to I) Peptide maps of albumin, tubulin, and alkaline phosphatase after digestion with 133 μg/ml, 67 μg/ml, 67 μg/ml, respectively, of chymotrypsin. (J to L) 2.5 μg each of undigested albumin, tubulin, and alkaline phosphatase, respectively. (M to O) Papain, *S. aureus* V8 protease, and chymotrypsin, respectively, at the highest amounts used in the digestion. (Reproduced from ref. 5 with permission.)

labelling, for example, is achieved by heating (e.g. 100 °C, 5 min) the peptide with 5% SDS in 0.15 M phosphate buffer pH 8.5, followed by cooling, and the addition of a solution of fluorescamine (1 mg/ml) in acetone, using 5 μl for every 100 μl of peptide solution. After shaking briefly, a small amount of bromophenol blue tracking dye (e.g. 5 μl, 5 mg/ml) is added and the sample applied directly to the sample well in the mapping gel. A very similar procedure is used with *o*-phthaldialdehyde except that a 1% solution in methanol is employed, and samples kept in the dark for 2 h before applying to the gel. With *o*-phthaldialdehyde, 2-mercaptoethanol is added to samples at the heating stage, but with fluorescamine labelling it is best added after the addition of fluorescamine.

The advantage of using a pre-labelling method is that there are no fixing and staining steps after the separation, which not only saves time but is very helpful if there are a number of small or very soluble peptides present which are poorly fixed and liable to be washed out of the gel and lost during staining and destaining. Another major advantage is that the progress of the peptide separation can be followed easily during the run. If dye pre-labelling is used, progress is followed by visual checking. If fluorescent pre-labelling has been employed, progress can be followed by examining the gel using a UV lamp. The main disadvantages of pre-labelling are that it tends to be less sensitive than post-separation staining, especially if dye pre-labelling is used, and that it can only be employed when SDS–PAGE is used to separate the peptides into a map and not with any other method.

Of much greater sensitivity is to use radioactively labelled proteins. There are many different ways of radioactive labelling (3). Proteins can be labelled *in vivo* using labelled amino acid precursors such as [^{3}H]leucine or [^{35}S]methionine, but this often gives low specific radioactivities. Labelling of either purified proteins or of mixtures in solution is also possible using a range of procedures.

It is, of course, also possible to separate unlabelled peptides into a map and then to transfer the whole pattern by capillary blotting (51) or electroblotting (52) to an immobilizing membrane for subsequent detection by the usual general protein detection methods for blots. Due to the large number of different peptides present, it is not usually practicable to use immunodetection methods. Nitrocellulose membrane is the least expensive immobilizing membrane and often will be satisfactory but it is worth remembering that small peptides are not strongly bound to it and may either pass through the membrane or be lost during staining, so the more strongly binding nylon or PVDF membranes may be preferable.

Yet another approach is to transfer the intact proteins after a preliminary separation onto an immobilizing nitrocellulose membrane and then to digest the bound protein while still on the membrane. Carrey and Hardie (53) successfully used Western blotting (electroblotting) to nitrocellulose membranes, trypsin treatment of the transferred proteins on strips of membrane, and IEF mapping of the peptides released (poorly bound to nitrocellulose).

Finally, it should be pointed out that a number of gel electrophoretic variations have been introduced specifically for peptide analysis and these have advantages over some of the earlier methods. They include improved procedures for PAGE (54), SDS–PAGE (55, 56), and IEF in immobilized pH gradients (57).

5. Peptide mapping of protein mixtures

Cutting zones out of a gel in which a preliminary separation has been done and then inserting the gel pieces into individual wells in a peptide mapping gel slab with a hydrolysis of proteins to peptides in between the two stages is essentially almost the same as a more conventional two-dimensional scheme. In fact, Bordier and Crettol-Jarvinen (58) and Saris *et al.* (15) carried this concept further and applied peptide mapping to a number of proteins in a whole strip of gel without isolation of the individual protein zones (see *Protocol 3* for detailed procedure). First the number of proteins in a heterogeneous sample are separated in the usual way on a normal slab gel. SDS–PAGE has been used for this but of course there is no reason why other techniques such as PAGE, IEF, agarose gel electrophoresis, etc, should not be used just as successfully. Once this separation has been completed, a strip of gel containing a complete lane of the separated components is cut out and the rest of the gel stained or autoradiographed in the usual way. The gel strip is then equilibrated for 30 min in about 50 ml of 0.125 M Tris–HCl pH 6.8, 0.1% SDS. A mapping gel is prepared which can be either a homogeneous (e.g. T = 15%) slab gel or one with an acrylamide concentration gradient. The stacking gel phase is poured without any well-forming sample comb, however, and the top 2–3 cm of the gel mould is left empty to accommodate the sample strip. The stacking gel mixture is overlaid with water, stacking gel buffer, or 10% ethanol before polymerization in order to obtain a flat surface. When the stacking gel has polymerized, the overlaying liquid is removed and the gel strip of sample proteins is added across the top of the stacking gel. The gel strip is sealed in place with melted 1% agarose in stacking gel buffer, taking care not to trap air bubbles. It may be easier to do this if the slab gel, mould, and agarose solutions are all kept warm (e.g. 55–60 °C) to keep the agarose molten, and if a little agarose solution is added above the polyacrylamide stacking gel before adding the sample strip. The sample gel strip should be covered with 2–3 mm of the agarose solution and the whole assembly allowed to cool, upon which the agarose sets to a gel. Alternatively, depending on the design of the apparatus and the thickness of the sample gel strip, the more direct approach of pushing the sample gel strip in between the glass plates and sealing it in with stacking gel mixture may be possible, but remember that there must be a depth of at least 1.5–2.0 cm of stacking gel beneath the sample strip and above the resolving gel to allow adequate stacking to occur.

The gel assembly is mounted in the electrophoresis apparatus which is then filled with reservoir buffer. Finally, 1 ml of the protease solution, as used in the usual Cleveland *et al.* method (21) in 0.125 M Tris–HCl pH 6.8, containing 0.1% SDS, 10% glycerol, and 0.001% bromophenol blue (i.e. stacking gel buffer plus tracking dye and glycerol to increase the density of the solution), is injected down through the buffer onto the top surface of the gel (*Protocol 3*). As in the original method (21), the electrical voltage is applied, and the sample and proteinase either allowed to migrate slowly through the stacking gel or, after the sample proteins and proteinase have stacked together and the bromophenol blue is close to the bottom of the stacking gel, the power is switched off for 20–30 min. In both approaches there is time for the sample proteins to be hydrolysed to a mixture of peptides which, when electrophoresis is resumed at full power, are then separated into a series of zones characteristic of both the protein substrate and of the proteinase used.

Protocol 3.   Peptide mapping of complex mixtures

Equipment and reagents

- Light box and scalpel
- Polyacrylamide gel slab containing duplicate lanes of each of the complex protein mixtures, already separated by gel electrophoresis
- Detection system: either Coomassie blue staining solution and destaining solution (see *Protocol 2*) or reagents for detecting radiolabelled proteins
- 0.125 M Tris–HCl pH 6.8, 0.1% SDS
- 1% agarose in 0.125 M Tris–HCl pH 6.8, 0.1% SDS
- Stacking gel mixture based on the Laemmli buffer system (Chapter 1, Section 5)

- Proteinase solution: 1–100 μg/ml proteinase in 0.125 M Tris–HCl pH 6.8, 0.1% SDS, 10% glycerol, 0.001% bromophenol blue
- Polyacrylamide peptide mapping slab gel: (either uniform concentration gel (e.g. 15%T, 2.6%C) or 5–20%T or 10–25%T gradient gel) prepared in the Laemmli buffer system (Chapter 1, Section 5) but without stacking gel in place
- Electrophoresis running buffer: Chapter 1, Section 5.2.4

Method

1. After electrophoresis of the first polyacrylamide slab gel to separate the components of the complex protein mixtures, for each sample, cut out a complete lane of the separated protein and equilibrate it for 30 min in 50 ml of 0.125 M Tris–HCl pH 6.8, 0.1% SDS. Stain the duplicate lane with Coomassie blue (see *Protocol 2*) or detect radiolabelled proteins by autoradiography or fluorography.

2. Pour stacking gel mixture on top of the polyacrylamide peptide mapping slab gel, leaving a 2–3 cm space at the top of the gel mould. The stacking gel must be at least 1.5–2.0 mm deep. Overlay with water.

3. After polymerization of the stacking gel, discard the water overlay and place the equilibrated strip of gel containing the sample proteins lengthwise on top of the stacking gel, making close contact with it.

4. Seal the sample gel strip in place using warm, melted 1% agarose solution in 0.125 M Tris–HCl pH 6.8, 0.1% SDS, removing trapped air bubbles as this is done, and allow to cool and gel. Alternatively use stacking gel mixture to seal the gel strip in place. In either case, ensure that the gel strip is covered with 2–3 mm of the sealing solution. Leave 5–6 mm clear above the gel between the plates to allow room for subsequent addition of proteinase solution. If stacking gel mixture is used as sealant, overlay it with water and allow it to polymerize.

5. Load the gel onto the slab gel apparatus and fill the reservoirs with electrophoresis running buffer.

6. Overlay the gel strip along its length with 1 ml of the proteinase solution, loading this through the running buffer.

Protocol 3. *Continued*

7. Begin electrophoresis. When the bromophenol blue is close to the bottom of the stacking gel, turn off the electrical current for 20–30 min to allow proteolysis to occur.

8. Continue electrophoresis until the bromophenol blue nears the bottom of the resolving gel then detect the separated peptides as usual by staining, autoradiography, or fluorography.

Typical results are shown in *Figure 4* in which erythrocyte membrane components are separated into about 30 zones in the preliminary SDS–PAGE separation, most of which give lanes of separated peptide zones in the second

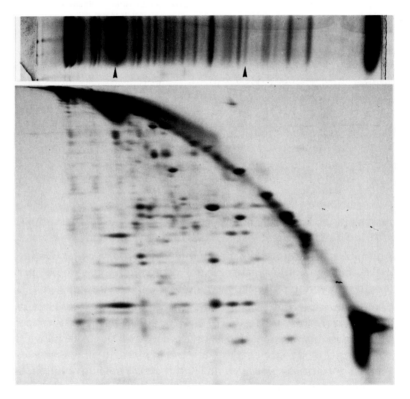

Figure 4. Peptide mapping of heterogeneous protein samples. Human erythrocyte membranes (100 μg protein) were electrophoresed on a linear 4–13% gradient slab gel and then the gel strip containing the sample was transferred to a uniform acrylamide gel (15% acrylamide) and overlayered with 1 μg of *S. aureus* V8 protease in sample buffer. The stained electrophoretogram of the same sample (50 μg) after one-dimensional electrophoresis is shown at the top of the two-dimensional gel. (Reproduced from ref. 9 with permission.)

dimension. Even with the relatively high sample loadings used in this example the pattern of zones revealed by Coomassie blue staining is faint, and silver staining could probably have been used with advantage. The poorly resolved curve of material extending from top left to bottom right represents largely unhydrolysed sample proteins and a more extensive hydrolysis (larger proportion of proteinase or longer time in the stacking gel stage) might have been beneficial. The method does have the advantage however that hitherto unsuspected homologies between different protein components in the sample may be revealed by similarities in peptide spot patterns (e.g. the two components arrowed in *Figure 4*).

6. Characterization of proteinases

When a substrate protein solution is incubated *in vitro* with a proteinase, it will hydrolyse the protein polypeptide chain at particular points depending upon the specificity of the enzyme concerned (e.g. see *Table 2*). The resulting mixture of peptides is therefore highly reproducible and characteristic of the particular enzyme although, of course, in all cases some bonds are more susceptible to attack than others and the large peptides formed initially are broken down to smaller peptides as time progresses. Peptide maps can then be constructed using any one- or two-dimensional electrophoretic separation technique, but SDS–PAGE and PAGE (e.g. *Figure 5*) are probably the most popular. The method is essentially the same as that described earlier for the peptide mapping of unknown purified proteins treated *in vitro* with added proteinases of known identity to generate the peptide mixture, but in this case the unknown is the proteinase and the substrate is the purified known component.

The activity of the unknown proteinase can also be explored by peptide mapping using what could be considered as the inverse of the Cleveland *et al.* (21) procedure. In this method, a proteinase is separated from other proteins, enzymes, or even other proteinases in a preliminary electrophoretic run by PAGE, IEF, etc. SDS–PAGE is often unsuitable for this initial separation since SDS is a potent denaturing agent. After the zones have separated, the gel can be stained briefly under non-denaturing conditions by immersion for a few seconds or minutes in 0.1% Coomassie blue G-250 in aqueous 10% methanol to reveal the separated components, and then immersed in the stacking gel buffer to be used in the peptide mapping gel. When the preliminary run has been performed at an acid pH (e.g. acid-gel PAGE or the acidic regions of an IEF gel) a 30 sec staining time is sufficient, but with basic gels slightly longer is needed (e.g. 5–10 min). When high loadings of sample are used, as is usually required in this mapping technique, the zones are generally visible immediately so the band of interest can be easily cut out. The advantage of using mild non-denaturing conditions for both the preliminary separation and the staining step is that enzyme activity is preserved

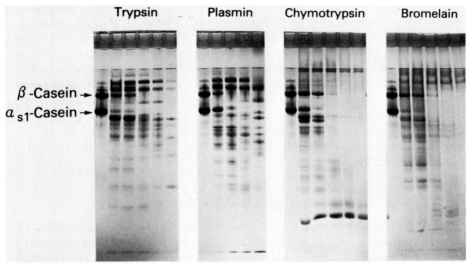

Figure 5. PAGE analysis on a $T = 12.5\%$, $C = 4\%$ slab gel containing 4.5 M urea of the hydrolysis of total casein by a number of different proteinases. Total casein (3 mg/ml) precipitated isoelectrically from bovine milk was dissolved in 0.1 M sodium phosphate buffer pH 7 (pH 5 for bromelain) and trypsin (2 μg/ml), plasmin; (20 μg/ml), chymotrypsin (3 μg/ml), or bromelain (20 μg/ml) added. Incubation times, *left* to *right* were 0, 2.5, 5, 20, 60, and 240 min for trypsin; 0, 10. 60, 80, and 1200 min for plasmin; 0, 5, 12, 40, 120, and 360 min for chymotrypsin; and 0, 2.5, 5, 12, 40, and 240 min for bromelain. At the end of each incubation, samples were heated at 100°C for 2 min to inactivate the proteinases and urea added to 8 M and 2-mercaptoethanol to 0.1 M. Portions of 25 μl were applied to the slab gel. Note that the peptide band patterns are very different for each enzyme and that there are even marked differences between the peptide maps given by enzymes as similar as trypsin and plasmin which have very similar bond specificity (Lys–, Arg–) but differ only in the rates at which individual bonds are cleaved. (Reproduced from ref. 5 with permission.)

to the maximum extent. As an alternative to a rapid staining, the guide strip method of locating enzyme bands can be used. Strips of gel are cut off each side of the slab gel, stained in the conventional way, and then aligned with the rest of the gel so that zones containing the sample component(s) of interest can be located and cut out of the central, unstained piece of gel.

The gel pieces containing the separated proteinase band are soaked for about 30 min in the same buffer as that used in the stacking gel phase of the peptide mapping gel. The mapping gel is the same as that described above and used in the standard protocol (*Protocol 1*) (i.e. a T = 15% gel or gradient gel). When equilibrated with stacking gel buffer, the gel pieces are cut to an appropriate size and inserted into the sample wells of the peptide mapping gel. Any gaps around, above, or below the gel pieces are filled by covering them with a few microlitres of stacking gel buffer containing 20% glycerol. The gel assembly is mounted in the electrophoretic chamber which is filled with electrophoresis buffer. Portions (15 μl) of the substrate protein solution are then added to the top of each well by injecting them down through the buffer into the wells. A suitable solution for most purposes will consist of 3–5 mg/ml of a protein such as purified α_{S1}-casein, β-casein, or haemoglobin dissolved in stacking gel buffer containing 10% glycerol or sorbitol and 0.001% bromophenol blue. Interpretation of the final peptide maps is usually easier if the substrate protein is purified and itself migrates as a single band during electrophoresis, but as long as control samples receiving no proteinase treatment are run on the same slab gel for comparison this is not an essential requirement. A 5 mg/ml solution of unfractionated total casein (with added glycerol and bromophenol blue, of course) is cheaper and may prove to be an acceptable alternative to purified individual caseins.

Electrophoretic running conditions, staining, and destaining are all performed in the usual way, with the power again being switched off for 20–30 min when the bromophenol blue is close to the bottom of the gel to allow time for the proteinase(s) in the sample to digest the casein substrate to peptides which are then separated into a map.

7. Interpretation of results

It has been implicit so far that similar band patterns produced by peptide digests of an unknown sample and a standard known protein, or between two unknowns, indicates a degree of identity. Obviously if the two band patterns are identical this is strong evidence that the two proteins are themselves identical, especially if the same result is obtained when the digests are examined on more than one peptide mapping gel run under different conditions or using different techniques. But what of the situation when some bands appear to coincide and others do not? In this case perhaps the most likely explanation is that the two proteins are related but not identical, as may occur, for example, in a precursor–product relationship or between a number of isozymes or when there are varying degrees of post-synthetic modification (e.g. glycosylation, phosphorylation, etc.). The more zones that there are in common, the more closely related the two proteins are. However, it is sometimes possible for two related proteins sharing close sequence homology to give rise to very different peptide maps on peptide hydrolysis, particularly if they differ slightly in molecular weight (11). In the example shown diagrammatically in *Figure 6*, protein A is slightly larger than protein B and points X and Y represent preferred major sites of cleavage by the proteinase, while 1–7 are secondary cleavage sites. If the principal cleavage occurs at X or at X *and* Y, then very similar peptide mixtures are formed, but if for some reason there is little or no cleavage at X while cleavage at the other sites continues, then many of the peptides formed will not coincide (e.g. 0–1, 0–2, 0–3, etc from A but 0′–1, 0′–2, 0′–3, etc from B). Although there are still likely to be a number of peptides that do coincide (e.g. 1–2, 1–3, 2–3, etc), these require two cleavages at secondary sites and may be relatively minor products at least in the initial

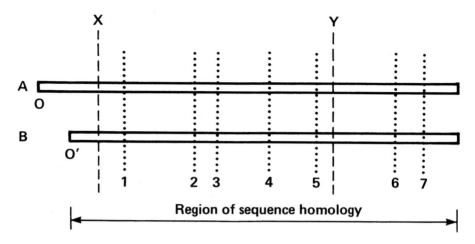

Figure 6. Hypothetical peptide mixtures that might be derived from two polypeptides, A and B, which have extensive primary amino acid sequence homology. A has a higher molecular weight than B, with an N-terminal extension to the sequence (as might occur with a signal sequence or a product–precursor relationship). If X and Y are major cleavage sites and 1–7 are minor sites and hydrolysis occurs at all of them, then most of the peptides generated from A will match those from B, but if little cleavage at X occurs, then many will be different.

stages of hydrolysis, so the peptide maps may look very different. Lam and Kasper (11) did in fact report that cleavage of two major nuclear envelope polypeptides with cyanogen bromide, hydroxylamine, and chymotrypsin gave peptide mixtures that led to similar maps for the two proteins, but cleavage with *S. aureus* proteinase gave two very different peptide maps. Thus it may be necessary to use several methods for generating peptides to establish a sequence homology unequivocally between two proteins.

If two lanes of zones on a peptide mapping gel have some zones of apparently identical mobility and some different, the degree of similarity can be expressed quantitatively as described by Calvert and Gratzer (59). For example, if the resolving power of the gel is such that bands 0.5 mm apart can be distinguished, then a gel could be considered as made up of a number of elements N, each of which may or may not contain a peptide band.

When comparing two lanes on a peptide map containing m and n bands respectively with x coincidences distributed over the N elements, the probability of these occurring by chance is given by:

$$P(x) = \frac{m!n! \, (N-m)! \, (N-n)!}{N!x! \, (m-x)! \, (n-x)! \, (N-m-n+x)!}$$

Thus, if x is large, $P(x)$ becomes very small, which means that it is very improbable that such an extent of coincidence could arise purely by chance, indicating that the two protein sequences are homologous.

As an example, Calvert and Gratzer (59) digested two spectrin subunits to peptides with papain and 7 cm long gels were used for mapping the peptides. Thus, $N = 140$ (7 cm gel; 0.5 mm resolution), $m = 23$, $n = 20$, $x = 11$ (the number of bands appearing to be identical in both maps) and therefore

$$P(x) = 1 \times 10^{-5}.$$

The usual criterion for a valid homology is $P(x) < 0.01$, so the value obtained excludes the possibility that this level of similarity in the maps occurred by chance and the two subunits are therefore at least partially homologous in amino acid sequence.

References

1. Ferguson, K. A. (1964). *Metabolism*, **13**, 985.
2. Hedrick, J. L. and Smith, A. J. (1968). *Arch. Biochem. Biophys.*, **126**, 155.
3. Andrews, A. T. (1986). *Electrophoresis: theory, techniques and biochemical and clinical applications*, 2nd edn. Oxford University Press, Oxford.
4. James, G. T. (1980). In *Methods of biochemical analysis* (ed. D. Glick), Vol. 26, p. 165. J. Wiley, New York.
5. Andrews, A. T. (1980). In *Methods of enzymatic analysis* (ed. H. U. Bergmeyer, J. Bergmeyer, and M. Grassl), Vol. V, p. 227. Verlag Chemie, GmbH, Weinheim.
6. Gooderham, K. (1987). *Sci. Tools*, **34**, 4.
7. Croft, L. R. (1980). *Handbook of protein sequence analysis*, 2nd edn, p. 19. J. Wiley, Chichester, UK.
8. Stark, G. R. (1977). In *Methods in enzymology*, (ed. C. H. W. Hirs and S. N. Timasheff), Vol. 47, p. 129. Academic Press, New York.
9. Mitchell, W. M. (1977). In *Methods in enzymology*, (ed. C. H. W. Hirs and S. N. Timasheff), Vol. 47, p. 165. Academic Press, New York.
10. Croft, L. R. (1980). *Handbook of protein sequence analysis*, 2nd edn, p. 9. J. Wiley, Chichester, UK.
11. Lam, K. S. and Kasper, C. B. (1980). *Anal. Biochem.*, **108**, 220.
12. Lonsdale-Eccles, J. D., Lynley, A. M., and Dale, B. A. (1981). *Biochem. J.*, **197**, 591.
13. Zingde, S. M., Shirsat, N. V., and Gothoskar, B. P. (1986). *Anal. Biochem.*, **155**, 10.
14. Mahboub, S., Richard, C., Delacourte, A., and Han, K-K. (1986). *Anal. Biochem.*, **154**, 171.
15. Saris, C. J. M., van Eenbergen, J., Jenks, B. G., and Bloemers, H. P. J. (1983). *Anal. Biochem.*, **132**, 54.
16. Detke, S. and Keller, J. M. (1982). *J. Biol. Chem.*, **257**, 3905.
17. Lischwe, M. A. and Ochs, D. (1982). *Anal. Biochem.*, **127**, 453.
18. Sonderegger, P., Jaussi, R., Gehring, H., Brunschweiler, K., and Christen, P. (1982). *Anal. Biochem.*, **122**, 298.
19. Macleod, A. R., Wong, N. C. W., and Dixon, G. H. (1977). *Eur. J. Biochem.*, **78**, 281.
20. Rittenhouse, J. and Marcus, F. (1984). *Anal. Biochem.*, **138**, 442.
21. Cleveland, D. W., Fischer, S. G., Kirschner, M. W., and Laemmli, U.K. (1977). *J. Biol. Chem.*, **252**, 1102.

22. Herman, H., Pytela, R., Dalton, J. M., and Wiche, G. (1984). *J. Biol. Chem.*, **259**, 612.
23. Mattick, J. S., Tsukamoto, Y., Nickless, J., and Wakil, S. J. (1983). *J. Biol. Chem.*, **258**, 15291.
24. Chen, Z., Agerberth, B., Gell, K., Andersson, M., Mutt, V., Östenson, C-G., *et al.* (1988). *Eur. J. Biochem.*, **174**, 239.
25. Tijssen, P. and Kurstak, E. (1983). *Anal. Biochem.*, **128**, 26.
26. Jekel, P., Weijer, W., and Beintema, J. (1983). *Anal. Biochem.*, **134**, 347.
27. Drapeau, G. R. (1980). *J. Biol. Chem.*, **255**, 839.
28. Levy, M., Fishman, L., and Shenkein, I. (1972). In *Methods in enzymology*, Vol. 19, p. 672. Academic Press, New York.
29. Colburn, J. C. (1992). In *Capillary electrophoresis, theory and practice* (ed. P. D. Grossman and J. C. Colburn), p. 237. Academic Press Inc., San Diego.
30. Smith, A., Kenny, J. W., and Ohms, J. I. (1992). In *Techniques in protein chemistry III* (ed. R. H. Angeletti), p. 113. Academic Press Inc., San Diego.
31. Hermann, P., Jannasch, R., and Lebl, M. (1986). *J. Chromatogr.*, **320**, 99.
32. Wong, S., Padua, A., and Henzel, W. J. (1992). In *Techniques in protein chemistry III* (ed. R. H. Angeletti), p. 3. Academic Press Inc., San Diego.
33. Stone, K. L., McNulty, E., LoPresti, M. L., Crawford, J. M., DeAngelis, R., and Williams, K. R. (1992). In *Techniques in protein chemistry III* (ed. R. H. Angeletti), p. 23. Academic Press Inc., San Diego.
34. Stone, K. L., LoPresti, M. B., Crawford, J. M., DeAngelis, R., and Williams, K. R. (1989). In *A practical guide to protein and peptide purification for microsequencing* (ed. P. T. Matsudaira), p. 31. Academic Press Inc., San Diego.
35. Le Gendre, N. and Matsudaira, P. T. (1989). In *A practical guide to protein and peptide purification for microsequencing* (ed. P. T. Matsudaira), p. 49. Academic Press Inc., San Diego.
36. Aebersold, R. (1989). In *A practical guide to protein and peptide purification for microsequencing* (ed. P. T. Matsudaira), p. 71. Academic Press Inc., San Diego.
37. Moritz, R. L., Ward, L. D., and Simpson, R. J. (1992). In *Techniques in protein chemistry III* (ed. R. H. Angeletti), p. 97. Academic Press Inc,., San Diego.
38. Aebersold, R., Patterson, S. D., and Hess, D. (1992). In *Techniques in protein chemistry III* (ed. E. H. Angeletti), p. 87. Academic Press Inc., San Diego.
39. Hancock, W. S. (1984). *CRC handbook of HPLC for the separation of amino acids, peptides and proteins*, Vol. I and II. CRC Press, Boca Raton, USA.
40. Hartman, P. A., Stodola, J. D., Harbour, G. C., and Hoogerheide, J. G. (1986). *J. Chromatogr.*, **360**, 385.
41. Scoble, H. A. (1989). In *A practical guide to protein and peptide purification for microsequencing* (ed. P. T. Matsudaira), p. 89. Academic Press Inc., San Diego.
42. Cohen, A. S. and Karger, B. L. (1987). *J. Chromatogr.*, **397**, 409.
43. Laemmli, U.K. (1970). *Nature*, **227**, 680.
44. Daverval, C., le Guilloux, M., Blaisonneau, J., and de Vienne, D. (1987). *Electrophoresis*, **8**, 158.
45. Griffith, I. P. (1972). *Anal. Biochem.*, **46**, 402.
46. Bosshard, H. F. and Datyner, A. (1977). *Anal. Biochem.*, **82**, 327.
47. Tzeng, M. C. (1983). *Anal. Biochem.*, **128**, 412.
48. Strottman, J. M., Robinson, J. B., and Stellwagen, N. C. (1983). *Anal. Biochem.*, **132**, 334.

49. Tijssen, P. and Kurstak, E. (1979). *Anal. Biochem.*, **99**, 97.
50. Weiderkamm, E., Wallach, D. F. H., and Flückiger, R. (1973). *Anal. Biochem.*, **54**, 102.
51. Southern, E. M. (1975). *J. Mol. Biol.*, **98**, 503.
52. Towbin, H., Staehelin, T., and Gordon, J. (1979). *Proc. Natl. Acad. Sci. USA*, **76**, 4350.
53. Carrey, E. A. and Hardie, D. G. (1986). *Anal. Biochem.*, **158**, 431.
54. West, M. H. P., Wu, R. S., and Bonner, W. M. (1984). *Electrophoresis*, **5**, 133.
55. Hashimoto, F., Horigome, T., Kanbayashi, M., Yoshida, K., and Sugano, H. (1983). *Anal. Biochem.*, **129**, 192.
56. De Wald, D. B., Adams, L. D., and Pearson, J. D. (1986). *Anal. Biochem.*, **154**, 502.
57. Gianazza, E., Chillemi, F., Duranti, M., and Righetti, P. G. (1983). *J. Biochem. Biophys. Methods*, **8**, 339.
58. Bordier, C. and Crettol-Järvinen, A. (1979). *J. Biol. Chem.*, **254**, 2565.
59. Calvert, R. and Gratzer, W. B. (1978). *FEBS Lett.*, **86**, 247.

8

Sequence analysis of gel-resolved proteins

RICHARD J. SIMPSON and GAVIN E. REID

1. Introduction

It is well recognized that polyacrylamide gel electrophoresis, particularly 2DE (1), is one of the most powerful techniques for rapidly analysing complex protein mixtures. Indeed, when the human genome is fully sequenced, predicted to occur by the turn of the century (2, 3), the attention of biologists will move away from *de novo* protein sequencing and towards establishing the function of mature proteins in terms of their post-transitional modifications, spatial and temporal characteristics. In this post-genome era, 2DE, undoubtedly, will play a key role in strategies aimed at analysing the full protein complement expressed by a cell or tissue type. Moreover, electrophoretic techniques will continue to play an important role in dissecting biological processes such as development, differentiation, and signal transduction which involve the coordinated expression of proteins that interact with one another in a regulated fashion.

Prior to the development of methodologies for identifying gel-separated proteins by direct N-terminal sequence analysis, protein identification was achieved by immunoblot analysis (4), co-migration of previously purified mature or recombinant proteins (5), amino acid compositional analysis (6, 7), or by comparative peptide mapping (8). While early electroelution procedures (9) for recovering proteins from SDS–polyacrylamide gels for sequence analysis yielded useful data, this approach was technically difficult due to the high concentrations of SDS and gel-related artefacts in the electroeluate which interfered with the available automated Edman degradation procedures. This latter problem could be overcome by recovering the protein from the electroeluate by 'inverse-gradient' reversed-phase HPLC (10) or by selective precipitation procedures (see ref. 11 for a review). A major impact in the ability to identify gel-separated proteins resulted from the development of electroblotting methods (12–14), followed not long afterwards by procedures for acquiring internal sequence information (15–18). In parallel with these developments were efforts directed towards the miniaturization of chromatographic columns for the high sensitivity HPLC separation of proteins and peptides (11).

In the early 1980s, two mass spectrometic ionization techniques, fast atom bombardment (19) and plasma desorption (20), pioneered the development of mass spectrometric-based procedures for protein characterization. While accurate molecular mass determination of peptides and small proteins was achievable at low picomole sensitivity, and sequence information on small peptides was possible by collision-induced dissociation (CID), these developments were restricted to only a few specialist mass spectrometry laboratories and major limitations were encountered in the analysis of intact proteins. The advent of two new ionization methods for biomolecules in the mid-1980s, electrospray ionization (ESI) (21) and matrix assisted laser desorption ionization (MALDI) (22), have now made routine affordable mass spectrometry (MS) available to biological researchers. These two MS techniques enable accurate molecular mass determinations of proteins with molecular masses exceeding 100 kDa at picomole to femtomole sensitivity methods (23–25). The sensitivity of MS is compatible with the protein quantities present in 2DE gels, and the mass accuracy for peptides, along with the capability of this technology to produce partial sequence information, facilitates unambiguous identification of gel-separated proteins provided they are listed in the available sequence databases.

This chapter describes the practicalities of obtaining sequence information from proteins fractionated by polyacrylamide gel electrophoresis. The methods described here are applicable to both one-dimensional as well as two-dimensional separations. The strategy outlined in *Figure 1* indicates an approach

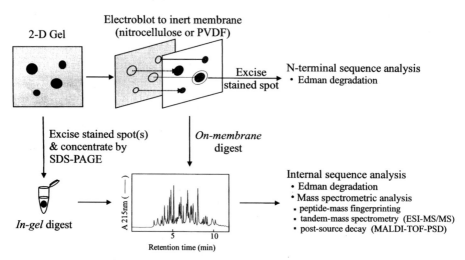

Figure 1. Experimental approaches for obtaining sequence information from gel-separated proteins.

for obtaining N-terminal sequence information, using Edman degradation procedures, for gel-separated proteins, as well as various approaches (both Edman degradation-based and MS-based) for acquiring internal sequence data.

2. Visualization and concentration of protein gel spots

2.1 Protein detection in primary gels

For the purpose of subsequent structural analysis, proteins can be visualized in one-dimensional and 2DE gels by a number of methods. The most widely used method is the conventional Coomassie staining procedure (26) which can detect proteins in the low microgram range ($\sim$ 0.5–1.0 µg). Care must be exercised with this staining method (see *Protocol 1*) since overstaining can interfere with subsequent in-gel proteolysis and peptide mapping steps. To minimize these potential problems, the gel is stained for 20 min only, and the gel background is destained overnight to visualize the proteins and remove excess Coomassie blue. After a protein spot has been excised, additional long washing steps with deionized water are necessary to remove Coomassie blue associated with the protein, and excess SDS.

Another category of protein visualization methods that is gaining wide acceptance is that of 'negative staining' which relies on selective precipitation of metal ions in gels, leaving protein bands unstained and readily amenable for structural analysis. Common negative staining methods include: potassium acetate/chloride (27, 28), sodium acetate (29), copper chloride (30), aurodye (31), zinc chloride (32), and zinc/imidazole (33, 34). Of these, the most widely used method, perhaps, is that of zinc/imidazole (see *Protocol 2*), which is approximately ten times more sensitive than Coomassie blue. Like Coomassie blue, this negative staining procedure does not result in detectable protein modification. Recently, a sensitive ($\sim$ 0.5 nanogram) procedure for the negative staining of proteins, based on the precipitation of methyltrichloroacetate in gels was reported (35). After separation, gels are incubated with 8% methyltrichloroacetate ester in 38% isopropanol and then washed in water to produce a negative image of colourless proteins against an opaque background (35). Owing to the reversibility of the process, gels can be restained (e.g. silver staining, see below) after rapid visualization.

Silver staining is the most widely used high sensitivity staining method for the detection of proteins in polyacrylamide gels (36), the sensitivity reported to be approximately 100 times better than Coomassie blue (see refs 37 and 38 for detailed reviews). Silver staining has not been used for protein microanalysis until recently since the detection limits (about 1–10 ng) were much too low for conventional sequence analysis by automated Edman degradation. However, with the improved sensitivity of protein structural analysis by mass spectrometric methods (low femtomole levels) (39), this staining method is gaining broader acceptance (40, 41). One critical parameter to heed in any silver staining protocol when using this visualization method for primary structural analysis is chemical modification of the protein. Such modifications are likely to occur during the pre-sensitizing step which involves pre-treatment of the gel with sensitizers (e.g. sulfosalicylic acid, sodium thiosulfate, DTT, glutaraldehyde, chelators, etc—for a review on various silver staining protocols see refs 37 and 38) in order to improve the contrast (and hence sensitivity) between the stained protein and the gel background. In a recent report (40) the 'acidic' silver staining method (37) was adapted for primary structural analysis by omission of the fixation/sensitization treatment with glutaraldehyde, without any compromise in sensitivity of protein detection (see *Protocol 3*). This silver staining method has been used recently for a number of biological applications (39, 41).

Protocol 1. Procedure for visualizing proteins in the primary gel by staining with Coomassie blue

Reagents

- Coomassie blue staining solution: mix 1 g Coomassie brilliant blue (Bio-Rad) in 1 litre of 50% (v/v) methanol (Mallinckrodt), 10% (v/v) acetic acid glacial (BDH), 40% deionized water—stir the solution for 3–4 h and then filter through Whatman filter paper.

- Destaining solution: 12% (v/v) methanol, 7% (v/v) acetic acid glacial, 81% deionized water

A. *Staining of gels*[a]

1. After electrophoresis, place the gel in a plastic container which has sufficient Coomassie blue staining solution to fully cover the gel. Place the container on a mechanical shaker and allow the gel to stain for 20 min at room temperature.

2. Discard the staining solution and add destain solution and three sheets of fine grade tissue paper (Kimwipe). Destain with shaking, replenishing the destain solution several times until the gel is fully destained. The tissue paper binds released stain.

B. *Storage of Coomassie blue stained gels*

1. For gels that are going to be processed in less than one month's time, store the gels in 200–250 ml of destain solution at 25°C in a sealed plastic container.

2. For long-term storage, place the gel between two sheets of cellophane (Hoefer) in a plastic gel frame (Hoefer), and air dry the gel in a fume-hood for ∼ 16 h.

3. Store dried gels wrapped in cling film at 25°C.

[a] Perform all the staining steps at room temperature unless otherwise indicated.

Protocol 2. Zinc/imidazole procedure for visualizing proteins in primary gels by negative staining[a]

Reagents

- Fixing solution: 50% (v/v) methanol, 5% (v/v) acetic acid glacial, 45% deionized water
- 0.2 M zinc sulfate solution: 28.7 g zinc sulfate (Merck) in 500 ml deionized water
- 0.2 M imidazole, 0.1% SDS solution: 6.8 g imidazole (Sigma), 0.5 g SDS, 500 ml deionized water

A. *Direct reverse staining with imidazole–SDS–zinc*

1. After electrophoresis, submerge the gel in fixing solution for 20 min with gentle shaking.
2. Discard the fixing solution and wash the gel twice with deionized water with gentle shaking for 15 min.
3. Incubate the gel in 0.2 M imidazole, 0.1% SDS solution for 15 min with gentle shaking.
4. Discard the imidazole–SDS solution, add 0.2 M zinc sulfate solution and agitate for 30–60 sec.
5. When satisfactory staining has been obtained, discard the zinc sulfate solution and add deionized water.

B. *Double staining of Coomassie blue stained polyacrylamide gels by imidazole–SDS–zinc*

1. After Coomassie blue staining (see *Protocol 1*), wash the destained gel twice in deionized water with gentle shaking for 15 min.
2. Incubate the gel in 0.2 M imidazole, 0.1% SDS solution for 15 min with gentle shaking.
3. Discard the imidazole–SDS solution, add 0.2 M zinc sulfate solution, and agitate for 30–60 sec.
4. When satisfactory staining has been obtained, discard the zinc sulfate solution and add deionized water.

[a] Procedure of Fernandez-Patron *et al.* (34).

Protocol 3. Procedure for visualizing proteins in primary gels by silver staining[a]

Reagents

- Fixing solution (see *Protocol 2*)
- Developing solution: 2% (w/v) sodium carbonate (Merck), 0.04% (v/v) formaldehyde (Merck) in deionized water
- 0.02% (w/v) sodium thiosulfate (Merck) in deionized water
- 0.1% (w/v) silver nitrate (Merck) in deionized water

Protocol 3. *Continued*

- Stopping solution: 5% (v/v) glacial acetic acid in deionized water
- 50% methanol
- 1% acetic acid in deionized water

Method

1. After electrophoresis, place the gel in a plastic container (it is best to use a disposable plastic container or a thoroughly washed container) and add sufficient fixing solution to cover the gel. Fix the gel for 20 min with gentle shaking.

2. Discard the fixing solution and add enough 50% methanol to fully cover the gel. Shake the gel gently for 10 min.

3. Replace the 50% methanol with deionized water and shake the gel gently for a further 10 min.

4. Discard the deionized water and soak the gel in 0.02% sodium thiosulfate for 1 min.

5. Rinse the gel twice with deionized water for 1 min.

6. Submerge the gel in chilled 0.1% silver nitrate solution and incubate for 20 min at 4°C.

7. Rinse the gel twice with deionized water for 1 min.

8. Submerge the gel in developing solution and shake intensively. Discard the developing solution if it turns a yellow colour and replenish it with fresh developing solution.[b]

9. Once the desired intensity of staining is achieved, terminate the development by replacing the developing solution with stopping solution.

10. Store the silver stained gel in 1% acetic acid until further use.

[a] Adapted from refs 37 and 40.
[b] It is critical that the development solution remains transparent during the development step.

2.2 Protein concentration

Empirical observations from a number of protein structural analysis laboratories indicate that pre-concentration of stained excised gel spots (or bands) prior to electroblotting onto PVDF is necessary in order to achieve high initial sequencing yields. Similarly, pre-concentration prior to in-gel proteolytic (or chemical) fragmentation is necessary to obtain high peptide yields. It has been stated that efficient cleavage of a protein is obtained only if the ratio of protein to polyacrylamide is kept high (18, 42), and that the lower limit from which useful peptide maps (for structural analysis using Edman degradation methods) can be obtained is 1 μg of protein per spot (43). Hence, to be able to proteolytically digest faint protein spots excised from multiple gels, the gel pieces must be concentrated into one single band prior to digestion. This can be readily achieved by re-electrophoresing the stained gel pieces in a second conventional SDS–polyacrylamide gel or, alternatively, in a secondary gel matrix in the tip of a conventional Pasteur pipette (44) as described in *Protocol 4* (and shown in *Figure 2*). For this procedure it is important to keep the total volume of combined gel pieces and equilibration buffer smaller than 0.5 ml (44). When handling larger volumes (e.g. 1.2 ml), this procedure can be performed in smaller batches (say, three 400 µl batches) and then these batches combined for further concentration in a single Pasteur pipette. Alternatively, larger volumes can be handled by similar means using a simple apparatus reported in refs 45 and 46.

An alternative approach to obtaining a critical protein concentration required for efficient digestion is to enrich the sample prior to loading a 2DE gel. This can be readily achieved by applying a total cell lysate onto a slab SDS–PAGE, excising a narrow M_r range containing the protein of interest, passively eluting the proteins, and then loading them onto a high-resolving 2DE for subsequent microsequence analysis (47). By these means, sequence data can be obtained from proteins in the submicrogram range.

Protocol 4. Procedure for concentrating polyacrylamide gel spots[a]

Equipment and reagents[b]

- Protean II xi 2D cell (Bio-Rad)
- 2 ml syringe
- 30% (w/v) acrylamide, 1%, *N,N'*-methylene-bisacrylamide (30%T, 3.3%C) (Bio-Rad) in deionized water
- 0.5 M Tris–HCl pH 6.8
- 5% (w/v) agarose
- 10% (w/v) SDS (Bio-Rad) in deionized water
- 5% stacking gel mixture: mix 2 ml 30% acrylamide, 3 ml 0.5 M Tris–HCl pH 6.8, 0.12 ml 10% SDS, 6.76 ml deionized water, 0.12 ml 10% ammonium persulfate, 6 µl TEMED (Bio-Rad).

- 10% (w/v) ammonium persulfate (Bio-Rad) in deionized water[c]
- Equilibration buffer: mix 12.5 ml 0.5 M Tris–HCl pH 6.8, 20 ml 10% SDS, 5 ml 2-mercaptoethanol (Sigma), 10 ml glycerol (Merck), and add a few grains of bromo-phenol blue (Bio-Rad)
- Coomassie blue staining solution (see *Protocol 1*)
- Destaining solution (see *Protocol 1*)
- Laemmli electrophoresis running buffer (see Chapter 1, Section 5.2.4)

Method

1. Excise the desired 2DE gel spots and place them in a 10 ml polypropylene tube. Rehydrate the dry spots in 9 ml deionized water.

2. Wash the rehydrated gel spots with 9 ml deionized water five times (~ 5 min per wash).

3. Discard the deionized water and replace it with equilibration solution (just enough to cover the gel pieces). Equilibrate the gel spots for at least 1 h at room temperature.

4. Cast the concentrating gel by pipetting the 5% stacking gel mixture into a Pasteur pipette.

Protocol 4. *Continued*

5. Quickly dip the tip of the Pasteur pipette into a solution of melted 5% agarose and hold the pipette for a short while until the agarose solidifies. Assemble the pipettes into the Protean II xi 2D cell with a tube gel adaptor, add 0.5 ml of water on top of each gel, and allow the gels to polymerize (see *Figure 2A*).

6. Remove the water from the top of the pipette and load the equilibrated gel spots and equilibration buffer (both from step 3) into the pipette (see *Figure 2B*). The total volume of combined gel pieces and equilibration buffer can be up to 0.5 ml.

7. Fill the remaining volume with Laemmli electrophoresis running buffer and perform electrophoresis in the Protean II xi 2D cell at 250 V at room temperature until the dye front reaches the end of the pipette.

8. Stop electrophoresis and remove the concentrating gel from the pipette by gently pressing from the bottom of the pipette using a 2 ml syringe.

9. Transfer the concentrating gel into a 50 ml tube filled with 30 ml of Coomassie blue staining solution and stain the gel for 10–20 min at room temperature.

10. Destain the gel with destaining solution.

[a] Adapted from ref. 45.
[b] Caution needs to be exercised when handling acrylamide, bisacrylamide, and TEMED, as these reagents are extremely toxic. These reagents should be weighed in a fume-hood, to avoid inhalation of the dust, and surgical gloves worn to avoid skin contact.
[c] This solution can be stored for two weeks at 4 °C.

3. N-terminal sequence analysis

Once an acceptable polyacrylamide gel-based protocol has been established for separating a particular mixture of biomolecules (see Chapters 1, 5, and 6), a number of critical decisions need to be made with respect to which strategy in *Figure 1* should be adopted for obtaining sequence information. A major consideration in this process relates to protein levels. The quantities of protein required for obtaining N-terminal sequence information using current commercially available automated Edman degradation instruments are in the order of low picomoles. (Note: for a protein of M_r 20 000, 1 µg is equivalent to 50 picomoles.) However, it should be stressed that this quantity relates to the yield of phenylthiohydantoin amino acid seen in the first cycle, and that this initial yield can vary from 25–80% of the *actual amount* of protein loaded onto the instrument.

A   Gel preparation   **B**   Electrophoresis

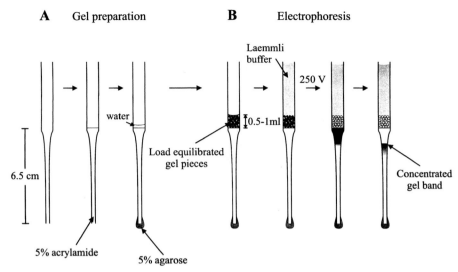

Figure 2. Design and operation of the Pasteur pipette concentration gel system (see *Protocol 4*).

Factors influencing the initial yield include:

(a) Losses incurred when electroblotting a protein from a gel onto an immobilizing matrix (e.g. polvinylidine difluoride, PVDF).

(b) Artefactual blocking of the protein's N-terminal α-NH$_2$ group (since a free α-NH$_2$ group is essential for the Edman degradation chemistry) (48) during the purification and handling stages.

(c) The optimization of the particular sequencing instrument being used for N-terminal sequence analysis.

Parameters influencing losses during electroblotting are the physical characteristics of the membrane (e.g. specific surface area, pore size distribution, pore volumes, and sequencing solvent/reagent permeability considerations) (49). For reviews of commercially available blotting membranes, the blotting process, and gel electrophoresis-induced protein modifications, see refs 11 and 50–53. *Protocols 5* and *6* describe the electrotransfer of proteins from gels to PDVF and their visualization.

N-terminal blocking may be due to formylation or acetylation of the protein's N-terminal amino acid or may occur by cyclization of N-terminal glutamine residues (i.e. formation of a pyroglutamyl group), particularly under acidic conditions (54, 55). While a number of elegant methods exist for deblocking electroblotted proteins (56–58), these methods are not recommended for low abundance proteins due to the poor efficiency of deblocking ($< 40\%$). Empirical estimates from a number of laboratories suggest that 40–50% of naturally occurring proteins resolved by 2DE are N-terminally blocked. For these reasons, one should err on the side of caution when estimating the quantities of protein required for sequence analysis by Edman degradation. An example of a sequence analysis of a relatively abundant gel-separated protein ('housekeeping protein'), heat shock protein 60, following electrotransfer onto PVDF membrane (see *Protocols 1, 5,* and *6*) is shown in *Figure 3*. For the above mentioned reasons, and taking into consideration the effort and cost involved in obtaining a highly purified protein, researchers need to decide whether to take the risk in attempting direct N-terminal sequence of an electroblotted protein. The alternative route indicated in *Figure 1*, internal sequence analysis, is more time-consuming but a safer option for a precious sample.

Protocol 5. Electrotransfer of proteins from gels to PVDF[a,b]

Equipment and reagents

- Electroblotting apparatus: e.g. Trans-Blot (Bio-Rad), Xcell II Blot Module (NOVEX), TE 22 Mighty Small Transphor Tank Transfer Unit (Hoefer), Transphor Tank Transfer Unit (Hoefer)
- 2.21% (w/v) CAPS (3-[cyclohexylamino]-1-propane-sulfonic acid) (Sigma) in deionized water—adjust to pH 11 with HCl.
- PVDF (polyvinylidene difluoride) membrane (Millipore)
- Plastic containers (e.g. domestic food storage containers)
- Transfer buffer: mix 1 part 2.2% CAPS, 1 part methanol, and 8 parts deionized water
- Methanol

Method

1. Immediately following electrophoresis, recover the gels from the electrophoresis apparatus and place them in the plastic container.

2. Add sufficient transfer buffer to the container to cover the gel(s) and let stand for 5–10 min at room temperature.

3. Cut out a piece of PVDF membrane to fit the gel size. Wet the PVDF membrane with neat methanol and then transfer it quickly into a plastic container containing transfer buffer, and soak for 5 min at room temperature.

4. Assemble the gel and PVDF membrane in the electroblotting apparatus according to the manufacturer's instructions.

5. Electrophorese at 4°C for 3 h at 500 mA.

[a] Adapted from ref. 59.
[b] The overall procedures for preparing SDS–PAGE and 2DE gels, sample loading, and electrophoresis are described in Chapters 1 and 6, and protocols contained therein.

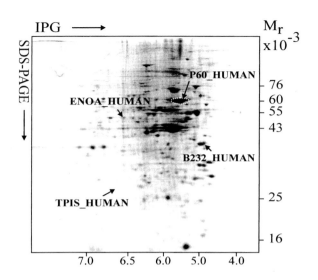

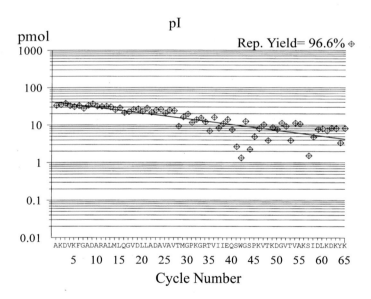

Figure 3. Amino acid sequence analysis of heat shock protein 60, from a reducing 2DE gel of the human colon carcinoma cell line LIM1215 (37). Coomassie blue stained protein spots from three identical 2DE gel/PVDF electroblots (one of which is displayed at the top) were combined for sequence analysis using a biphasic column sequencer (Hewlett-Packard model G1200). The phenylthiohydantoin amino acids are indicated in the lower diagram (above the Cycle number) by the one letter notation used for amino acids. (Reproduced with permission from ref. 59.)

Protocol 6. Visualization of proteins on PVDF[a]

Reagents

- Coomassie brilliant blue staining solution: dissolve 1 g Coomassie brilliant blue in aqueous 50% methanol, stir the solution for 3–4 h, and then filter through Whatman filter paper.

- Destaining solution: 50% methanol, 10% acetic acid, 40% water[b]

Method

1. Immediately following electrotransfer, place the PVDF membrane in a plastic container which contains enough Coomassie brilliant blue staining solution to cover the membrane. Stain the membrane for 5–10 min at room temperature.

2. Discard the staining solution and replace with destaining solution. Change the destain solution until the PVDF membrane is fully destained.

3. Wash the PVDF membrane with deionized water for 10–15 min (repeat five times).

[a] Adapted from ref. 60.
[b] To avoid N-terminal blockage of proteins, only high quality acetic acid (i.e. free of peroxides, aldehydes, etc.) should be used during the staining and destaining procedures.

---

4. Internal sequence analysis

4.1 General considerations

Internal protein sequence information is invaluable for a number of reasons, not the least since many eukaroytic proteins have blocked N-termini (refs 54 and 55; see also Section 3). Other reasons include:

- the construction of oligonucleotide probes for molecular cloning
- confirming the integrity of recombinant proteins
- disulfide bond linkages
- epitope mapping
- identifying post-translational modifications, which is becoming increasingly important

Several methods have been reported for obtaining peptides from proteins separated by polyacrylamide gel electrophoresis. Earlier approaches involved the electrophoretic or passive extraction of proteins from the gel followed by proteolytic digestion in solution (9, 10, 17). These approaches were often technically difficult due to the problem of removing detergents which interfere with both proteolytic digestion and chromatographic fractionation of peptide mixtures (10) and have now been superseded by two general techniques:

(a) In-gel digestion and extraction of the resultant peptides (17, 18, 61–64).

(b) *In situ* cleavage of proteins electroblotted onto nitrocellulose (15) or PVDF (65).

Methods for chemical cleavage of proteins (e.g. cyanogen bromide) have been reported for both in-gel (66) and on-membrane (67) approaches. While each of these procedures have some disadvantages, the recent impetus has been towards the in-gel approach. This has been due mainly to protein losses incurred by overblotting as well as inefficient transfer to the membrane (particularly of high M_r proteins). A comparison of peptide maps of proteins digested in-gel and on-membrane usually reveals fewer peptides for the on-membrane digested proteins. Anecdotal observations suggest that this is due to the protease having limited access to the substrate embedded in the nitrocellulose/PVDF. Several attempts have been made to improve digests of membrane bound proteins. These include the use of a cationic PVDF membrane which allows high recovery of peptides (68) and the use of the detergent Zwittergent 3–16 (69). For reviews of in-gel and on-membrane digestion procedures see refs 53 and 70.

4.2 Choice of proteolytic enzyme

Most 'sequencing grade' endoproteinases that are used in conventional solution digests (e.g. trypsin, *Staphylococcus* protease V8, endoproteinase Lys–C, and endoproteinase Asp–N) give comparable peptide maps for in-gel digests. It is generally considered that of the above mentioned proteases, endoproteinase Lys–C (*Achromobacter* protease I) gives considerably higher in-gel cleavage efficiencies. Also, it should be noted that Lys–C retains full activity in the presence of 0.1% SDS while other proteases are inhibited to varying extents by this detergent, trypsin being the most sensitive. A serious disadvantage of Lys–C is that several of the generated peptides will contain internal arginine residues which complicate the interpretation of mass spectrometric collision-induced dissociation fragmentation patterns. This, in turn, limits the usefulness of some mass spectrometric-based peptide identification procedures (see Section 5). For these reasons, trypsin is recommended for methods relying on mass spectrometric-based identification procedures, as long as procedures are in place to quantitatively remove SDS prior to digestion (see below).

4.3 Reduction/alkylation and proteolytic digestion of protein spots

Crucial to many of the peptide-based methods for identifying proteins separated by 2DE is the ability to achieve efficient digestion of proteins immobilized in the acrylamide gel matrix or upon a blotting membrane. It is well recognized that attempts to digest mature proteins composed of disulfide bonded structures (in contrast to non-disulfide bonded proteins) are often unsuccessful, regardless of whether the digest was performed on-membrane or in-gel. Moreover, for disulfide bonded proteins that do digest, the peptide maps are often extremely complex. For these reasons, it is useful to reduce and alkylate gel-resolved proteins prior to proteolytic digestion. This additional manipulation step prior to performing in-gel digestion has the additional advantage of lowering the level of background artefacts (71), an important consideration for any subsequent mass spectrometric analysis. A number of procedures for *in situ* reduction and alkylation of proteins on PVDF blots (43, 72, 73) and in-gel digestion protocols (71) have been described. A procedure for the *S*-pyridylethylation of proteins in intact polyacrylamide gels that is compatible with subsequent mass spectrometric analysis techniques is given in *Protocol 7* and *Figure 4*. Procedures for performing in-gel and on-membrane digestions of gel-separated proteins are described in *Protocols 8* and *9*, respectively.

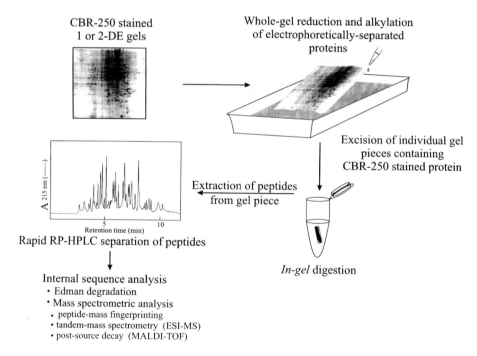

Figure 4. Flow diagram for the whole gel reduction, *S*-pyridylethylation, *in situ* digestion, and identification of acrylamide gel-resolved proteins. (Reproduced with permission from ref. 73.)

Protocol 7. In-gel *S*-pyridylethylation of gel-resolved proteins

Equipment and reagents
- Centrifugal lyophilizer (e.g. Savant) (optional)
- 4-vinylpyridine (Aldrich)
- Reduction buffer: 0.2 M Tris–HCl pH 8.4, 10 mM DTT (Calbiochem), 2 mM EDTA
- 2-mercaptoethanol (Sigma)

A. *Whole gel reduction and S-pyridylethylation*

1. Ensure that the gel has been appropriately stained and destained (see *Protocols 1–3*).

2. Wash the intact gel extensively in large volumes of deionized water, typically 400–500 ml, with three changes, over approx. 1.5 h. This step removes any acetic acid resulting from the destain protocol which would otherwise adversely affect the pH of the reduction.

3. Transfer the intact gel to a clean container.

4. Immerse the gel in 50 ml of reduction buffer (the volume depends on the size of both the gel and the container). Incubate the gel at 40°C for 2 h.

5. Add 4-vinylpyridine to the reduction buffer to a final concentration of 2% (v/v). Incubate the gel in the dark for 1 h at room temperature.

6. Halt alkylation by the addition of excess 2-mercaptoethanol (2%, v/v final concentration).

7. Wash the gel extensively in a large volume of deionized water, typically 400–500 ml, with three changes, over approx. 1.5 h. This step is included to remove 2-mercaptoethanol which would otherwise interfere with proteolysis.

8. If the protein spots are no longer visible, repeat the staining and destaining procedures (see *Protocols 1–3*).

B. *Individual gel band reduction and S-pyridylethylation*

1. Ensure that the gel was appropriately stained and destained (see *Protocols 1–3*).

2. Place the excised gel piece in a clean 1.5 ml polypropylene microcentrifuge tube (e.g. Eppendorf).

3. Wash the excised gel piece extensively with deionized water, 1 ml, with three changes, over approx. 1.5 h. Alternatively, completely dehydrate the gel by centrifugal lyophilization (e.g. using a Savant lyophilizer).

4. Remove deionized water from the microcentrifuge tube.

5. Add enough reduction buffer to completely cover the gel piece(s),

Protocol 7. *Continued*

typically 100–200 µl. Dehydrated gel pieces will swell as they absorb the reduction buffer, so ensure that there is enough buffer to accommodate this.

6. Incubate the gel at 40°C for 1 h.

7. Add 4-vinylpyridine to the reduction buffer to a final concentration of 2% (v/v) and incubate the gel in the dark for 1 h at room temperature.

8. Halt alkylation by the addition of excess 2-mercaptoethanol (2%, v/v final concentration).

9. Wash the gel pieces extensively in large volumes of deionized water, typically 1 ml, with three changes, over approx. 1.5 h.

Protocol 8. In-gel proteolytic digestion procedure and extraction of peptides

Equipment and reagents

- Centrifugal lyophilizer (e.g. Savant)
- 1% (v/v) TFA (Pierce, Sequanal grade) in deionized water
- 0.1% (v/v) TFA, 60% CH_3CN

For *in situ* trypsin cleavage
- Wash buffer: 0.1 M NH_4CO_3, 50% CH_3CN
- Digestion buffer: 0.2 M NH_4HCO_3, 0.5 mM $CaCl_2$
- 0.1 mg/ml trypsin in digestion buffer

For *in situ Achromobacter lyticus* protease I (Lys–C cleavage)
- Wash buffer: 0.05 M Tris–HCl pH 9.3, 50% CH_3CN
- Digestion buffer: 0.1 M Tris–HCl pH 9.3
- 0.1 mg/ml *A. lyticus* protease I in digestion buffer

Method

1. Excise the protein gel spots of interest as described in *Protocol 7* and place in a microcentrifuge tube (Eppendorf).

2. Remove excess Coomassie blue stain by washing with 1 ml of either 0.1 M NH_4HCO_3, 50% CH_3CN (for trypsin) or 0.05 M Tris–HCl pH 9.3, 50% CH_3CN (for Lys–C) twice, for 30 min each time, at 30°C.

3. Dry each gel piece completely by centrifugal lyophilization. The gel piece should *not* stick to the walls of the Eppendorf tube when completely dry.

4. Rehydrate the gel piece by adding 5 µl of the relevant digestion buffer containing 0.5 µg of the relevant protease directly onto the gel piece.

5. Wait until the solution has been absorbed by the gel piece (~ 5–10 min).

6. If necessary, repeat steps 4 and 5 to fully swell the gel piece.

7. Add 200 µl of digestion buffer to fully immerse the gel piece.

8. Incubate for ~ 16 h at 37 °C.

9. Carefully remove the digestion buffer and place into a separate, clean Eppendorf tube. (The digestion buffer contains > 80% of extractable peptides.)

10. Add 200 μl of 1% TFA to the Eppendorf tube containing the gel piece.

11. Sonicate for 30 min.

12. Carefully recover the 1% TFA and add it to the tube containing the digestion buffer.

13. Repeat steps 10–12 substituting 0.1% TFA, 60% CH₃CN, for the 1% TFA solution.

14. Add the 0.1% TFA, 60% CH₃CN extract to the extract from step 12.

15. Reduce the volume of the pooled extracts by centrifugal lyophilization. Do not dry the pooled extracts completely otherwise sample loss may result. The objective of this step is to remove CH₃CN from the pooled extracts and reduce the volume for subsequent dilution and injection onto a capillary RP-HPLC system.

Protocol 9. On-membrane proteolytic digestion of electroblotted proteins[a]

Equipment and reagents

- Electroblotting apparatus (see *Protocol 5*)
- 0.1% Amido black 10B dye (Sigma) in water: acetic acid:methanol (45:10:45, by vol.)
- 0.1% Ponceau S dye (Sigma) in 1% aqueous acetic acid
- 1% aqueous acetic acid
- 0.5% (w/v) PVP-40[b] (Sigma) in 100 mM acetic acid

- 200 mM NaOH
- Digestion buffer: for trypsin (bovine or porcine), chymotrypsin, Lys–C, and Asp–N use 100 mM NH₄HCO₃ pH 7.9, 10% (v/v) acetonitrile, at 37 °C overnight (use an enzyme concentration of approx. 100 ng enzyme/μl digestion buffer, irrespective of the amount of substrate).

A. *Amido black staining (74)*

1. Electroblot the proteins from the gel onto a nitrocellulose or PVDF membrane.[c]

2. Immerse the nitrocellulose membrane in 0.1% Amido black 10B for 1–3 min.

3. Rapidly destain with several washes of water:acetic acid:methanol.

4. Thoroughly rinse the destained blots with deionized water to remove any excess acetic acid.

5. Cut out the stained protein bands (for 2DE gel spots, up to 40 spots from identical gels may be required) and transfer these to 1.5 ml Eppendorf tubes for immediate processing or storage at −20 °C.

6. Add 1.2 ml of 0.5% (w/v) PVP-40[b] in 100 mM acetic acid to each tube.

7. Incubate the tube for 30 min at 37 °C.

Protocol 9. *Continued*

   8. Centrifuge the tube (approx. 1000 *g*) for 5 min.
   9. Remove the supernatant solution and discard.
   10. Add approx. 1 ml of water to the tube.[d]
   11. Vortex the tube for 5 sec.
   12. Repeat steps 8 and 9.
   13. Repeat steps 10–12 (five times).
   14. Cut the nitrocellulose strips into small pieces ($\sim 1 \times 1$ mm) and place them in a fresh tube (typically, a 0.5 ml or a 0.32 ml microcentrifuge tube).
   15. Add the minimal quantity of digestion buffer (10–20 μl) to submerge the nitrocellulose pieces.
   16. After digestion, load the total reaction mixture onto an appropriate RP-HPLC column for peptide fractionation (or store the peptide mixture at –20 °C until use).

B. *Ponceau S staining (74)*

   1. Electroblot the proteins from the gel onto a nitrocellulose or PVDF membrane.[c]
   2. Immerse the nitrocellulose membrane in 0.1% Ponceau S for 1 min.
   3. Remove excess stain from the blot by gentle agitation for 1–3 min in 1% aqueous acetic acid.
   4. Cut out the protein bands of interest and transfer them to Eppendorf tubes.
   5. Destain the protein bands by washing the membrane pieces with 200 mM NaOH for 1–2 min.
   6. Wash the membrane pieces with deionized water and process immediately or store wet at –20 °C (avoiding excessive drying).
   7. Perform on-membrane digestion as described in part A, steps 6–16.

[a] Procedure of Aebersold (52).
[b] PVP-40 (Sigma) is used to prevent absorption of the protease to the nitrocellulose during digestion. This detergent can be replaced by 1% Zwittergent 3–16 (Calbiochem) (69) which is fully compatible with downstream processing methods such as RP-HPLC (Zwittergent 3–16 is UV_{214} transparent) and MALDI-TOF mass spectrometry. Caution: Zwittergent 3–16 is not compatible with nanoseparation techniques such as capillary RP-HPLC since the detergent competes with the peptides for the binding sites on the reversed-phase column.
[c] Typically, 2 h for 0.5 mm thick gels. For proteins that are difficult to transfer, add up to 0.005% SDS to the transfer buffer. Nitrocellulose membranes are preferred to PVDF membranes due to the hydrophobic surface of PVDF limiting the recovery of peptide fragments and the higher yields of recovered peptides from nitrocellulose. It is not possible to perform on-membrane digestion (PVDF) after N-terminal sequence analysis of the protein. Since nitrocellulose is not inert to Edman chemistry reagents and solvents, this membrane can not be used for direct N-terminal sequence analysis. Both Amido black and Ponceau S staining procedures do not interfere with proteolytic digestion, the extraction of peptides from the membrane, or subsequent RP-HPLC analysis of peptides.
[d] It is essential to remove excess PVP-40 before RP-HPLC peptide mapping due to the strong UV absorbance of this detergent. Moreover, breakdown products of PVP-40 also produce major contaminant peaks in ESI-MS.

4.4 Reverse-phase HPLC of peptides

The isolation of pure peptides from a complex digest mixture usually requires a number of chromatographic steps employing various column and solvent-mediated selective steps (for a review see ref. 11, and references therein). For peptide mixtures derived from abundant SDS–PAGE or 2DE resolved proteins, an initial separation on a narrow bore (2.1 mm i.d. × 30 mm) Brownlee RP-300 (Perkin-Elmer), or Vydac C4 (Vydac, Hesperio, CA), reversed-phase column using a trifluoroacetic acid (TFA)/acetonitrile gradient system is recommended. A typical solvent system is: solvent A, 0.1% (v/v) aqueous TFA (HPLC/Spectro grade Cat. No. 28901, Pierce); solvent B, 60% acetonitrile containing 0.085% (v/v) aqueous TFA. The column is usually developed with a linear 60 min gradient from 0–100% solvent B at a flow rate of 100 μl/min. Peptides can be detected at UV_{214nm}. The tandem use of photodiode array detection (11) allows the rapid, highly sensitive identification of aromatic amino acid containing peptides (particularly tryptophan-, tyrosine- and *S*-pyridylethylated-cysteine containing peptides) (71) from the characteristic ultraviolet spectra available for each peptide. Peptides are collected into 1.5 ml tubes according to UV absorbance at 214 nm, and immediately capped to prevent loss of organic solvent. The use of Eppendorf brand tubes is recommended due to the low levels of contaminants which otherwise interfere with RP-HPLC and mass spectrometry. For high sensitivity work, it is recommended that tubes are rinsed with solvent B prior to sample collection.

In cases where the collected peptide is clearly impure (as judged by peak shape and/or UV spectrum across the peak), the collected sample is re-chromatographed using a separation system with a different selectivity (11). Solvent systems that can be used for the second chromatographic system include: aqueous 1% (w/v) NaCl/acetonitrile; aqueous 0.1% (v/v) TFA/methanol; 50 mM (w/v) sodium phosphate/acetonitrile. Using these solvent systems, collected samples can be applied directly to conventional Edman chemistry sequencers.

To avoid sample loss, it is recommended that peptide fractions never be dried between chromatographic steps. Since peptide fractions usually contain sufficient quantities of secondary solvent to prevent their retention on a similar interactive support, it is necessary to dilute them (one- to twofold) with the primary solvent to facilitate trace enrichment in the subsequent chromatographic step. Dilution can be readily accomplished in a large volume (1.5 ml) sample loading syringe immediately prior to injection. An illustration of the selective effects achieved with various mobile phase conditions is given in *Figure 5*.

For the separation of peptide mixtures from trace-abundant proteins, it is recommended that microbore columns (~ 1 mm i.d.) or capillary columns (< 0.5 mm i.d.) be employed. While microbore columns can be operated on conventional HPLC systems that can deliver accurate, low flow rates

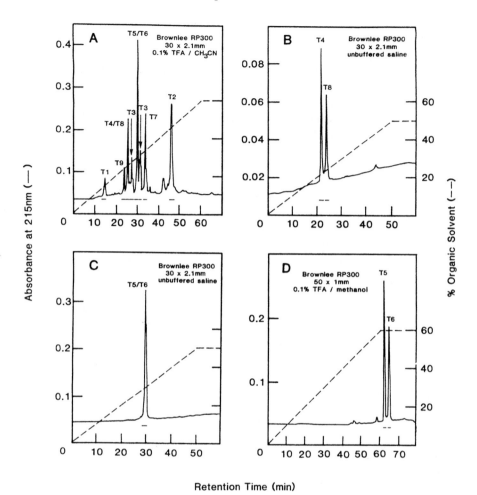

Retention Time (min)

Figure 5. Microbore HPLC purification of peptides from murine interleukin-6 (IL-6). (A) Fractionation of tryptic peptides of IL-6 (~ 14 μg) by RP-HPLC. Column: Brownlee RP-300, 30 × 2.1 mm internal diameter equilibrated with 0.1% TFA. A linear 60 min gradient was used from 0–100% B where solvent A was 0.1% TFA (aq) and solvent B was 60% acetonitrile/40% water containing 0.09% TFA. Flow rate was 100 μl/min. Column temperature was 45°C. Peak fractions T4/T8 and T5/T6 from (A) were re-chromatographed on the same column as in (A) but employing unbuffered 1% NaCl as the mobile phase. Peak T4/T8 was resolved into two peptide fractions (B) but peak fraction T5/T6 re-chromatographed as a single peak (C). Using a third chromatographic step, peak fraction T5/T6 from (C) was resolved into two homogeneous peptides T5 and T6 (D) when chromatographed on an ODS-Hypersil reversed-phase support (12 nm pore size, 5 μm particle diameter C18) (Hypersil, Runcorn, Cheshire, UK) packed into a glass lined stainless steel column (50 × 1.0 mm internal diameter). (Reproduced with permission from ref. 87.)

($\sim$ 50 μl/min) (11), as yet there are no commercially available HPLC systems for capillary column usage. However, conventional HPLC systems can be modified to provide the accurate low flow rates (0.4–4 μl/min) and gradients necessary to operate capillary columns (75–79). Such columns are essential for the preparation of samples for a number of mass spectrometric methods for peptide identification (see below). Detailed procedures for the facile fabrication of < 0.32 mm i.d. polyimide coated fused-silica columns which allow the detection of 500 pg amounts of protein can be found in refs 78 and 79.

5. Mass spectrometric methods for identifying gel-separated proteins

5.1 General concepts

Dramatic advances in mass spectrometric methodologies over the past five years have had a great impact on the rapid identification and characterization of small quantities (submicrogram) of gel-separated proteins (25, 50, 53, 80–86). The development of ESI and MALDI ionization methods (21, 22) has lead to the introduction of low cost (relatively) commercially available mass spectrometers for routine use in protein characterization.

MALDI-TOF mass spectrometry (22) is a technique where analyte molecules, embedded in an excess of specific wavelength (usually UV_{337nm}) absorbing matrix, are protonated and desorbed into the gas phase following a laser pulse. The mass-to-charge (m/z) ratio of the analyte ions is then determined by their time of flight. Mass accuracy determinations vary from $\pm$ 0.01% to 0.1% depending on the sample preparation technique and the method used for mass calibration (for a routine sample preparation procedure, see *Protocol 10*).

Electrospray ionization mass spectrometry involves the introduction of an aqueous solution of analyte molecules via a capillary at high voltage (21). A spray of charged droplets is formed which, upon desolvation, leads to gas phase ions. Typically, ions formed by ESI are analysed by either quadrupole or ion-trap mass spectrometers. Mass accuracy is routinely $\pm$ 0.01% (i.e. $\pm$ 2 Da for a 20 000 Da protein).

Amino acid sequence information can be obtained by interpretation of the fragmentation pattern of peptide ions resulting from fragmentation along the peptide backbone of selected peptide ions (for a recent review see ref. 25, and references therein). In MALDI-TOF MS, such fragmentation patterns result as a consequence of post-source decay (PSD) processes occurring in the field-free region of a reflectron time of flight mass spectrometer (88). In the case of ESI-MS, selected ions undergo collision-induced dissociation (CID) by fragmentation with a heavy inert gas such as argon (89). The resultant fragment ion series in each case provides redundant information regarding the amino acid sequence as fragment ions can retain charge on either the N- (*a,b,c* type), or C-terminal (*x,y,z* type) series of ions, where the mass difference between ions in each series corresponds to the residue mass of the amino acid. The success of this approach depends on the equal fragmentation of each of the amide bonds along the peptide backbone. This however does not commonly occur, due to the contribution of such factors as the proton affinities and intramolecular side chain interactions within the peptide (90, 91). Routine interpretation of peptide fragment ion spectra is therefore, particularly sequence-dependent. For a recent review on the interpretation of the tandem mass spectra of peptides see ref. 92.

Concomitant with the above mentioned advances in hardware developments has been the development, since 1993, of a variety of computer algorithms for the identification of known proteins using mass spectrometric-derived information. These approaches now allow the identification of proteins in mixtures at levels significantly lower than those achievable by Edman degradation methods. The most widely used mass spectrometric methods for identifying proteins include:

(a) Peptide mass fingerprinting searches (93–97) (see also *Table 1* for URL addresses of World Wide Web-based search algorithms).

(b) Peptide 'sequence tag' searches (98) (see *Table 1*).

(c) Peptide fragment ion searches (99–101) (see *Table 1*).

Table 1. World Wide Web-based search algorithms[a]

Peptide–mass fingerprinting search algorithms

Ms-Fit	http://rafael.ucsf.edu/msfit.htm
Peptide Search	http://www.mann.embl-heidelberg.de/Services/PeptideSearch
ProFound	http://128.122.10.5/cgi-bin/prot-id
Mass Search	http://cbrg.inf.ethz.ch/MassSearch.html
Peptide Mass Search	http://www.mdc-berlin.de/~emu/peptide_mass.html
Mowse	http://gservl.dl.ac.uk/SEQNET/mowse.html
Peptide Mass	http://www.expasy.ch/www/tools.html

Peptide sequence tag search algorithms

Peptide Search	http://www.mann.embl-heidelberg.de/Services/PeptideSearch

Peptide fragment ion search algorithms

SEQUEST	http://thompson.mbt.washington.edu/sequest.html
MS-Tag	http://rafael.ucsf.edu/mstag.htm
PepFrag	http://128.122.10.5/cgi-bin/prot-id-frag

[a] URL addresses were current at time of manuscript submission (April 1997). For updated links to WWW-based search algorithms, see http://www.ludwig.edu.au/www/jpsl/jpslhome.html.

All of these methods utilize mass spectrometric-derived information, be it peptide mass or peptide fragment ion masses, to probe the available protein sequence databases in order to identify the protein of interest. These methods will be discussed in more detail in Sections 5.3–5.5.

5.2 Sample preparation

While peptide detection is in the femtomole range for MALDI-TOF MS, and the presence of salts and other contaminants can be tolerated to some degree, the presence of buffer salts, detergents, and other gel-related artefacts may result in poor quality spectra and reduced peak intensities. Significant ion suppression may also be observed, limiting the usefulness of the technique for the analysis of complex mixtures. For these reasons, it is recommended that samples be cleaned up prior to MALDI-TOF MS analysis. On-slide rinsing procedures to remove salts and other contaminants have been suggested (102–104). Alternatively, samples may be desalted by rapid chromatographic techniques prior to sample application (59, 79, 105, 106). An added advantage of any chromatographic step is that while samples are desalted, trace enrichment occurs, leading to improved sensitivity.

ESI-MS is less tolerant to buffer salts and detergents than MALDI-TOF MS. The requirement for aqueous sample introduction makes ESI particularly suited for on-line coupling to HPLC. Trace enrichment of samples onto a reversed-phase support enables removal of buffer salts prior to sample elution (either stepwise elution or via a linear gradient). As ESI-MS sensitivity is concentration-dependent, the application of capillary RP-HPLC (59, 71, 75–79), or low flow sample (e.g. nano/microspray) introduction is recommended for the analysis of trace abundant samples (107–109), particularly for applications requiring subpicomole level sequence analysis.

Approaches for the sample preparation of peptide mixtures for mass spectrometric determination are outlined in *Figure 6*.

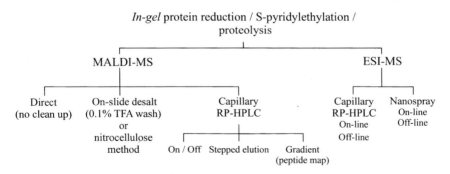

Figure 6. Approaches for sample preparation for mass spectrometric determination of peptide mixtures following in-gel reduction, *S*-pyridylethylation, and proteolysis.

Protocol 10. Preparing samples for MALDI analysis

Reagents

- Matrix solution: 20 mg/ml α-cyano-4-hydroxycinnamic acid (Sigma), in 60% acetonitrile, 0.1% TFA[a]
- Calibration standards: 1–10 pmol/μl in 60% acetonitrile, 0.1% TFA[b]
- Unknown samples: if these are lyophilized, dissolve them in 60% acetonitrile, 0.1% TFA at a concentration of 1–10 pmol/μl, or use HPLC purified fractions directly

Method

1. Pipette 0.5 μl of the matrix solution (using a 2 μl pipette for accuracy) into a well on the metal sample slide used for MALDI analysis.
2. Quickly add 0.5 μl of the standard (or sample) to the matrix before it dries.
3. Allow the solvent to evaporate and the samples to dry either in the atmosphere or in a low temperature (40°C) oven.
4. Transfer the metal sample slide to the vacuum chamber of the mass spectrometer.
5. Aquire an initial mass spectrum at a laser power well above the ionization threshold to 'warm up' the calibration standard.
6. Decrease the laser power until a good spectrum is obtained.[d]
7. Obtain a linear external two point calibration using a matrix-derived ion (e.g. $(M + H)^+ - OH$, m/z 173.2) and the singly charged ion of the appropriate calibration standard.
8. Repeat steps 5 and 6 for unknown samples and compare with the calibration values to obtain accurate mass.[e]

[a] Under these conditions α-cyano-4-hydroxycinnamic acid is a saturated solution and must be centrifuged prior to use. For uniform sample/matrix formation, the use of undissolved matrix crystals should be avoided.

[b] Choose well-characterized proteins or peptides of known molecular mass, close to the molecular mass of the unknown sample, as calibration standards (see instrument manufacturers instructions for preferred standards). This will ensure a linear calibration curve. The concentration of the calibration standards must be similar to that of the unknown sample (range 1–10 pmol/μl) to ensure accurate results.

[c] Samples containing contaminating salts need to be desalted prior to analysis (37, 80, 91, 92). Check the low mass region of the spectrum for major ions indicative of contaminating salts. Minor quantities of sodium (m/z 23) and potassium (m/z 39) ions, readily generated via laser ionization, can cause significant ion suppression.

[d] For optimal results, run the unknown samples at approximately the same laser power as the calibration standard. Whilst a high laser power may give a good signal, resolution may be compromised due to peak broadening. The use of a low laser power at the ionization threshold may give high resolution, but result in poor signal-to-noise ratios. Although the sample appears to be uniform, it is recommended that different regions of the spot be examined to find 'sweet spots', i.e. regions of the sample spot that give superior signal-to-noise ratios. Check that different 'sweet spots' for the same sample give similar spectra. Also, ensure that those ions corresponding to multiply charged species of the same sample compute to give the same molecular mass as the singly charged ion.

[e] The use of an internal calibrant may be preferred for samples requiring more precise mass determinations.

5.3 Peptide–mass fingerprinting search algorithms

Peptide–mass fingerprinting search algorithms (93–97) (see also *Table 1*) compare a set of experimentally determined peptide masses (generated by specific enzymatic digestion) with the theoretical masses of the enzymatically-derived peptides of all the amino acid sequences in a protein sequence database (see *Figure 7*). The identified proteins are arranged as a list that match, to varying degrees, the input data. The search specificity can be enhanced by including additional supplementary information about the intact protein such as mass, isoelectric point, species, amino acid composition, and possible post-translational modifications. Simple heterogeneous protein mixtures can be tolerated provided that sufficient peptide masses are obtained to unambiguously identify each component. This limits the analysis of protein digests present as mixtures to those containing approximately equimolar amounts of peptides in the digest.

5.4 Peptide sequence tag search algorithms

It is often relatively easy (particularly for tryptic peptides) to rapidly assign several residues of an amino acid sequence following tandem mass spectrometric analysis. The mass of the peptide, together with the partially interpreted amino acid sequence, the masses of the 'flanking' regions (i.e. the N- and C-terminal masses either side of the interpreted sequence), and the enzyme cleavage specificity can be used to generate a peptide 'sequence tag' (98) (see *Figure 8* and also *Table 1*). This data, along with any other supplementary information (see Section 5.1) can be used to probe the protein sequence databases to locate proteolytically-derived peptides of the same mass, partial sequence, and flanking regions. In this manner, protein identification can be achieved from a single peptide when only a short stretch of

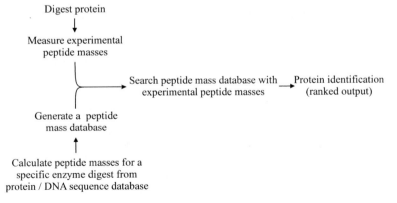

Figure 7. Strategy for the identification of known proteins using peptide–mass fingerprinting search algorithms. For World Wide Web accessible tools, see *Table 1.*

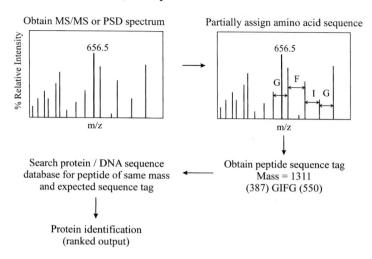

Figure 8. Strategy for the identification of known proteins using peptide sequence tag search algorithms. For World Wide Web accessible tools, see *Table 1*.

amino acid sequence has been interpreted. Identification can still be achieved even when errors in the protein sequence database exist, where cross-species matching is required or when amino acids are post-translationally modified, by allowing errors in either one or two regions of the sequence tag. Additionally, results from several peptide sequence tag searches from the same digest can be compiled to give further evidence for a protein identification. Confidence in the results can be enhanced by manual comparison of the experimental data with those predicted from the expected fragment ions of the matched sequences.

5.5 Peptide fragment ion search algorithms

Automated search routines have been developed where prior identification of the peptide fragment ion spectra is not required (99–101) (see also *Table 1*). The mass of the unknown peptide is used to search the protein sequence database to find all peptides of the same mass. Then, the peptide fragment ion spectra of the unknown is compared with the predicted peptide fragment ion spectra of each matching peptide in the database (see *Figure 9*). Factors such as enzyme cleavage specificity, partial amino acid sequence, or partial amino acid composition may be incorporated to increase the search specificity. The potential presence of post-translationally modified residues may also be incorporated into the search routine. A correlation function is applied to the search algorithm to generate a ranked output of peptide sequences. These search algorithms, when coupled to LC-MS/MS systems, allow fully automated sequence analysis and identification of protein spots isolated from 1D or 2D gels provided the protein sequence is listed in the protein sequence database.

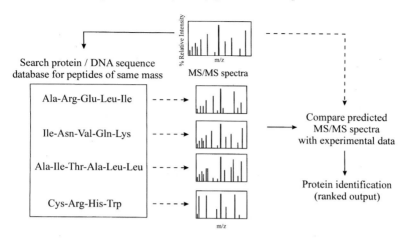

Figure 9. Strategy for the identification of known proteins using peptide fragment ion search algorithms. For World Wide Web accessible tools, see *Table 1*.

As mentioned above, manual interpretation of the experimental data with that predicted by the search algorithm and compilation of several different data sets from within the same digest provide a good way of increasing the confidence in the search results.

Acknowledgements

Thanks are due to Hong Ji, Lisa Zugaro, Axel Ducret, and James Eddes for critical reading of many of the protocols in this manuscript.

References

1. O'Farrell, P. H. (1975). *J. Biol. Chem.*, **250**, 4007.
2. Lander, E. S. (1996). *Science*, **274**, 536.
3. Chait, B. T. (1996). *Nat. Biotechnol.*, **14**, 101.
4. Towbin, H., Staehelin, T., and Gordon, J. (1979). *Proc. Natl. Acad. Sci. USA*, **76**, 4350.
5. Celis, J. E., Rasmussen, H. H., Olsen, E., Madsen, P., Leffers, H., Honoré, B., *et al.* (1993). *Electrophoresis*, **14**, 1091.
6. Cornish-Bowden, A. (1980). *Anal. Biochem.*, **105**, 233.
7. Eckerskorn, C., Jungblut, P., Mewes, W., Klose, J., and Lottspeich, F. (1988). *Electrophoresis*, **9**, 830.
8. Cleveland, D. W., Fischer, S. G., Kirschner, M.W., and Laemmli, U.K. (1977). *J. Biol. Chem.*, **252**, 1102.
9. Hunkapiller, M. W., Luhan, E., Ostrander, F., and Hood, L. E. (1982). In *Methods in enzymology* (ed. C. H. W. Hirs and S. N. Timasheff), Vol. 91, p. 227–36. Academic Press, San Diego.

10. Simpson, R. J., Moritz, R. L., Nice, E. C., and Grego, B. (1987). *Eur. J. Biochem.*, **165**, 21.
11. Simpson, R. J., Moritz, R. L., Begg, G. S., Rubira, M. R., and Nice, E. C. (1989). *Anal. Biochem.*, **177**, 221.
12. Vandekerckhove, J., Bauw, G., Puype, M., Van Damme, J., and Van Montagu, M. (1985). *Eur. J. Biochem.*, **152**, 9.
13. Aebersold, R. H., Teplow, D. B., Hood, L. E., and Kent, S. B. H. (1986). *J. Biol. Chem.*, **261**, 4229.
14. Matsudaira, P. (1987). *J. Biol. Chem.*, **262**, 10035.
15. Aebersold, R. H., Leavitt, J., Saavedra, R. A., Hood, L. E., and Kent, S. B. H. (1987). *Proc. Natl. Acad. Sci. USA*, **84**, 6970.
16. Bauw, G., Van Damme, J., Puype, M., Vandekerckhove, J., Gesser, B., Ratz, G. P., *et al.* (1989). *Proc. Natl. Acad. Sci. USA*, **86**, 7701.
17. Ward, L. D., Reid, G. E., Moritz, R. L., and Simpson, R. J. (1990). *J. Chromatogr.*, **519**, 199.
18. Eckerskorn, C. and Lottspeich, F. (1989). *Chromatographia*, **28**, 92.
19. Barber, M., Bordoli, R. S., Elliot, G., Sedwick, R. D., and Tyler, A. N. (1982). *Anal. Chem.*, **54**, 645A.
20. Macfarlane, R. D. and Torgerson, D. F. (1976). *Science*, **191**, 920.
21. Fenn, J. B., Mann, M., Meng, C. K., Wong, S. F., and Whitehouse, C. M. (1989). *Science*, **246**, 64.
22. Karas, M. and Hillenkamp, F. (1988). *Anal. Chem.*, **60**, 2299.
23. Hillenkamp, F., Karas, M., Beavis, R. C., and Chait, B. T. (1991). *Anal. Chem.*, **63**, 1193.
24. Burlingame, A. L., Baillie, T. A., and Russell, D. H. (1992). *Anal. Chem.*, **64**, 634R.
25. Burlingame, A. L., Boyd, R. K., and Gaskell, S. J. (1996). *Anal. Chem.*, **68**, 599R.
26. Wilson, C. M. (1983). In *Methods in enzymology* (ed. C. H. W. Hirs and S. N. Timasheff, Vol. 91, p. 236. Academic Press, San Diego.
27. Nelles, L. P. and Bamburg, J. R. (1976). *Anal. Biochem.*, **73**, 522.
28. Bergman, T. and Jörnvall, H. (1987). *Eur. J. Biochem.*, **169**, 9.
29. Higgins, R. C. and Darmus, M. E. (1979). *Anal. Biochem.*, **93**, 257.
30. Lee, C., Levin, A., and Branton, D. (1987). *Anal. Biochem.*, **166**, 308.
31. Casero, P., Del Campo, G. B., and Righetti, P. G. (1985). *Electrophoresis*, **6**, 362.
32. Lonnie, D. A. and Weaver, K. M. (1990). *Appl. Theor. Electrophoresis*, **1**, 279.
33. Fernandez-Patron, C., Calero, M., Collazo, P. R., Garcia, J. R., Madrazo, J., Musacchio, A., *et al.* (1995). *Anal. Biochem.*, **224**, 203.
34. Fernandez-Patron, C., Hardy, E., Sosa, A., Seoane, J., and Castellanon, L. (1995). *Anal. Biochem.*, **224**, 263.
35. Candiano, G., Porotto, M., Lanciotto, M., and Ghiggeri, G. M. (1996). *Anal. Biochem.*, **234**, 245.
36. Switzer, R. C., Merril, C. R., and Shifrin, S. (1979). *Anal. Biochem.*, **98**, 231.
37. Rabilloud, T. (1990). *Electrophoresis*, **11**, 785.
38. Rabilloud, T., Vuillard, L., Gilly, C., and Lawrence, J-J. (1994). *Cell Mol. Biol.*, **40**, 57.
39. McCormack, A. L., Schieltz, D. M., Goode, B., Yang, S., Barnes, G., Drubin, D., *et al.* (1997). *Anal. Biochem.*, **69**, 767.

40. Shevchenko, A., Wilm, M., Vorm, O., and Mann, M. (1996). *Anal. Chem.*, **68**, 850.
41. Muzio, M., Chinnaiyan, A. M., Kischkel, F. C., O'Rourke, K., Shevchenko, A., Ni, J., *et al.* (1996). *Cell*, **85**, 817.
42. Tempst, P., Link, A. J., Riviere, L. R., Fleming, M., and Elicone, C. (1990). *Electrophoresis*, **11**, 537.
43. Jenö, P., Mini, T., Moes, S., Hintermann, E., and Horst, M. (1995). *Anal. Biochem.*, **224**, 75.
44. Gevaert, K., Verschelde, J-L., Puype, M., Van Damme, J., De Boeck, S., and Vandekerckhove, J. (1996). *Electrophoresis*, **17**, 918.
45. Gevaert, K., Rider, M. H., Puype, M., Van Damme, J., De Boeck, S., and Vandekerckhove, J. (1995). In *Methods in protein structure analysis* (ed. M.Z. Atassi and E. Appella), p. 15. Plenum Press, New York, USA.
46. Dainese, P., Staudenmann, W., Quadroni, M., Korostensky, C., Gonnet, G., Kertesz, M., *et al.* (1997). *Electrophoresis*, **18**, 432.
47. Ji, H., Baldwin, G. S., Burgess, A. W., Moritz, R. L., Ward, L. D., and Simpson, R. J. (1993). *J. Biol. Chem.*, **268**, 13396.
48. Inglis, A. S., Reid, G. E., and Simpson, R. J. (1995). In *Interface between chemistry and biochemistry* (ed. P. Jollés and H. Jörnvall), p. 141. Birkhäuser Verlag, Basel, Switzerland.
49. Eckerskorn, C. and Lottspeich, F. (1993). *Electrophoresis*, **14**, 831.
50. Patterson, S. C. (1994). *Anal. Biochem.*, **221**, 1.
51. Eckerskorn, C. (1994). In *Microcharacterisation of proteins* (ed. R. Kellner, F. Lottspeich, and H. E. Meyer), p. 75. VCH Publishers Inc., New York, USA.
52. Aebersold, R. (1993). In *A practical guide to protein and peptide purification for microsequencing* (ed. P. Matsudaira), 2nd edn, p. 103. Academic Press, Inc., San Diego, USA.
53. Patterson, S. C. and Aebersold, R. (1995). *Electrophoresis*, **16**, 1791.
54. Brown, J. L. and Roberts, W. K. (1976). *J. Biol. Chem.*, **251**, 1009.
55. Brown, J. L. (1979). *J. Biol. Chem.*, **254**, 1447.
56. Wellner, D., Panneerselvam, C., and Horecker, B. L. (1990). *Proc. Natl. Acad. Sci. USA*, **87**, 1947.
57. Hirano, H., Komatsu, S., Takakura, H., Sakiyama, F., and Tsunasawa, S. (1992). *J. Biochem.*, **111**, 754.
58. Hirano, H., Komatsu, S., Kajiwara, H., Takagi, Y., and Tsunasawa, S. (1993). *Electrophoresis*, **14**, 839.
59. Ji, H., Reid, G. E., Moritz, R. L., Eddes, J. S., Burgess, A. W., and Simpson, R. J. (1997). *Electrophoresis*, **18**, 605.
60. Ward, L. D., Ji, H., Whitehead, R. H., and Simpson, R. J. (1990). *Electrophoresis*, **11**, 883.
61. Kawasaki, H., Emori, Y., and Suzuki, K. (1990). *Anal. Biochem.*, **191**, 332.
62. Rosenfeld, J., Capdevielle, J., Guillemot, J., and Ferrara, P. (1992). *Anal. Biochem.*, **203**, 173.
63. Hellman, U., Wernstedt, C., Gonez, J., and Heldin, C-H. (1995). *Anal. Biochem.*, **224**, 451.
64. Moritz, R. L., Eddes, J. S., Ji, H., Reid, G. E., and Simpson, R. J. (1995). In *Techniques in protein chemistry VI* (ed. J. W. Crabb), p. 311. Academic Press, San Diego, USA.

65. Fernandez, J., DeMott, M., Atherton, D., and Mische, S. M. (1992). *Anal. Biochem.*, **201**, 255.
66. Jahnen, W., Ward, L. D., Reid, G. E., Moritz, R. L., and Simpson, R. J. (1990). *Biochem. Biophys. Res. Commun.*, **166**, 139.
67. Stone, K. L., McNulty, D. E., LoPresti, M. L., Crawford, J. M., DeAngelis, R., and Williams, K. R. (1992). In *Techniques in protein chemistry III* (ed. R. Angeletti), p. 23. Academic Press, San Diego, USA.
68. Patterson, S. D., Hess, D., Yungwirth, T., and Aebersold, R. (1992). *Anal. Biochem.*, **202**, 193.
69. Lui, M., Tempst, P., and Erdjument-Bromage, H. (1996). *Anal. Biochem.*, **241**, 156.
70. Williams, K., Kobayashi, R., Lane, W., and Tempst, P. (1993). *Assoc. Biomol. Res. Fac. News*, **4 (4)**, 7.
71. Moritz, R. L., Eddes, J. S., Reid, G. E., and Simpson, R. J. (1996). *Electrophoresis*, **17**, 907.
72. Iwamatsu, A. (1992). *Electrophoresis*, **13**, 142.
73. Henzel, W. J., Billeci, T. M., Stults, J. T., Wong, S. C., Grimley, C., and Watanabe, C. (1993). *Proc. Natl. Acad. Sci. USA*, **90**, 5011.
74. Schaffner, W. and Weissman, C. (1973). *Anal. Biochem.*, **56**, 502.
75. Moritz, R. L. and Simpson, R. J. (1992). *J. Chromatogr.*, **599**, 119.
76. Moritz, R. L. and Simpson, R. J. (1992). *J. Microcol. Sep.*, **4**, 485.
77. Moritz, R. J. and Simpson, R. J. (1993). In *Methods in protein sequence analysis* (ed. K. Imahori and F. Sakiyama), p. 3. Plenum Press, New York, USA.
78. Moritz, R. L., Reid, G. E., Ward, L. D., and Simpson, R. J. (1994). *Methods: a companion to methods in enzymology*, **6**, 213.
79. Tong, D., Moritz, R. L., Eddes, J. S., Reid, G. E., Rasmussen, R. K., Dorow, D. S., *et al.* (1997). *J. Protein Chem.*, **16**, 425.
80. Patterson, S. D. (1997). *Biochem. Soc. Trans.*, **25**, 255.
81. Shevchenko, A., Jensen, O. N., Podtelejnikov, A. V., Sagliocco, F., Wilm, M., Vorm, O., *et al.* (1996). *Proc. Natl. Acad. Sci. USA*, **93**, 14440.
82. Roepstorff, P. (1997). *Curr. Opin. Biotechnol.*, **8**, 6.
83. Yates, J. R., McCormack, A. L., Link, A. J., Schieltz, D., Eng, J., and Hays, L. (1996). *Analyst*, **121**, 65R.
84. Yates, J. R., McCormack, A. L., and Eng, J. (1996). *Anal. Chem.*, **68**, 534A.
85. Epstein, L. B., Smith, D. M., Matsui, N. M., Tran, H. N., Sullivan, C., Raineri, I., *et al.* (1996). *Electrophoresis*, **17**, 1655.
86. Patterson, S. D., Thomas, D., and Bradshaw, R. A. (1996). *Electrophoresis*, **17**, 877.
87. Simpson, R. J., Moritz, R. L., Rubira, M. R., and Van-Snick, J. (1988). *Eur. J. Biochem.*, **176**, 187.
88. Spengler, B., Kirsch, D., Kaufmann, R., and Jaeger, E. (1992). *Rapid Commun. Mass Spectrom.*, **6**, 105.
89. Hunt, D. F., Yates, J. R., Shabanowitz, J., Winston, S., and Hauer, C. R. (1986). *Proc. Natl. Acad. Sci. USA*, **83**, 6233.
90. Morgan, D. G. and Bursey, M. M. (1994). *Org. Mass Spectrom.*, **29**, 354.
91. Burlet, O., Yang, C-Y., and Gaskell, S. J. (1992). *J. Am. Soc. Mass Spectrom.*, **3**, 337.
92. Papayannopoulos, I. A. (1995). *Mass Spectrom. Rev.*, **14**, 49.
93. Yates, J. R., Griffin, P. R., Speicher, S., and Hunkapiller, T. (1993). *Anal. Biochem.*, **214**, 397.

94. Henzel, W., Billeci, T., Stults, J., Wond, S., Grimley, C., and Watanabe, C. (1993). *Proc. Natl. Acad. Sci. USA*, **90**, 5011.
95. James, P., Qaudroni, M., Carafoli, E., and Gonnet, G. (1993). *Biochem. Biophys. Res. Commun.*, **195**, 58.
96. Pappin, D., Hojrup, P., and Bleasby, A. (1993). *Curr. Biol.*, **3**, 327.
97. Mann, M., Hojrup, P., and Roepstorff, P. (1993). *Biol. Mass Spectrom.*, **22**, 338.
98. Mann, M. and Wilm, M. (1994). *Anal. Chem.*, **66**, 4390.
99. Eng, J. K., McCormack, A. L., and Yates, J. R. (1994). *J. Am. Soc. Mass Spectrom.*, **5**, 976.
100. Yates, J. R., Eng, J. K., McCormack, A. L., and Schieltz, D. (1995). *Anal. Biochem.*, **67**, 1426.
101. McCormack, A. L., Schieltz, D., Goode, B., Yang, S., Barnes, G., Drubin, D., *et al.* (1997). *Anal. Chem.*, **69**, 767.
102. Vorm, O., Roepstorff, P., and Mann, M. (1994). *Anal. Chem.*, **66**, 3281.
103. Shevchenko, A., Wilm, M., Vorm, O., and Mann, M. (1996). *Anal. Chem.*, **68**, 850.
104. Cohen, S. L. and Chait, B. T. (1996). *Anal. Chem.*, **68**, 31.
105. Zhang, H., Andren, P. E., and Caprioli, R. M. (1995). *J. Mass Spectrom.*, **30**, 1768.
106. Courchesne, P. L. and Patterson, S. D. (1997). *BioTechniques*, **22**, 246.
107. Davis, M. T., Stahl, D. C., Hefta, S. A., and Lee, T. D. (1995). *Anal. Chem.*, **67**, 4549.
108. Wilm, M., Shevchenko, A., Houthaeve, T., Breit, S., Schweigerer, L., Fotsis, T., *et al.* (1996). *Nature*, **379**, 466.
109. Wilm, M. and Mann, M. (1996). *Anal. Chem.*, **68**, 1.

9

Fluorophore-labelled saccharide electrophoresis for the analysis of glycoproteins

PETER JACKSON

1. Introduction

It has become apparent that the carbohydrate moieties of glycoproteins (glycones) have important and varied functions (1–3) and this knowledge has stimulated considerable advances in methods for their purification and analysis. However these procedures are not routine. Current analytical strategies usually involve the release of the free saccharide (glycan) from the glycoprotein or glycopeptide followed by chromatographic isolation and structural determination using a variety of techniques including compositional analysis, chemical and enzymatic degradation, lectin binding, mass spectrometry, and nuclear magnetic resonance (4–10).

Polyacrylamide gel electrophoresis of fluorophore-labelled saccharides (PAGEFS), which is known variously in its commercial formats as fluorophore assisted carbohydrate electrophoresis (FACE) and GlycoMap, is a method in which reducing saccharides, including mono- and oligosaccharides and complex glycans, are labelled at their reducing terminals by reductive amination with a fluorophore and the fluorescent saccharide derivatives separated by polyacrylamide gel electrophoresis (11–29). The fluorophore allows the labelled saccharide to be detected in the polyacrylamide gel with sub-picomolar sensitivity. Several fluorophores have been applied to PAGEFS (11, 13, 29, 30) but 8-aminonaphthalene-1,3,6-trisulfonic acid (ANTS) and 2-aminoacridone (AMAC) (*Figure 1*), which were introduced with the original work (11, 13), have been most widely used and will be the subject of this chapter. ANTS carries negative charges that enable the electrophoresis of ANTS labelled neutral saccharides as well as acidic saccharides so producing a fluorescent band pattern (oligosaccharide profile; OSP) of the saccharides. AMAC carries no charge at the pH of the electrophoretic system used and can be applied in analyses that complement those of ANTS.

PAGEFS provides a convenient method for comparing and assessing

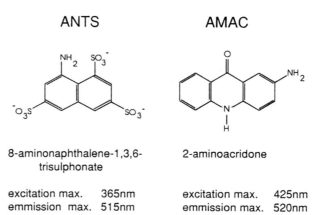

ANTS   AMAC

8-aminonaphthalene-1,3,6-   2-aminoacridone
      trisulphonate

excitation max.   365nm   excitation max.   425nm
emmission max. 515nm   emmission max. 520nm

Figure 1. The chemical structures of ANTS and AMAC. (From ref. 21, with permission.)

changes in the glycosylation of glycopeptides or glycoproteins, for example when proteins are expressed in different cell lines. The degree of complexity of the glycosylation is revealed together with information on the types of saccharide present. The method is sensitive, having been used on as little as 2 μg of protein. PAGEFS enables multiple samples to be electrophoresed in parallel so that, for instance, it can be used to analyse chromatographic column fractions. PAGEFS has also been used in conjunction with lectins to measure binding constants (28) and it has proved ideal for the enzymatic structural analysis of purified glycans. Therefore PAGEFS is a method that can provide researchers in biological sciences, who are not specialists in carbohydrate analysis, with a means of obtaining limited, important information about the glycosylation of glycoproteins rapidly and inexpensively. It should be emphasized that the method is technically easy to carry out and requires only standard biochemical laboratory equipment. The reagents, other than the enzymes, are relatively inexpensive when obtained from the suppliers indicated (see below). Consequently, PAGEFS can provide a readily accessible launching-pad to more sophisticated analyses.

This chapter describes methods for releasing saccharides from glycoproteins, and their labelling , electrophoresis, and recording of the resultant electrofluorograms.

2. The release of oligosaccharides from glycoconjugates

Most saccharides of interest exist in covalently bound form as glycones and must be released from the glycoproteins, either enzymatically or chemically, to yield the corresponding glycans before they are labelled with the fluorophore.

2.1 Enzymatic release of glycans

The release of asparagine-linked glycans (*N*-glycans) is achieved most conveniently by using one of several enzymes, the applications of which have been well described (31, 32). These can be obtained, together with good documentation on their use, from one of the numerous suppliers such as: Boehringer Mannheim, Calbiochem, Dextra Laboratories, New England Biolabs, Oxford Glycosystems, and Seikagaku. The glycopeptidase, peptide-N^4-(*N*-acetyl-β-glucosaminyl)asparagine amidase (PNGase F), is particularly useful since it releases the whole oligosaccharide (leaving an aspartyl residue on the protein). PNGase F has the additional advantage that it has a broad specificity, being able to cleave all types of *N*-linked glycone except when the glycosylated asparagine is either amino or carboxy terminal or on very small peptides. Endoglycosidases can also be used: these cleave within the glycone, the structure of which influences their specificity, leaving residual saccharide attached to the asparagine. The glycans released by both PNGase F and the endoglycosidases have a single reactive reducing terminal carbon (carbon 1) on the GlcNAc adjacent to the cleaved bond.

Before releasing the saccharides, the glycoproteins are precipitated with an organic solvent such as ethanol (*Protocol 1*) to free them from substances that might inhibit the cleaving enzymes or cause artefacts on the electrofluorograms. Some glycoproteins may not be precipitated efficiently by ethanol when dissolved in water or dilute salt solutions but phosphate-buffered saline (PBS) has proved to be an effective solvent for a variety of glycoproteins. The precipitation of glycoproteins that are in low concentrations can be enhanced by the addition of bovine serum albumin (BSA) (1 mg/ml) as a carrier. However, each batch of BSA should be first tested alone for the presence of material that may introduce artefactual bands into the electrofluorogram.

For the saccharide release step, PNGase F works most effectively on glycoproteins that have been denatured, a state that can be achieved using SDS and 2-mercaptoethanol (see *Protocol 1*). The SDS is subsequently sequestered by adding Nonidet P-40 (a non-ionic detergent) to prevent it denaturing the PNGase F. When using endoglycosidases, the denaturation steps (*Protocol 1*, steps 6–10) are omitted. In general, consult suppliers' or review data for digestion conditions but avoid digestion buffers that contain primary amines since they may interfere with the subsequent fluorophore labelling.

The oligosaccharides of glycopeptides can also be released using *Protocol 1* after they have been purified in a salt-free medium. Isolation by HPLC using reverse-phase chromatography in a volatile eluent is a convenient method.

Only one enzyme, *O*-glycosidase, is available for the release of serine- or threonine-linked glycans (*O*-glycans). However, it has limited applicability, since it is capable of cleaving only the oligosaccharide Glcβ1–4GalNAc. *O*-glycans are usually more complex than this simple structure and need to be degraded enzymically *in situ* before the *O*-glycosidase is effective.

Protocol 1. Enzymatic release of *N*-glycans from 50–200 μg[a] of glycoprotein using PNGase F

Equipment and reagents

All reagents should be of analytical grade.

- Centrifugal vacuum evaporator (e.g. Jouan Ltd.) and associated vacuum pump and cold trap
- Microcentrifuge
- Oven at 37°C
- Ethanol: containing no denaturant (Merck Ltd.)
- Phosphate-buffered saline (PBS): 10 mM sodium phosphate buffer pH 7.2, 0.15 M NaCl
- Denaturing solution A: 1% (w/v) SDS, 0.1 M EDTA, 0.5 M 2-mercaptoethanol

- Denaturing solution B: 0.5% (w/v) SDS, 0.05 M EDTA, 0.25 M 2-mercaptoethanol
- Incubation buffer: 0.2 mM sodium phosphate buffer pH 8.6 at 37°C
- 7.5% (v/v) Nonidet P-40 (Calbiochem-Novabiochem Ltd.) solution in water
- PNGase F (EC 3.2.2.18) from *Flavobacterium meningosepticum*: recombinant form expressed in *E. coli* (Boehringer Mannheim) in solution; 1 U/5 μl (supplier's definition)

Method

1. Dry a volume of solution containing the glycoprotein (50–200 μg) in a 1.5 ml microcentrifuge tube using the centrifugal vacuum evaporator.

2. Dissolve or suspend the sample in 50 μl PBS.

3. Add 500 μl of ice-cold ethanol, vortex mix well, then leave on ice for at least 1 h.

4. Pellet the precipitated glycoprotein by centrifugation at 10 000 *g* for 2 min. Remove and discard the supernatant.

5. Dry the pellet in the centrifugal vacuum evaporator for 15 min.

6. Add 5 μl of denaturing solution A to the dry pellet and vortex mix well.[b]

7. Centrifuge briefly in the microcentrifuge to collect all the liquid at the tip of the tube.

8. Incubate at room temperature (22°C) for 30 min to allow contact with the concentrated denaturing solution.

9. Add 40 μl of incubation buffer and vortex mix well.

10. Heat the sample at 100°C for 5 min to fully denature the protein then allow to cool to room temperature.

11. Centrifuge briefly to collect the liquid at the tip of the tube, add 5 μl of 7.5% Nonidet P-40 solution, and vortex mix. Disregard the fact that the protein may not be fully in solution.

12. Add 5 μl (1 U; supplier's definition) of PNGase F solution, vortex mix, and centrifuge briefly to collect the suspension at the tip of the tube.

13. Agitate the tube gently to resuspend the protein as completely as possible whilst keeping the suspension at the tip of the tube. Incubate in an oven at 37°C for 18–20 h.

14. Add 165 μl (3 vol.) cold ethanol and leave on ice for at least 1 h to precipitate protein.

15. Centrifuge at 10 000 g for 2 min in a microcentrifuge and remove the supernatant, which contains the released glycans. Save the pellet for further analysis if required.

16. Dry the supernatant in the centrifugal vacuum evaporator. Do not heat above 45 °C to avoid degradation of the saccharides. The dried glycans are now ready for derivatization with fluorophore using the standard method (see *Protocol 3*). They can be stored at −20 °C.

---

[a] For small quantities of glycoproteins (1–50 μg) precipitate the protein from 20 μl of PBS, use 2 μl of denaturing solution B, and 0.2 of the volumes given above for the incubation buffer, NP-40, and PNGase F solutions. Use the reduced volume method for the fluorophore labelling (*Protocol 3*, footnote a).

[b] For digestion with an endoglycosidase, omit the denaturation steps 6–11, and substitute the appropriate digestion buffer and enzyme.

2.2 Release of saccharides using hydrazinolysis

Hydrazinolysis is the only chemical releasing method that produces glycans with a suitable reactive terminal saccharide. Chemical cleavage methods that generate non-reducing glycans, such as alkaline borohydride cleavage (which yields alditols) are not applicable to PAGEFS. Hydrazine can release from glycoproteins either *O*-glycans alone or both *N*- and *O*-glycans depending on the conditions (32–34). However, it requires a more careful preparation of the glycoconjugate before treatment compared with enzymatic release since the cleavage reaction is inhibited by the presence of either salts or water. The hydrazine and the reaction mixture must be handled so as to maintain anhydrous conditions and care is required since hydrazine is both toxic and volatile. However, it has been reported that glycan release with hydrazinolysis is nearer completion than when using enzymatic release (31–34). A disadvantage of hydrazinolysis is that during the procedure the acetyl group of the acetyl aminosaccharide residues is released and so it is necessary to re-acetylate the released glycans before fluorophore labelling. Hydrazinolysis is described in *Protocol 2*; (see also refs 31–34). The reducing glycan that is obtained is dried in a centrifugal vacuum evaporator and fluorophore-labelled as described in *Protocol 3*.

The glycans are released as acetohydrazide derivatives, that is, linked to hydrazine by a glycosylamine bond. During the re-*N*-acetylation the glycosylamine bond is cleaved and the glycans are mainly converted to the aldehydic reducing form which can then be labelled by the fluorophore. However, a small proportion of the released glycans may remain as the acetohydrazide derivative. Complete conversion to the aldehydic form can be ensured by incubating the dry released glycan residue (obtained in *Protocol 2*, step 10) in approx. 0.5 ml of 1 mM cupric (CuII) acetate in 1 mM acetic acid solution for

1 h. The volume of this solution is not critical. The copper ions are removed by passing the reaction mixture over a 0.5 ml Bio-Rad AG50-X12 cation exchange column (H$^+$ form) and the hydrazide-free glycans collected by repeating *Protocol 2*, steps 9 and 10.

Protocol 2. Release of glycans using hydrazinolysis

Equipment and reagents

Reagents should be of analytical grade.

- Centrifugal vacuum evaporator (e.g. Jouan Ltd.) and associated vacuum pump and cold trap
- Microcentrifuge
- Oven at 100 °C
- Screw-capped glass vial (1–2.5 ml having a Teflon cap liner), washed, preferably by sonication in 2 M nitric acid, rinsed with water, and dried in an oven
- Glass microsyringe with a Teflon piston (Hamilton), rinsed with dry methanol, and dried in an oven
- Hydrazine, in sealed vials (Sigma Chemical Co. Ltd.)
- Ethanol: containing no denaturant (Merck Ltd.)

- Dessicator containing phosphorus pentoxide
- Methanol, anhydrous (Aldrich Chemical Co. Ltd.)
- Toluene, anhydrous (Aldrich Chemical Co. Ltd.)
- 0.1% (w/v) trifluoroacetic acid solution
- 1% (v/v) acetic acid solution
- Saturated sodium bicarbonate solution, ice-cold
- Acetic anhydride (Sigma Chemical Co., Ltd.)
- Phosphate-buffered saline (PBS): 10 mM sodium phosphate buffer pH 7, 0.15 M NaCl
- Dry nitrogen gas (Merck Ltd.)
- Bio-Rad AG 50W-X12 cation exchange resin (H$^+$ form) (Bio-Rad Laboratories Ltd.)

Method

NB: hydrazinolysis according to the following method requires the use of hazardous reagents. It is therefore essential that the correct safety procedures be consulted by the researcher and followed carefully.

1. Dialyse the glycoprotein against either water, or 0.1% trifluoroacetic acid, or 1% acetic acid, at 4 °C and place the required volume, containing up to 0.1–20 mg of glycoprotein, in a screw-capped glass vial. Alternatively, using a similar glass vial, precipitate the glycoprotein twice from an aqueous solution with ethanol as described in *Protocol 1*, steps 2–4.

2. Dry the pellet in a desiccator containing phosphorous pentoxide for at least 24 h using a vacuum of 50 mm Hg or higher.

3. Gas the tube with nitrogen and add rapidly a minimum of 50 µl hydrazine from a freshly opened vial using a dry glass microsyringe. Do not allow the sample to be in contact with the atmosphere for more than 2 min. Gas the tube with nitrogen again and seal. The concentration of glycoprotein should be in the range 2–25 mg/ml hydrazine.[a]

4. Heat in an oven at 100 °C for 8–12 h.[b] For release of *O*-glycans only incubate at 60 °C for 5 h.

5. Open the reaction vial, place in a desiccator containing concentrated sulfuric acid, and remove the hydrazine by vacuum for 16 h. If any

remaining hydrazine is visible, add 100 μl of dry toluene and repeat the evaporation. Repeat this step if necessary.

6. Cool the reaction tube on ice, add 0.1 ml ice-cold saturated sodium bicarbonate solution (liquid phase) per mg original glycoprotein,[c] and dissolve the residue.

7. Acetylate all available amino groups by adding 50 μl acetic anhydride[d]/ ml of sodium bicarbonate solution, mix gently, and incubate for 10 min at room temperature. Repeat the addition of the acetic anhydride and the incubation a further three times to ensure complete acetylation.

8. Pass the re-acetylation reaction mixture down a column of Bio-Rad AG 50W-X12 cation exchange resin (H[+] form). Use a column volume equal to five times the volume of sodium bicarbonate.

9. Wash the column with five bed-volumes of water and combine the washings which contain the released glycans.

10. Dry the washings in the centrifugal vacuum evaporator.

11. Label the released glycans as described in *Protocol 3*.

[a] The presence of an excess of hydrazine should not cause any problems.
[b] The rate of glycan release may vary from sample to sample but by 10 h most glycans will have been released.
[c] Use a minimum of 0.1 ml saturated sodium bicarbonate.
[d] 50 μl of acetic anhydride is approx. 0.5 mmole which is approx. five times the molar quantity of amino groups on the amino acids generated by the degradation of 10 mg of glycoprotein by the hydrazine.

3. Labelling of saccharides with fluorophores

The initial step in the reductive amination of the saccharide with the fluorophore leads to the formation of a Schiff's base in a reaction that is reversible. The Schiff's base is therefore reduced with sodium cyanoborohydride to a stable secondary amine. The method shown in *Protocol 3* is for labelling with either one of the fluorophores ANTS or AMAC. The procedure for labelling with AMAC varies only slightly from that used for ANTS but the variation is critical and must be adhered to. For labelling quantities of less than 100 pmol of saccharides, use a variation of the standard protocol (11, 13, 22) as described in footnote a of *Protocol 3*.

The ANTS- and AMAC-labelling reactions have been shown to be virtually quantitative for glucose in the range 0.39–100 nmol per reaction (11, 13). No rigorous data has been published on the efficiency of labelling *N*-glycans. However, a detailed study (29) of the reductive amination of saccharides with anthranilic acid (available from Oxford Glycosystems Ltd.) (see also Section 5) indicates that this fluorophore will label 77–89% of the total *N*-glycans. In addition, Guttman *et al.* (35) have shown that labelling of GlcNAc with 1-aminopyrene-3,6,8-trisulfonate is virtually quantitative when using similar conditions for those described above for ANTS-labelling.

Protocol 3. The standard method[a] for either ANTS or AMAC labelling

Equipment and reagents

- Centrifugal vacuum evaporator (e.g. Jouan Ltd.) and associated vacuum pump and cold trap
- 0.15 M ANTS (disodium salt, Ubichem Ltd.) in acetic acid:water solution (3:17, v/v): dissolve the ANTS in water with warming (60°C), then add the required volume of acetic acid. Store at –70°C.
- Reducing solution for ANTS-labelling: 1 M sodium cyanoborohydride (NaCNBH₃) (Aldrich Chemical Co.) in DMSO—this solution is made immediately before use (within a few minutes). Do not store.

- Microcentrifuge
- Oven at 37°C
- Glycerol:water (1:4, v/v)
- 0.1 M AMAC (Lambda Fluoreszenzetechnologie Ges.mbh) in acetic acid:DMSO (3:17, v/v)—this solution can be stored at –70°C but appears to degrade at –20°C.
- Reducing solution for AMAC-labelling: 1 M sodium cyanoborohydride (NaCNBH₃) (Aldrich Chemical Co.) in water—make this solution immediately before use (within a few minutes). Do not store.
- DMSO:glycerol:water (2:1:7, by vol.)

Method

1. Place a suitable volume of saccharide solution in a microcentrifuge tube (1.5 ml or 0.5 ml capacity) and dry in a centrifugal vacuum evaporator. Do not heat above 45°C.

2. Add 5 μl of 0.15 M ANTS (or 0.1 M AMAC[b]) solution and mix.

3. Add 5 μl of the appropriate reducing solution (in DMSO for ANTS and water for AMAC[c]), mix, and centrifuge briefly to bring the reactants to the tip of the tube.

4. Incubate for 16 h at 37°C.

5. Dry the reaction mix in the centrifugal vacuum evaporator for about 1–2 h. Some heating will be required to evaporate the DMSO but avoid temperatures greater than 45°C. Store at –70°C.[d]

6. To prepare samples for electrophoresis, dissolve ANTS labelled samples in a suitable volume of glycerol:water solution[e] and store at –70°C. Dissolve AMAC-labelled samples in DMSO:glycerol:water (2:1:7, by vol.) and store at –70°C.

---

[a] This method can be used for quantities of saccharide in the range 0.1–100 nmol. For smaller amounts, use a 0.5 ml microcentrifuge tube, reduce the volume of both the ANTS (or AMAC) and NaCNBH₃ solutions to 1 μl, otherwise proceed as above. This is known as the reduced volume method (see *Protocol 1*, footnote a).

[b] It is important to ensure that the AMAC is made in acetic acid:DMSO.

[c] The reducing solution used for the AMAC-labelling **must** be made in water and is used without delay.

[d] It may not be possible to remove all of the DMSO completely but samples appear to be stable for several months when stored at –70°C. AMAC-labelled samples should be kept at room temperature as little as possible. These are noticeably unstable when stored at –20°C and artefactual bands become visible on the electrofluorograms.

[e] The volume of sample buffer should be adjusted to suit the experiment but 1–10 μl is typical. As a general guide, good fluorescent band patterns can usually be obtained from samples in which each band contains labelled saccharide in the range 5–400 pmol although with complex patterns, bands containing smaller quantities may be resolved more easily.

4. Electrophoretic standards

The electrophoretic standard that is used most commonly for PAGEFS is an ANTS-labelled, partial α-amylase digest of heat hydrolysed wheat starch (11, 27). This produces a ladder of fluorescent bands on an electrofluorogram each differing one from another by one glucose residue (162 Da). A method for the preparation of this standard is given in *Protocol 4* (see also ref. 11) and a typical result is shown in *Figure 2*. It is convenient to prepare a few aliquots simultaneously which will provide sufficient standard for several hundred gels.

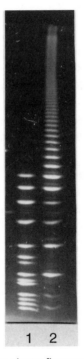

Figure 2. A photographic image of an electrofluorogram showing ANTS-labelled saccharide standards. The gel was a 20–40% linear acrylamide gradient and a discontinuous buffer system was used. Lane 1: a standard mixture of 14 ANTS-labelled saccharides chosen to give a wide spread of well-resolved bands. In order of decreasing mobilities, the order of the bands is: 6-deoxyglucose, glucose, galactose, *N*-acetylgalactosamine, galactosylgalactose, lactose, maltose, galactobiose, cellotriose, maltotriose, maltotetraose, maltopentaose, maltohexaose, and maltoheptaose. Lane 2: malto-oligosaccharide ladder generated through the partial digestion of wheat starch with α-amylase and labelled subsequently with ANTS. The degree of polymerization represented by each band can be determined from the matching positions of the standards in lane 1. Each band in the ladder is different from its immediate neighbours by a single glucose residue. (Adapted from ref. 11, with permission.)

Protocol 4. Preparation of fluorescent electrophoretic mobility
standard

Equipment and reagents

All reagents should be of analytical grade.

- Centrifugal vacuum evaporator (e.g. Jouan Ltd.) and associated vacuum pump and cold trap
- Oven at 37°C
- Ethanol: containing no denaturants (Merck Ltd.)
- Enzyme incubation buffer: 0.1 M ammonium acetate buffer pH 5.5 at 37°C
- Heat hydrolysed wheat starch (Sigma Chemical Co. Ltd.)
- α-amylase (EC 3.2.1.1) from *Bacillus amyloliquefaciens* (Boehringer Mannheim UK Ltd.)
- Glycerol:water solution (1:4, v/v)

Method

1. Suspend the heat hydrolysed wheat starch in the ammonium acetate buffer (10 mg/ml) by vortex mixing.

2. Dissolve the α-amylase in enzyme incubation buffer at 0.75 mg protein/ml.

3. Place 50 μl of heat hydrolysed wheat starch suspension in a 1.5 ml microcentrifuge tube, add 5 μl of α-amylase solution, and mix.

4. Incubate at 37°C for 30 min.

5. Stop the reaction by adding 1 ml of ice-cold ethanol. Dry in the centrifugal vacuum evaporator.

6. Use the standard conditions to label the dry digest with ANTS (see *Protocol 3*) but add 20 μl of both the ANTS solution and the sodium cyanoborohydride solutions.

7. Dry the reaction mixture in the centrifugal vacuum evaporator and store at −20°C. The dry product is stable for at least 12 months.

8. For electrophoresis, dissolve the digest in 250 μl of glycerol:water solution (1:4, v/v). A typical gel loading is 2 μl per lane. This solution can be stored at −20°C for short periods but is better kept at −70°C for the longer-term (months).

One should be aware that the α-amylase digestion can be variable and it is prudent to set-up a series of digestions either having varying digestion times or varying amounts of α-amylase. The aim should be to produce a mixture of glucose oligomers (malto-oligosaccharides) with a range of degrees of polymerization from 2 (maltose) to 20. Higher degrees of polymerization can sometimes be obtained if the digestion conditions are controlled sufficiently carefully.

The identities of each band (i.e. its degree of polymerization) can be determined by electrophoresing in an adjacent gel lane a sample of a known purified malto-oligosaccharide: a range from maltose to maltoheptaose is obtainable in purified form from Sigma Chemical Co. Ltd. The main bands of the ANTS-labelled wheat starch digest ladder have a yellowish fluorescence but sometimes blue fluorescent bands are displayed between the main bands. This phenomenon is probably owing to insufficient reduction of the initial labelling product (Schiff's base) and can be eliminated by doubling the volume of the labelling reagents. This problem may also occur if the sodium cyanoborohydride is not fully effective, for instance if the solution has not been made immediately before use.

Another standard, which consists of a mixture of 14 relatively simple saccharides, is also shown in *Figure 2*. This standard is useful for monitoring the effectiveness and correct operation of the electrophoretic system.

5. Polyacrylamide gel electrophoresis of ANTS- and AMAC-labelled saccharides

5.1 Electrophoresis of ANTS-labelled neutral and acidic saccharides and AMAC-labelled acidic saccharides

The standard method for PAGE of ANTS-labelled saccharides is described in *Protocol 5*. This method is also used for the analysis of acidic saccharides that have been labelled with AMAC but it should be noted that neutral saccharides that have been labelled with AMAC will not electrophorese in this system since the derivatives carry no charge at the pH of the buffer (however, see Section 5.2). The method shown is derived from the alkaline discontinuous electrophoresis buffer system described by Laemmli (36) with the SDS omitted throughout as it serves no purpose in the separation of the relatively small and readily water soluble fluorophore-labelled saccharides. The other major difference from methods for the separation of proteins is the relatively high acrylamide concentrations. Uniform gel concentrations in the range 20% to 40% (w/v) acrylamide are typical and a linear gradient from 20–40% was used successfully in the initial work (11). A method for a typical 30% (w/v) acrylamide uniform gel is given in *Protocol 5*. Lower gel concentrations may give better resolution of lower mobility saccharides and vice versa. When using AMAC labelling, it is advisable to use acrylamide concentrations of either 20% or 25% (w/v) since higher concentrations lead to very low electrophoretic mobilities of labelled saccharides especially when analysing neutral saccharides in the presence of borate buffers (see *Protocol 6*).

Standard electrophoretic equipment is used for the analysis but higher than normal voltages and current are used in order to obtain relatively rapid movement of the labelled saccharides through the dense gel. In order to minimize joule heating, the gel is kept relatively thin, typically 0.5 mm, and efficient cooling is used. It is important that the equipment ensures even cooling of both glass plates of the gel cassette. If one side of the gel is warmer than the other then loss of resolution can occur. The Hoefer SE600 apparatus described in *Protocol 5* is one of the few available that provides for even cooling of the gel plates and has the advantage of having some flexibility in format since both long (160 mm) and short (80 mm) glass plates are available. Commercial minigel equipment can be used but the resolution may not be optimal unless even cooling of the plates can be achieved.

Protocol 5. Electrophoresis of ANTS-labelled neutral and acidic saccharides and AMAC-labelled acidic saccharides

Equipment and reagents

- Electrophoresis apparatus: Hoefer SE600 (Pharmacia Biotech Ltd.) (the glass plates must be scrupulously cleaned in detergent and high purity water and polished with ethanol using a lint-free tissue since dust particles will fluoresce strongly).
- Glass plates for the gel cassette are made preferably of Pyrex or other low fluorescence glass (e.g. glass used for microscope slides) since this allows sensitive imaging of the gels without the need to remove them from the cassette. Normal window glass as supplied with the Hoefer apparatus is relatively fluorescent but can still be used. Pyrex glass cut to size can be obtained from Soham Scientific.
- Recirculating water cooler to maintain the electrophoresis tank buffer in the range 5–10°C.
- Power supply unit: this should be capable of delivering a minimum of 1000 V and 250 mA with either constant voltage or constant current.
- Disposable gloves: these must be powder-free
- 60% (w/v) acrylamide, 1.6% (w/v) *N,N'*-methylene-bisacrylamide

- Microsyringe (10 μl capacity) (Hamilton)
- Pipettor with flat-ended disposable tips (Sigma Chemical Co. Ltd.)
- Electrophoresis reagents from Merck Ltd. (use 'Electran' or analytical grade): store all the long-term stock solutions at –20°C or short-term stocks (days) in the dark at 4°C.
- 10% (w/v) ammonium persulfate: either made freshly or from a frozen stock.
- TEMED
- 4 × resolving gel buffer (1.5 M Tris–HCl buffer pH 8.8): mix 1.5 M Tris base (Trizma base, Sigma Chemical Co. Ltd.) and 1.5 M Tris–HCl (Trizma–HCl, Sigma Chemical Co. Ltd.) until pH 8.8 is obtained at 22°C.
- 10 × electrode buffer: 1.92 M glycine, 0.25 M Tris base pH 8.5 at 22°C.
- Electrophoresis marker dyes: bromophenol blue, xylene cyanol FF, thorin 1, direct red 75 (Aldrich Chemical Co. Ltd.)—use these dyes together in a mixture of 1 mg/ml of each in glycerol:water (1:4, v/v).
- Electrophoretic standard: ANTS labelled partial α-amylase digest of heat hydrolysed wheat starch (see Section 4).

Method

NB. Wear powder-free disposable gloves for all steps.

1. Dilute the stock electrode buffer tenfold to prepare a 1 × working solution and place the requisite volume in the anode tank of the apparatus. Cool to 5–10°C.

2. Set-up the gel cassette/mould using spacers 0.5 mm thick.

3. Prepare the resolving gel solution (final concentration is 30% (w/v) acrylamide) as follows. For 10 ml of solution, mix 2.5 ml resolving gel buffer (4 × concentrated), 5 ml 60% acrylamide, 1.6% bisacrylamide solution, 50 μl ammonium persulfate solution, and 2.5 ml water. Degas the solution under vacuum briefly. Then add 10 μl TEMED and mix well.

4. Pour the resolving gel mixture into the gel mould immediately after adding the TEMED. The height of the gel is typically 6–14 cm.[a] Leave 2 cm above the resolving gel for the stacking gel and sample wells.

5. Overlay the gel solution with water to a depth of about 1 cm but avoid, as far as possible, allowing the water to mix with the acrylamide. Allow the solution to polymerize (about 10 min).

6. Prepare the stacking solution (final concentration is 3.78% (w/v) acrylamide) as follows. For 10 ml of solution, mix 2.5 ml resolving gel buffer (4 × concentrated) (a special stacking gel buffer is not required), 0.63 ml stock 60% acrylamide, 1.6% bisacrylamide solution, 100 μl ammonium persulfate solution, and water to a total volume of 10 ml. Add 10 μl TEMED and mix well. Degassing is not necessary.

7. Within 20 min of the resolving gel setting, pour off all of the water and pour the stacking gel solution into the gel mould so as to fill all the space above the resolving gel. Delay at this step will cause bubbles to form at the interface between the two gels as the stacking gel polymerizes.[b]

8. Generate the sample wells by inserting into the stacking gel solution a suitable comb containing typically 8–12 teeth for each 8 cm of gel width. As far as possible, avoid trapping air bubbles; remove them by tapping the glass or agitating the comb.

9. Let the stacking gel acrylamide polymerize for at least 30 min. Remove the comb immediately prior to sample loading.

10. Fill the sample wells with 1 × electrode buffer.

11. Load the samples under the buffer in the wells using either a microsyringe or a small volume disposable pipette tip. Typically load 1–10 μl per well. The sample solutions will sink onto the gel.

12. Load 1–2 μl of both the electrophoretic standard and the marker dyes together in the same outer wells at both sides of the gel.

13. Assemble the electrophoretic apparatus according to the supplier's instructions. Ensure that there is even cooling of each gel plate by stirring the anodic electrode buffer.

14. Add the diluted stock electrode buffer to the cathode compartment taking care not to disturb the samples.

15. Turn on the current to start the electrophoresis. For a gel with the dimensions 0.5 mm × 140 mm × 140 mm, use 100 V for 30 min, then increase to 500 V for 30 min, and finally 1000 V for approx. 120 min. Hold all the voltages constant.

16. When the thorin 1 (orange in colour) marker dye has reached the base

Protocol 5. *Continued*

of the gel, turn off the current and view the electrofluorogram. When AMAC-labelled saccharides are being analysed, the excess AMAC in the reaction mixture remains at the origin in the well and is strongly fluorescent. Therefore this must be removed by flushing each well with water using a syringe before imaging.

[a] In general the longer the gel the greater the resolution.
[b] The resolving gel may peel away from the glass at the edges of the cassette. This can be ignored and the central unblemished area used. For small gels (0.5 mm × 140 mm × 80 mm) (thickness × width × length) the solutions are not degassed before polymerization but for large gels (140 mm long) degassing will improve the adhesion of the gel to the glass. Equipment that allows the use of thin glass plates instead of the 3 mm glass used in the Hoefer apparatus tends to reduce the peeling of the gel from the glass.

The precise duration of the electrophoretic run will depend on the gel length, buffer system, acrylamide concentration, and voltage used but is typically in the range 1–4 h. Thorin 1 has a similar mobility to unreacted ANTS which appears as the strongest fluorescent band moving through the gel ahead of the labelled saccharides. ANTS-labelled maltotetraose has a mobility slightly greater than the bromophenol blue. Oxford Glycosystems Ltd. has recently introduced another PAGEFS fluorophore, anthranilic acid (29), in their commercially available Glyco Map method which apparently has electrophoretic separations at least as good as ANTS. This system requires no external cooling equipment.

In general, the charge-to-mass ratio of labelled saccharide determines its mobility but other factors also have an effect since various isomers such as either ANTS- or AMAC-labelled maltose (Glcα1–4Glc), isomaltose (Glcα1–6Glc), and cellobiose (Glcβ1–4Glc) can be separated. It appears that either the effective sizes of the labelled saccharides differ or there is some interaction between the saccharides and the gel which affects their mobilities (11). This latter seems to be the most likely explanation for the separation of the epimers glucose and galactose.

5.2 Electrophoresis of AMAC-labelled neutral saccharides

In order to analyse AMAC-labelled neutral saccharides, a buffer is required that contains borate ions that associate with the saccharides and confers on them negative charges (13). This method is described in *Protocol 6* using a 20% (w/v) acrylamide gel.

In this system good resolution of small neutral saccharide isomers can be obtained but they electrophorese much more slowly than AMAC-labelled acidic saccharides which will also move in the system. The bromophenol blue marker will move ahead of most acidic N-linked glycans. Sialic acids can be electrophoresed in the system. However, AMAC-labelling of N-acetylneuraminic acid (NeuAc) generates more than one fluorescent band one of which appears close to neutral monosaccharide bands; the exact identities of these products have not yet been reported. NeuAc shows no significant reaction with ANTS.

Protocol 6. Separation of neutral and acidic saccharides labelled with AMAC

Equipment and reagents

- For equipment see *Protocol 5*
- 10 × electrode buffer: 1 M Tris base (Trizma base, Sigma Chemical Co. Ltd.), boric acid pH 8.3—add solid boric acid (approx. 60 g/litre) to approx. 1.5 M Tris base solution until the pH is 8.3, then dilute the solution to give a final concentration of 1 M Tris; store at 4°C.
- Electrophoresis marker dye: 1 mg/ml bromophenol blue (Aldrich Chemical Co. Ltd.) in glycerol:water (1:4, v/v)
- For other reagents see *Protocol 5*

Method

1. Follow the instructions in *Protocol 5* but *do not* use the buffers described in that protocol. Instead use a final concentration of 0.1 M Tris base–borate/boric acid pH 8.3 for both electrode compartments, and for both the stacking gel and the resolving gel.

2. Use a uniform concentration 20% (w/v) acrylamide resolving gel and a 4% (w/v) acrylamide 'stacking gel'.

3. When loading the samples, the volumes should be no greater than 2 μl in an 8 mm wide sample well.[a]

4. Electrophorese the samples until the bromophenol blue reaches the base of the gel. For an 80 mm long resolving gel typical electrophoretic conditions are 250 V for 30 min followed by 500 V for 120 min, although much longer times (several hours) are sometimes needed for some separations particularly of the large neutral saccharides; the voltages are held constant.

5. Rinse out the sample wells with water from a syringe at the end of the electrophoresis to remove unreacted AMAC that remains in the wells during the electrophoresis.

6. View the electrofluorogram as described in Section 7.

[a] Keep the sample volume to a minimum since this method does not use a discontinuous buffer system and the resolution is dependent on the depth of the sample in the wells. The low concentration, so-called 'stacking gel', serves to inhibit band distortion.

5.3 Oligosaccharide profiles

The method described in *Protocol 5* has been shown to be ideal for producing OSPs of ANTS-labelled acidic and neutral glycans and acidic AMAC-labelled glycans. It is of particular use for the analysis of multiple samples which can be electrophoresed simultaneously. Examples of the oligosaccharide profiles of ANTS-labelled, and acidic AMAC-labelled saccharides are shown in *Figures 3* and *4* respectively. The method is highly sensitive, requiring glycans released from as little as 2 μg of glycoprotein, although it is more typical to analyse quantities of glycoprotein in the region of 20 μg. The separations are notable for their high resolution: for instance two isomeric sialylated triantennary *N*-glycans, that differ only in the linkage position (position 3 instead of 6) of one of the sialic acid residues, can be separated after ANTS labelling (23).

Since ANTS and AMAC have some different properties, they can be used in a complementary way. ANTS is used for producing total OSPs since both neutral and acidic *N*-glycans are detected. AMAC can be used to distinguish

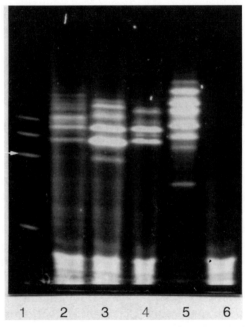

Figure 3. Graphics display image of a gel of ANTS-labelled oligosaccharides cleaved from various glycoproteins using PNGase F. The gel was imaged for 10 sec using a CCD imager. The length of the resolving gel was approx. 85 mm and the position of maltopentaose in the malto-oligosaccharide standard ladder is indicated by an *arrow* on the left-hand side. Lane 1: malto-oligosaccharide standard derived from hydrolysed wheat starch. Lane 2: ovalbumin. Lane 3: ribonuclease B (bovine pancreas). Lane 4: apo-transferrin (human). Lane 5: trypsin inhibitor (ovomucoid from chicken egg white). Lane 6: control. Each lane contains oligosaccharides derived from 10 mg of glycoprotein except for trypsin inhibitor in which 4 mg was analysed and the control in which no glycoprotein was added. A greater number of bands can be revealed in each sample and in the malto-oligosaccharide standard by adjustment of the grey scaling of the image on the graphics display. (From ref. 22, with permission.)

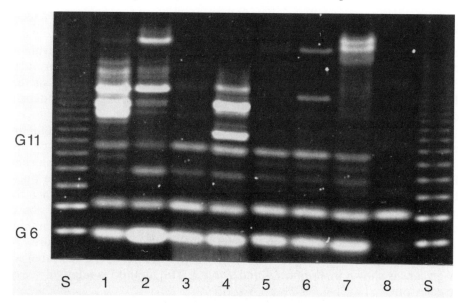

Figure 4. A composite image of two electrofluorograms showing the separation of the AMAC derivatives of glycans released from various glycoproteins. Lanes S: standard fluorescent, ANTS-labelled wheat starch digest marker as described in *Figure 2*. Lane 1: α1-acid glycoprotein (human). Lane 2: α1-antitrypsin (human plasma). Lane 3: BSA. Lane 4: fetuin (fetal calf serum). Lane 5: ribonuclease B (bovine pancreas). Lane 6: apo-transferrin (human). Lane 7: trypsin inhibitor (ovomucoid from chicken egg white). Lane 8: control without protein. The glycans from approx. 10 mg of protein were analysed in each lane. The gel was imaged for 10 sec. The bands with mobilities greater than G11 appear to be artefactual. (From ref. 25, with permission.)

unequivocally between acidic and neutral saccharides since AMAC labelled neutral saccharides do not electrophorese in the buffer system described in *Protocol 5*. AMAC-labelled neutral and acidic saccharides of similar size tend to separate well in the borate buffer described in *Protocol 6*.

5.4 Preparative elution of fluorophore-labelled saccharides from gels

In order to carry out further analyses, labelled saccharides may be eluted preparatively from gels. Fluorescent bands of interest are excised from the gel while viewing using a low intensity UV lamp (Mineralite; UVP International Ltd.). The gels are boken with a spatula or similar instrument and eluted by shaking in the dark with either water for 3 h or longer, for ANTS-labelled saccharides, or DMSO:water (1:4, v/v) for AMAC-labelled saccharides (26, 27). One should be aware that high intensity UV light tends to cause the fluorescence to fade and appears to cause the fluorophore-labelled saccharides to cross-link to the gel.

6. Enzymatic structural analysis of *N*-glycans

Structural analysis of *N*-glycans can be carried out by sequential enzymatic degradation of both fluorophore-labelled and unlabelled *N*-glycans (21, 23–27). Considerable structural data may be deduced from analysis of the PAGEFS oligosaccharide profiles obtained from the resulting enzyme degradation products.

Individual glycans that are to be analysed are purified either by chromatographic methods, either before or after fluorophore labelling, or by elution from a preparative scale PAGEFS gel after excision of the relevant fluorescent band (see Section 5.4). Small quantities (picomoles) of a purified fluorophore-labelled or unlabelled glycan can be degraded either by any suitable exoglycosidase acting individually or by mixtures of these enzymes. However, the selection of the enzymes and the order in which they are used requires some assumptions concerning the glycan structure. A degradation scheme designed for commonly occurring ANTS-labelled *N*-glycans is shown in *Protocol 7* and *Table 1*. The method is relatively easy to carry out.

Table 1. Format for the multiple sequential enzymic digestion of an ANTS-labelled *N*-glycan

Enzyme	Reaction tube					
	A	**B**	**C**	**D**	**E**	**F**
Neuraminidase	–	+	+	+	+	+
β-galactosidase	–	–	–	+	+	+
N-acetylhexosaminidase	–	–	–	–	+	+
α-mannosidase	–	–	–	–	–	+

Protocol 7. Enzymatic structural analysis of ANTS-labelled *N*-glycans

Equipment and reagents

All reagents should be analytical grade.

- Centrifugal vacuum evaporator (e.g. Jouan Ltd.) and associated vacuum pump and cold trap
- Oven or heating block at 37 °C
- Microcentrifuge
- Neuraminidase[a] (*Arthrobacter*; Calbiochem-Novabiochem Ltd.): 5 mU/ml in 0.2% (w/v) bovine serum albumin solution
- β-galactosidase[a] (Jack bean; Seikagaku Corporation): 10 mU/ml in water

- *N*-acetylhexosaminidase[a] (Jack bean; Dextra Laboratories Ltd.): 10 mU/ml, use without dilution
- α-Mannosidase[a] (Jack bean; Dextra Laboratories Ltd.): 168 mU/ml, use without dilution
- Enzyme incubation buffer: 0.2 M sodium citrate buffer pH 4.5 at 37 °C
- Glycerol:water solution (1:4, v/v)
- ANTS-labelled purified *N*-glycan sample

Method

1. Dry a solution of 5 pmol or more of the ANTS-labelled purified *N*-glycan to be analysed in each of six, 0.5 ml microcentrifuge tubes (labelled A–F) using the centrifugal vacuum evaporator.

2. Add 45 μl of enzyme incubation buffer to each tube and mix.

3. Dilute each stock enzyme solution, where necessary, to the required concentration.

4. Add 1 μl of each enzyme solution to the reaction tubes as indicated in *Table 1*. Vortex mix and centrifuge briefly to collect the reactants at the tip of each tube.

5. Incubate at 37 °C for 16–20 h then dry in the centrifugal vacuum evaporator.

6. Add 3 vol. of ethanol to each sample and leave on ice for 1 h to precipitate the protein.

7. Centrifuge at 10 000 *g* for 2 min. Recover the supernatant which contains the labelled glycans and cleaved unlabelled monosaccharides.

8. Dry the supernatant in the centrifugal vacuum evaporator.

9. Add glycerol:water solution (1:4, v/v) to bring the concentration of labelled saccharide to 1–10 pmol/μl and mix by vortexing.

10. Centrifuge briefly to collect the solution at the tip of the tube.

11. Apply 2.5 μl of the sample containing, preferably, at least 5 pmol of glycan to the electrophoresis gel.

12. Electrophorese as described in *Protocol 5* for ANTS labelled saccharides.[b,c]

[a] These and other glycosidases can be obtained from other suppliers such as Oxford Glycosystems Ltd., Boehringer Mannheim Gmbh, and New England BioLabs Inc.
[b] AMAC-labelled saccharides can also be analysed. Use *Protocol 5* for electrophoresis of AMAC-labelled acidic glycans, and *Protocol 6* for AMAC-labelled neutral glycans.
[c] If the sample solution is very viscous, add 3 vol. of ethanol to each enzyme digest to stop the reaction and also precipitate the protein. Leave on ice for 1 h, centrifuge at 10 000 *g* for 2 min, remove the supernatant which contains the digested saccharides, and dry this in the centrifugal vacuum evaporator. Dissolve the residue in glycerol:water (1:4, v/v) for electrophoresis as described in *Protocol 5*.

An electrofluorogram of an enzymatic structural analysis of a typical ANTS labelled *N*-glycan is shown in *Figure 5*. The glycan was eluted previously from a preparative gel and concentrated in a centrifugal vacuum evaporator prior to the analysis. Any other suitable purification method can be used providing that any substance present in the purified glycan solution is compatible with the enzyme activity. No problems have been found due either to impure enzymes or incomplete digestion, although the latter occasionally occurs for α-mannosidase which is relatively slow in its cleavage of the Manα1–3Man–chitoboise produced as a partial product of the cleavage of Manα1–6

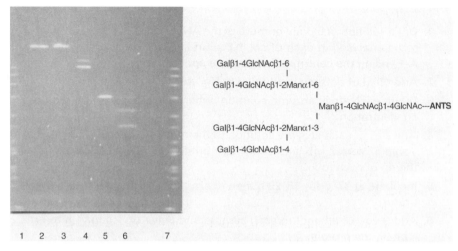

Galβ1-4GlcNAcβ1-6
   |
Galβ1-4GlcNAcβ1-2Manα1-6
         |
            Manβ1-4GlcNAcβ1-4GlcNAc---**ANTS**
        |
Galβ1-4GlcNAcβ1-2Manα1-3
       |
Galβ1-4GlcNAcβ1-4

Figure 5. An image of an electrofluorogram showing analytical enzymic digestions of the *N*-glycan, the structure of which is shown on the *right*. Lanes 1 and 7 show a series of ANTS-labelled saccharide markers. Lanes 2–6 each contain approx. 10 pmol of the ANTS-labelled oligosaccharide which had been treated with a series of various exoglycosidases as follows: lane 2, no enzyme added; lane 3, neuraminidase; lane 4, neuraminidase and β-galactosidase; lane 5, neuraminidase, β-galactosidase, and *N*-acetylhexosaminidase; lane 6, neuraminidase, β-galactosidase, *N*-acetylhexosaminidase, and α-mannosidase. (Adapted from ref. 21, with permission.)

(Manα1–3)Man–chitobiose. Two products may therefore occur, Manα1–3Man–chitoboise and Man–chitobiose, which lead to the two observed bands in lane 6 of the electrofluorogram in *Figure 5*. The cleaved monosaccharides were not detected in this study since there is no post-degradation labelling. The structure is determined in part from a knowledge of the specificity of the enzymes present.

The number of glycan antennae can be determined by comparison of the mobility change caused by the digestion with the mobility change occurring in known glycan standards. Glycan standards can be obtained from Oxford Glycosystems Ltd. and Dextra Laboratories Ltd. A method has been described for the limited enzymatic analysis of mixtures of AMAC-labelled acidic glycans in a non-borate containing buffer system (25). However, after the removal of any sialic acid residues the fluorophore-labelled glycans are no longer charged and so further analysis would require the use of a borate buffer system (see *Protocol 6*).

Alternatively, glycans can be fluorophore-labelled after the enzyme digestion (11, 12). In this case the released monosaccharides are labelled simultaneously with the glycan. Both the glycan and the saccharide are visible on the electrofluorogram and their quantitative ratios can be determined to give data on the number of antennae in the glycan.

7. Viewing and imaging

The electrofluorograms of both ANTS- and AMAC-labelled saccharides are viewed or recorded photographically or electronically when illuminated on a UV transilluminator with maximum emission at wavelengths of either 254 nm, 302 nm, or 365 nm. UV transilluminators are available from numerous suppliers; a model with the maximum linear dimensions available should be used to facilitate accurate quantitation (see below). Gels can be recorded either whilst still in their glass electrophoresis cassettes or after their removal. Viewing with the gel in the cassette is very simple and convenient and obviates the hazard of handling the gels. However, the glass supplied with most equipment is relatively fluorescent, especially when viewed at 254 nm or 302 nm, and it is preferable to use low fluorescence glass (see *Protocol 5*).

The Polaroid MP4 camera system and Polaroid type 55 film will give not only a positive photographic print but also a 4 × 5 inch, high quality negative that can be used for densitometric quantitation of bands. A typical exposure is 60 sec at f4.5 when using a transilluminator with a power of 7 mW/cm^2. There is a consequent fading of the fluorescence of about 10% when exposing for 60 sec. Workers must ensure that they are properly protected during the photography. As little as 1 pmol per band can be detected and quantitation in the range 10–500 pmol is possible (11).

Gels can also be imaged using electronic cameras based on charge-coupled devices (CCD). The highest sensitivity is obtained using integrating, cooled, CCD cameras with slow scan readout of the image. There are numerous suppliers of these cameras such as AstroCam Ltd., Digital Pixel Ltd., and Wright Instruments Ltd. These cameras are designed to have low levels of noise and a wide dynamic range and can be supplied with sophisticated software. However, they are expensive. A cheaper but effective alternative is one of the gel documentation systems that are commonly used for imaging ethidium bromide stained DNA gels. These are available with easy-to-use quantitation analysis software. It is advisable to use a camera with the largest CCD available in order to optimize resolution and therefore picture quality. Typical systems are available from Flowgen Instruments Ltd., and Pharmacia Biotech Ltd. These have a narrower dynamic range than the more expensive cameras but can operate either in integrating mode or in video mode. In video mode the sensitivity and image quality are inferior but it is a useful facility for focusing the image. CCD cameras that operate only in video mode should be avoided. As little as 0.2 pmol per band can be detected with the most sensitive CCD cameras (11, 13).

When photographing the electrofluorograms, use a Wratten (Kodak) gelatin filter; either number 8 or alternatively a combination of Wratten 2A and 12. When imaging using a CCD camera it is advisable to use an interference filter designed to allow optimum transmission at the maximum emission wavelengths of the ANTS (515 nm) and AMAC (520 nm). Such a filter will reduce the background to a minimum and will provide the highest sensitivity when used with the cooled, slow scan CCD cameras. Interference filters designed for the imaging of fluorescein are available commonly and can be used for both ANTS and AMAC. Information on suitable interference filters can be obtained either from the camera suppliers, or from either Omega Optical or Glen Spectra Ltd., who supply such filters.

It is important to ensure that the illumination of the gels is as uniform as possible especially when photographic negatives are to be scanned densitometrically. All transilluminators are less intense towards their peripheries than at the centre so that bands containing the same quantity of fluorescent material will fluoresce more strongly if they are near the centre of the transilluminator as compared with similar bands at the periphery. To diminish this effect to acceptable levels, use a transilluminator that has linear dimensions that are at least twice those of the gel. When imaging electronically, the uniformity of the exciting light does not need to be so carefully controlled since it is possible to correct for non-uniformity by a process called flat-fielding which has the effect of simulating a uniform illumination across the gel. Flat-fielding should not be confused with background subtraction; the latter allows for fluorescence in the gel but not for non-uniformity of the transilluminator. Flat-fielding programs are usually available with the more expensive imaging systems.

For quantitative analysis, known quantities of glucose (or lactose or maltopentaose), are included in separate gel lanes so that a calibration curve can be drawn (11, 13). It has been shown that ANTS-labelling of glucose in the range 0.39–100 nmol per reaction tube is quantitative when using the standard conditions described in *Protocol 3*. It should be noted that the photographic response is non-linear but the electrophoresis of suitable standards enables good quantitation over the range 10–500 pmol per band (11, 13). The CCD response of the AstroCam camera is linear over the range 12–500 pmol per band (11).

In summary there are advantages and disadvantages for both photography and electronic imaging (see *Table 2*) but on balance the ease of electronic data

Table 2. Comparison of digital (CCD) and photographic methods for imaging electrofluorograms[a]

Digital camera (CCD)	Photography
Greater sensitivity (five- to tenfold)	Higher resolution
Easier to obtain quantitative results	Less expensive to buy
Wider linear dynamic range	Available commonly
Easy electronic image storage and retrieval	
Easy image enhancement	
Less expensive to record individual images	

[a] From ref. 26, with permission.

storage, image manipulation, and quantitation makes CCD-based cameras the method of choice.

References

1. Rademacher, T.W., Parekh, R.B., and Dwek, R.A. (1988). *Annu. Rev. Biochem.*, **57**, 755.
2. Varki, A. (1993). *Glycobiology*, **2**, 97.
3. Kobata, A. (1992). *Eur. J. Biochem.*, **209**, 483.
4. Welply, J. K. (1989). *Trends Biotechnol.*, **7**, 5.
5. Laine, R.A. (1990). In *Methods in enzymology* (ed. J.A. McCloskey), Vol. 193, p. 539. Academic Press, London.
6. Spellman, M.W. (1990). *Anal. Chem.*, **62**, 1714.
7. Vapnek, D. (1990). *Glycobiology*, **1**, 3.
8. Varki, A. (1991). *FASEB J.*, **5**, 226.
9. Kobata, A. and Takasaki, S. (1994). In *Glycobiology: a practucal approach* (ed. M. Fukuda and A. Kobata), p. 165. IRL Press, Oxford.
10. Dell, A., Khoo, K-H., Panico, M., McDowell, R.A., Etienne, A.T., Reason, A.J., *et al.* (1994). In *Glycobiology: a practical approach* (ed. M. Fukuda and A. Kobata), p. 187. IRL Press, Oxford.
11. Jackson, P. (1990). *Biochem. J.*, **270**, 705.
12. Jackson, P. and Williams, G.R. (1990). *Electrophoresis*, **12**, 94.
13. Jackson, P. (1991). *Anal. Biochem.*, **196**, 238.
14. Lee, K-B., Al-Hakim, A., Loganathan, D., and Linhardt, R.J. (1991). *Carbohydr. Res.*, **214**, 155.
15. Stack, R.J. and Sullivan, M.T. (1992). *Glycobiology*, **2**, 85.
16. Jackson, P. and Williams, G.R. (1988). *Analysis of carbohydrates*. Patent Publication No. WO88/10422.
17. Jackson, P. (1991). *Analysis of carbohydrates*. Patent Publication No. WO91/05256.
18. Jackson, P. (1991). *Treatment of carbohydrates*. Patent Publication No. WO91/05265.
19. Jackson, P. (1992). *Analysis of carbohydrates*. Patent Publication No. WO92/11531.
20. Jackson, P. (1993). *Analysis of carbohydrates*. Patent Publication No. WO93/02356.
21. Jackson, P. (1993). *Biochem. Soc. Trans.*, **21**, 121.
22. Jackson, P. (1994). In *Methods in enzymology* (ed. W.J. Lennarz and G.W. Hart), Vol. 230, p. 250. Academic Press, London.
23. Jackson, P. (1994). *Anal. Biochem.*, **216**, 243.
24. Jackson, P. (1994). In *Advances in electrophoresis* (ed. A. Chrambach, M.J. Dunn, and B. Radola), Vol. 7, p. 225. VCH, Weinheim.
25. Jackson, P. (1994). *Electrophoresis*, **15**, 896.
26. Jackson, P. (1996). *Mol. Biotechnol.*, **5**, 101.
27. Jackson, P. (1997) In *A laboratory guide to glycocojugate analysis* (ed. P. Jackson and J.T. Gallagher), p. 113. Birkhauser Verlag, Basel.
28. Hu, G-F. and Vallee, B.L. (1994). *Anal. Biochem.*, **218**, 185.

29. Bigge, J.C., Patel, T.P., Bruce, J.A., Goulding, P.N., Charles, S.M., and Parekh, R.B. (1995). *Anal. Biochem.*, **230**, 229.
30. O'Shea, M.G. and Morell, M.K. (1996). *Electrophoresis*, **17**, 681.
31. Tarentino, A.L. and Plummer, T.H., Jr. (1994). In *Methods in enzymology* (ed. W.J. Lennarz and G.W. Hart), Vol. 230, p. 44. Academic Press, London.
32. Kobata, A. and Endo, T. (1993). In *Glycobiology: a practical approach* (ed. M. Fukuda and A. Kobata), p. 79. IRL Press, Oxford.
33. Patel, T.P. and Parekh, R. (1994). In *Methods in enzymology* (ed. W.J. Lennarz and G.W. Hart), Vol. 230, p. 57. Academic Press, London.
34. Patel, T., Bruce, J., Merry, A., Bigge, C., Wormald, M., Jaques, A., *et al.* (1993). *Biochemistry*, **32**, 679.
35. Guttman, A., Chen, F-T.A., Evangelista, R.A., and Cooke, N. (1996). *Anal. Biochem.*, **233**, 234.
36. Laemmli, U.K. (1970). *Nature*, **227**, 680.

10

Analysis of protein:protein interactions by gel electrophoresis

VINCENT M. COGHLAN

1. Introduction

With the application of modern techniques for ultrastructural imaging and macromolecular characterization has come an appreciation for the high degree of structural organization found within cells. An emerging theme in cellular organization is that for many metabolic pathways, cells utilize large multienzyme complexes that are maintained through specific protein:protein interactions (1). Defining these interactions has become, and will remain, an essential part of the functional characterization of proteins.

Several methods exist for the analysis of protein interactions, each with distinct advantages and disadvantages (2). Interaction cloning methods such as the yeast two-hybrid system are well suited for initial identification of binding partners, but require secondary methods to verify complex formation and to determine the strength of the interaction. More quantitative methods such as equilibrium dialysis, fluorescence polarization, and surface plasmon resonance are useful for measuring binding affinities, but typically require purified or modified proteins and elaborate (expensive) equipment.

Some of the most useful methods for characterizing multiprotein complexes employ gel electrophoresis and these are the subject of this chapter. In general, gel-based methods are rapid, sensitive, and can be performed using equipment commonly found in most biochemical laboratories. Several protocols are described for isolating protein complexes, identifying binding partners, and characterizing protein interactions. The success of a particular protocol depends largely on the nature of the complex under investigation and quite often rules of trial and error prevail. Indeed, some complexes associate weakly or transiently and are not amenable to analysis by gel electrophoresis. Another factor to consider is the availability of specific reagents: some methods require purified antibodies or recombinant protein. An attempt has been made to provide protocols that can be adapted to many different experimental situations from the characterization of large multiprotein complexes to the study of intramolecular interactions.

2. Isolation of protein complexes

Several approaches are available for isolating protein complexes from solutions. Most techniques involve specific purification of one component of a complex under conditions where associated proteins co-purify. This is often achieved using antibodies or affinity chromatography to isolate the target protein. To ensure that binding partners remain associated throughout the purification protocol, chemical cross-linking is often used to stabilize complexes. Each technique relies on gel electrophoresis as a means to monitor the purification procedure and to characterize the components that make up the protein complex.

2.1 Co-immunoprecipitation

In cases where antibodies to one component are available, specific complexes can be immunoprecipitated from protein solutions. In a typical experiment (e.g. *Protocol 1*), antibody against one of the proteins of a complex is incubated with a crude cellular extract. Protein A linked to agarose beads is then added to precipitate the antibody with bound antigen and antigen-containing complexes. The precipitates are washed to remove non-specific contaminants, eluted, and subjected to denaturing gel electrophoresis to separate complexes into individual components that are characterized by molecular weight or identified by Western blotting. For this technique to be successful, the following comments should be considered:

(a) The starting material should be chosen carefully. When using extracts of animal tissue for immunoprecipitations, be aware of potential species cross-reactivity of antibodies and secondary reagents. For cultured cells, it is advantageous to radiolabel the proteins metabolically (e.g. using [^{35}S]methionine) to facilitate subsequent analysis. If one component is being overexpressed in cells and antibodies are not available, the incorporation of an epitope tag into the construct provides a means for specific precipitation. If the cellular compartment of a complex is known, or can be inferred, the use of purified fractions may increase yields. When known proteins are suspected to associate, recombinant preparations can be mixed and used as starting material. This method has an advantage in that one protein can be radiolabelled or affinity-tagged for subsequent identification. Keep in mind, however, that important cellular factors or post-translational modifications may be absent in recombinant preparations. Due to non-specific aggregation, the use of *in vitro* translated proteins for co-immunoprecipitation experiments is not recommended.

(b) The extract must be prepared carefully. While buffered saline solutions work well in many immunoprecipitation experiments, the stability or accessibility of some complexes may be affected by divalent cations, detergents, or other factors. For example, the addition of Triton X-100 (up to 1%) may help solubilize some complexes while disrupting others. Buffers should contain protease inhibitors to help minimize degradation. Irrespective of the extraction buffer used, it is critical to avoid excessive dilution of the extract. Significant dissociation of biologically relevant complexes can occur at dilutions greater than twofold (3). In some cases, the presence of a volume-excluding polymer (e.g. polyethylene glycol) may increase yields through molecular crowding (4). The method of extraction (e.g. sonication, freeze–thaw, etc.) can also affect the stability of protein complexes.

(c) The primary antibody must be specific. Affinity purified polyclonal antibodies work well for most immunoprecipitations. Monoclonal antibodies can also be used, but be aware that some epitopes may be unavailable for reaction when the antigen is bound with other proteins in a complex. In the most straightforward approach, the antibodies are directly coupled to agarose beads. Alternatively, primary antibodies can be used in combination with commercially available secondary precipitating reagents. These usually consist of anti-IgG antibodies, protein A, or protein G covalently linked to agarose or magnetic beads. In all cases, control experiments should be performed to demonstrate that the complex is not precipitated using pre-absorbed or pre-immune sera.

(d) The detection method must be sensitive. The amount of complex precipitated may be too small to detect by conventional gel staining methods. While results can vary, it has been estimated that for a standard 50 kDa protein, silver staining (Chapter 2) is practical for detection of complexes with dissociation constants in the nanomolar range (2). Radiographic methods can increase sensitivity 1000-fold and should be used where feasible (e.g. metabolically labelled cells). For identification of specific components, Western blot analysis is quite useful, but keep in mind that secondary antibodies may cross-react with the IgG used for immunoprecipitation, resulting in large background bands in the 50–55 kDa range (IgG heavy chain; 70–75 kDa for IgM). This problem can be circumvented by using species-specific secondary antibodies or labelled primary antibodies for Western blots. When immunoprecipitations are performed using primary antibodies directly coupled to beads, elutions should be performed in the absence of reducing agents to eliminate heavy and light chain contamination. Finally, combining immunoprecipitation with chemical cross-linking may increase yields by stabilizing some complexes (see Section 2.3).

Protocol 1. Immunoprecipitation of protein complexes from cultured cells

Equipment and reagents

- Orbital mixer (Nutator, Becton-Dickinson)
- Extraction buffer: 50 mM Tris–HCl pH 7.5, 100 mM NaCl, 0.1% Triton X-100, 1 μg/ml leupeptin, 1 μg/ml pepstatin, 1 mM benzamidine, 0.1 mM AEBSF (2-aminoethylbenzenesulfonyl fluoride; Calbiochem).
- Elution buffer: 50 mM Tris–HCl pH 7.5, 1 mM EDTA, 1% SDS
- Protein A–agarose beads (Sigma)
- SDS–PAGE sample buffer and SDS–polyacrylamide slab gel (see Chapter 1)

Method

1. Grow the cells in the presence of [^{35}S]methionine under conditions where the protein of interest is maximally expressed. Prepare a lysate using $\sim 5 \times 10^7$ cells/ml of extraction buffer by repeated cycles of freezing and thawing. Spin in microcentrifuge at 16 000 g for 30 min (4°C) to remove insoluble material.[a] *Caution*: all fractions are radioactive and should be appropriately handled, stored, and disposed.

2. After removing a small aliquot (e.g. 10 μl), divide the supernatant (< 1 ml) into two microcentrifuge tubes. Add 0.1–2 μg of antigen-specific antibody to one tube and an equivalent amount of control antibody to the other. Incubate both tubes on an orbital mixer for 4 h at 4°C.

3. Wash 0.2 ml of protein A–agarose beads twice with extraction buffer, then add 0.1 ml to each tube containing the protein/antibody solution, and incubate for 2 h at 4°C with mixing.[b]

4. Collect the beads by gentle centrifugation (1000 g for 5 min) and wash using an equal volume of extraction buffer. Repeat wash three times.

5. Resuspend each tube of beads in 0.1 ml of elution buffer and incubate at 65°C for 10 min. Spin for 1 min in microcentrifuge and remove the supernatant to a fresh tube.[c]

6. Add SDS–PAGE sample buffer to aliquots of the crude lysate, washes, and elutions. Boil samples for 5 min and load onto an SDS–polyacrylamide slab gel (see earlier chapters in this book).

7. After the marker dye front has reached the gel bottom, remove the gel and fix in 25% methanol, 10% acetic acid. Dry the gel and expose it to X-ray film to detect radiolabelled protein bands.

[a] The pellet should be saved, dissolved in SDS–PAGE sample buffer, and included on gels with other samples when determining optimal extraction conditions.

[b] Protein G should be substituted when using antibodies from species other than rabbit, guinea-pig, or cat.

[c] The pellets can be re-extracted by boiling for 5 min in SDS–PAGE sample buffer if the complex is not solubilized in extraction buffer.

2.2 Affinity purification

Many protein complexes can be isolated through affinity purification of one of the components. There are a number of ways to demonstrate co-purification of associated proteins, each with specific utility. The most straightforward and least physiological approach is to covalently attach a recombinant protein to agarose beads and use the beads to affinity purify binding proteins from crude extracts. In cases where binding activity is lost upon covalent attachment to beads, a recombinant fusion protein containing an affinity tag (e.g. poly-histidine as described in *Protocol 2*) can be incubated with the protein. After affinity purification of the fusion or 'bait' protein (e.g. on a Ni^{2+} or glutathione column), associated proteins are identified on stained polyacrylamide gels (*Protocol 2*) or Western blots (5). When recombinant proteins are unavailable or do not retain full binding activity, purified 'native' protein can be covalently modified (e.g. biotinylated) or directly linked to agarose or magnetic beads and used to purify binding proteins from solutions. A more physiological approach is to overexpress a fusion protein in transfected cells and co-purify endo-genously formed complexes using an affinity tag. Sensitivity can often be increased in these experiments by culturing the cells in the presence of [^{35}S]methionine to allow detection of the radiolabelled purified protein. Finally, purification schemes based on the natural affinity of specific protein compo-nents should also be considered. For example, a ternary complex containing the cAMP-dependent protein kinase, AKAP75, and protein phosphatase 2B can be purified from brain extracts using either calmodulin–Sepharose (to bind the phosphatase) or cAMP–agarose (to bind the kinase) (6, 7).

In any affinity purification protocol, it is important to demonstrate that co-purified proteins require the bait protein for purification and do not, by them-selves, bind the affinity media. This can be accomplished using purified proteins or extracts from cells not expressing the bait protein. Slab gels are usually used to assay the purified fractions since multiple samples (e.g. starting material, flow-through, washes, elution, and controls) can be directly compared on the same gel.

Protocol 2. Affinity purification of protein complexes using a recombinant, poly-histidine-tagged fusion protein

Equipment and reagents

- Sonic cell disrupter
- Orbital mixer (Nutator, Becton-Dickinson)
- Extraction buffer (see *Protocol 1*)
- His-tag binding resin (Novagen)
- Ni^{2+} charging buffer: 50 mM $NiSO_4$
- SDS–PAGE sample buffer and 10% discon-tinuous SDS–polyacrylamide slab gel (see Chapter 1)

- Binding buffer: 25 mM Hepes, 250 mM NaCl pH 7.8
- Wash buffer: 25 mM Hepes, 500 mM NaCl, 30 mM imidazole pH 7.8
- Elution buffer: 25 mM Hepes, 250 mM NaCl, 250 mM imidazole pH 7.8
- Coomassie gel stain and destain (see Chapter 2)

Protocol 2. *Continued*

Method

1. Prepare a lysate using ~ 5 × 10⁶ cells/ml in extraction buffer by sonication on ice (three times, 2 min each at maximum microprobe setting).

2. Quick freeze the sample at –80°C, then thaw.

3. Spin in the microcentrifuge at 16 000 g for 30 min (4°C) to remove insoluble material.[a]

4. Split the supernatant into two fresh microcentrifuge tubes. To one tube, add 1–10 μg/ml (final concentration) of recombinant histidine-tagged binding protein. Do not add any protein to the other tube (control tube).[b] Incubate both tubes on mixer for 4 h at 4°C.

5. Prepare two small (1–2 ml) columns of His-tag binding resin, charge these by adding 2 ml of Ni²⁺ charge buffer, and wash with 5 ml of binding buffer. Add the histidine-tagged protein sample to one column and the unsupplemented control supernatant to the other column. Wash both columns with 5 ml of wash buffer.

6. Elute proteins from both columns using elution buffer. Collect 0.25 ml fractions.

7. Load equal amounts (50 μl) of the column eluates onto a 10% discontinuous SDS–polyacrylamide slab gel (see Chapter 1) and electrophorese until the tracking dye reaches the bottom of the gel.

8. Stain the gel with Coomasie stain and destain the gel as described in Chapter 2.[c]

9. Compare the binding patterns from the two columns to identify proteins within the extract that specifically bind the histidine-tagged fusion and those that bind the affinity column by themselves (false positives).

[a] The pellet should be saved, dissolved in SDS–PAGE sample buffer, and included on gels with other samples when determining optimal extraction conditions by Western blot analysis.
[b] An alternative control is to add an equivalent amount of bait protein that is missing the His-tag sequence.
[c] For interactions of known proteins to which antibodies are available, omit steps 8 and 9 and then transfer the slab gel to nitrocellulose or PVDF membranes for Western blot analysis as described in Chapter 2, Section 2.

2.3 Chemical cross-linking

Protein complexes can be stabilized by chemical cross-linking, allowing for the study of weak or transient complexes that may otherwise be difficult to detect. For this reason, cross-linking is often used in combination with immunoprecipitation or affinity purification techniques (Sections 2.1 and 2.2). Various bifunctional cross-linking reagents are available that differ with

Table 1. Properties of bifunctional cross-linking reagents

Cross-linker	Reactive towards	Spacer length	Features	References
Homobifunctional				
Disuccimidyl tartate (DST)	$-NH_2$	6.4 Å	c, e	19
Disuccimidyl glutarate (DSG)	$-NH_2$	7.7 Å		20
Disuccimidyl suberate (DSS)	$-NH_2$	11.4 Å	e	20
Dithiobis(succimidyl propionate) (DSP)	$-NH_2$	12.0 Å	b, e	8
Ethyleneglycolbis(succimidylsuccinate) (EGS)	$-NH_2$	16.1 Å	d, e	21
Bis-(4-azidosalicylamido)ethyl disulfide (BASED)	Non-specific	–	a, b	22
Bismaleimidohexane (BMH)	$-SH$	16.1 Å	c	23
Heterobifunctional				
N-Succimidyl-3-(2-pyridyldithio)-propionate (SPDP)	$-SH$, $-NH_2$	6.8 Å	b, e	24
Succimidyl 4-(N-maleimidomethyl)-cyclohexane-1-carboxylate (SMCC)	$-SH$, $-NH_2$	11.6 Å	e	25
Succimidyl 4-(p-maleimidophenyl)-butyrate (SMPB)	$-SH$, $-NH_2$	14.5 Å	e	26
Succimidyl-6-[3-(2-pyridyldithio)-propionamido] hexanoate (LC-SPDP)	$-SH$, $-NH_2$	15.6 Å	b, e	24

[a] Photoactivated.
[b] Cleavable by disulfide reducing reagents (e.g. 2-mercaptoethanol).
[c] Cleavable by oxidizing reagents (e.g. periodate).
[d] Cleavable by hydroxylamine.
[e] Also available in a sulfonated, water soluble form.

respect to their reactivity, length, reversibility, and solubility (see *Table 1*). *Protocol 3* describes a method for cross-linking proteins in intact cells using dithiobissuccinimidyl propionate (DSP), a membrane permeable cross-linking reagent that forms intermolecular covalent bonds through ε-amino groups of lysine side chains. One advantage to using DSP is that it is thiol-cleavable, making it possible to directly compare unlinked and linked proteins separated on gels run with or without reducing agents, respectively.

Another use for chemical cross-linking is for stabilizing complexes formed *in vitro*. In this type of experiment, the bait protein is radiolabelled and incubated with a crude cellular lysate. After cross-linking, the formed complexes are separated by SDS–PAGE and those specifically containing the bait protein (the only labelled species) are visualized by autoradiography.

An inherent problem with all chemical cross-linking experiments is a lack of specificity. Any protein in the immediate vicinity has the potential to cross-link even if not directly associated with the target protein. This 'nearest neighbour' problem may be controlled by experimenting with cross-linkers of different length and chemical reactivity (see *Table 1*) and by confirming results using an independent method.

Protocol 3. Chemical cross-linking intact cells[a]

Equipment and reagents

- 10% discontinuous SDS–polyacrylamide gel prepared as described in Chapter 1, Section 5.
- Tris–glycine running buffer (Chapter 1, Section 5.2.4)
- Coomassie stain/destain solutions or reagents for Western blotting (Chapter 2)
- KRH buffer: 25 mM Hepes pH 7.4, 125 mM NaCl, 4.8 mM KCl, 2.6 mM CaCl$_2$, 1.2 mM MgSO$_4$, 5.6 mM glucose

- 25 mM DSP solution (dithiobissuccinimidyl proprionate; Pierce) in DMSO
- TBS buffer: 50 mM Tris–HCl, 137 mM NaCl, 2.7 mM KCl pH 7.5
- 4 × non-reducing sample buffer: 125 mM Tris–HCl pH 6.8, 4% SDS, 20% glycerol, 0.05% bromophenol blue

Method

1. Wash approx. 2 × 10^6 cultured cells twice with KRH buffer. To half the cells, add DSP to a final concentration of 1 mM and incubate at 20°C for 5–30 min. Leave the other cells untreated.

2. Wash both sets of cells three times with TBS buffer and collect them by scraping (adherent cells) or centrifugation (suspended cells).

3. Lyse the cells by repeated cycles of freezing and thawing.

4. Spin the lysate in a microcentrifuge at 16000 *g* for 30 min (4°C) to separate the extract into soluble and particulate fractions.

5. Mix approx. 20 μg of each sample with 0.25 vol. non-reducing sample buffer and load onto a 10% SDS–polyacrylamide gel.[b] Mix another 20 μg of each sample with 0.25 vol. sample buffer containing 0.2 M dithiothreitol and load onto the gel, leaving an empty lane between reduced and non-reduced samples.

6. Carry out electrophoresis (Chapter 1, Section 5.5).

7. When the tracking dye has reached the bottom, remove the gel and stain it, or, alternatively, transfer the separated proteins to a membrane for Western blot analysis (Chapter 2).

[a] Adapted from ref. 8.
[b] Alternatively, the samples can be immunoprecipitated (*Protocol 1*) prior to loading.

---

3. Identifying binding partners

With many of the techniques described above, protein complexes are purified and the individual components are then separated and characterized by denaturing gel electrophoresis. When combined with Western blotting, unequivocal identification of individual components is achieved. It is also possible, however, to use gel electrophoresis and blotting techniques to identify binding partners directly in crude mixtures or within large complexes.

3.1 Protein overlay

Also called 'ligand-', 'affinity-', or 'far-Western' blotting, the protein overlay method is a very powerful technique for confirming suspected interactions and for identifying new binding partners. The method is similar to traditional Western blot analysis where proteins are separated by gel electrophoresis then transferred to the membrane. Instead of using antibodies, however, a protein of interest is labelled and used to probe the blots for binding partners. An example of this method, where the regulatory subunit of the cAMP-dependent protein kinase was used to probe various crude extracts to identify specific binding proteins called AKAPs (9), is shown in *Figure 1*. It is notable that the overlay technique can also be used for identifying intramolecular interactions (e.g. dimerization), presumably through an exchange reaction (10). For known interactions, the overlay method can be used semi-quantitatively (Section 4.2) or to assay a series of deletion or point mutants for identification

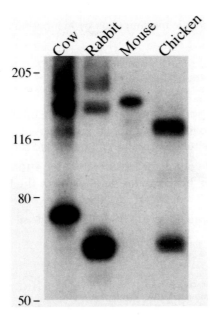

Figure 1. Protein overlay. ^{32}P-labelled regulatory subunit of the cAMP-dependent protein kinase was used as a probe to detect specific binding proteins (called AKAPs) within crude brain extracts from the animal sources indicated. An autoradiograph after 16 h exposure is shown. Numbers represent the size of molecular weight markers (in kDa) run in parallel lanes (not shown).

of binding determinants (11). *Protocol 4* describes the use of overlays using a radiolabelled protein probe whilst *Protocol 5* describes a non-radioactive protein overlay.

Protocol 4.   Radioactive protein overlay

Equipment and reagents

- Orbital or rocking shaker
- Liquid scintillation counter
- X-ray film
- Desalting column (e.g. Excellulose GF-5; Pierce)
- PVDF membrane (e.g. Immobilon; Millipore)
- [γ-³²P]ATP (> 3000 Ci/mmol; Dupont)

- Protein kinase (e.g. PKA catalytic subunit; Promega)
- 10 × kinase buffer: 0.5 M Mops, 0.5 M NaCl, 20 mM MgCl₂, 10 mM DTT pH 7
- TBS buffer (see *Protocol 3*)
- TBSA: 50 mM Tris–HCl, 137 mM NaCl, 2.7 mM KCl pH 7.5, 1% bovine serum albumin, 0.02% sodium azide

Method

1. Separate the protein samples by SDS–PAGE (see Chapter 1).[a]

2. Transfer the separated proteins to PVDF membrane by Western blotting (see Chapter 2), marking the position of molecular weight markers with a pen.

3. Block the non-specific binding sites by incubating the blot in TBSA for 2 h (25°C).[b]

4. Prepare 0.1–2 μg of protein probe by *in vitro* phosphorylation using 20–100 U of protein kinase and 10–50 μCi of [³²P]ATP in 1 × kinase buffer.[c]

5. Remove unincorporated nucleotide using a 2 ml desalting column run in TBS buffer. Collect 150 μl fractions and count 1 μl aliquots of each fraction. Pool the tubes containing the labelled protein.[d]

6. Using a minimal volume of TBSA (~ 8 ml for a 10 × 15 cm blot), dilute the probe to 5 × 10⁴ c.p.m./ml and incubate the blot in this solution with shaking for 2 h at 25°C.[e]

7. Wash the blot three times in TBSA (using ~ 20 ml for each 10 × 15 cm blot).

8. Dry the blot and expose it to X-ray film.

[a] Alternatively, native gels can be used.
[b] The blots can also be stored in TBSA overnight at 4°C.
[c] For known PKA substrates where phosphorylation does not affect binding activity. Other protein kinases may be substituted. Handle radioactive samples with proper protection and care.
[d] The labelled protein should come off the column within the void volume (~ 0.67 ml for 2 ml column).
[e] Incubations can also be performed overnight at 4°C.

Protocol 5. Non-radioactive protein overlay

Equipment and reagents

- Rocking or platform shaker
- X-ray film
- PVDF membrane (e.g. Immobilon, Millipore)
- HRP-conjugated secondary antibody (e.g. HRP-goat anti-rabbit IgG; Amersham)
- Chemilumenescence reagents (e.g. ECL, Amersham)
- TBS buffer (see *Protocol 3*)
- TBSA (see *Protocol 4*)

Method

1. Separate the protein samples by SDS–PAGE (see Chapter 1).[a]

2. Transfer the separated proteins to PVDF membrane by Western blotting (see Chapter 2), marking the position of molecular weight markers with a pen.

3. Block the non-specific binding sites by incubating the blot in TBSA for 2 h (25°C).[b]

4. Discard the TBSA blocking solution and add a minimal volume of fresh TBSA (~ 8 ml for a 10 × 15 cm blot). Add 0.01–1 µg/ml final concentration[c] of binding protein and incubate the blot in this solution for 2 h on a rocking or platform shaker at 25°C.

5. Wash the blot three times in TBSA (using ~ 20 ml for each 10 × 15 cm blot).

6. Incubate the blot in TBSA containing an appropriate dilution[d] of binding protein-specific antibody for 1 h on the shaker at 25°C.

7. Wash the blot as described in step 5.

8. Incubate the blot in TBSA containing HRP-conjugated secondary antibody (at 1:20 000 dilution) for 1 h on the shaker at 25°C.

9. Wash as described in step 5.

10. Incubate the blot in chemilumenescence detection reagents for 1 min, and then expose to film (1–5 min.).

[a] Alternatively, native gels can be used.
[b] The blots can also be stored in TBSA overnight at 4°C.
[c] The optimum amount of binding protein added depends on a number of factors including binding affinity and should be determined empirically. Step 4 is omitted in control blots, which are probed with antibodies alone (steps 6–9).
[d] The optimal dilution of antibody can vary and is usually similar to the amount that is used on Western blots (e.g. 1:2000).

Several points should be considered to ensure success with this technique:

(a) The protein overlay method requires that the binding partners associate with the probe after being subjected to electrophoresis and immobilized onto the membrane. Often, binding proteins retain this ability even after separation on denaturing gels because the recognition sites are relatively short sequences, or never totally denature, or renature on the membrane. Some proteins may be renatured *in situ* by treating the membrane with guanidine hydrochloride in a manner similar to that used for detecting DNA binding proteins using Southwestern blots (12). Alternatively, non-denaturing polyacrylamide gels can also be used and are preferred for complexes involving multiple proteins or where binding activity is easily lost. In some cases, the overlay is non-reciprocal, i.e. it works only when one binding partner is always used as the probe and the other is always immobilized on the membrane.

(b) Either nitrocellulose or PVDF membranes can be used for blotting proteins. In the author's experience, PVDF provides greater sensitivity with lower background in most instances. To minimize denaturation of the proteins on the membrane, mild staining techniques (e.g. Ponceau S, see Chapter 2) should be used. However, staining is usually unnecessary if pre-stained molecular weight markers are used.

(c) The membranes can be blocked using a variety of reagents. Buffers containing BSA or milk are the most popular, but can interfere with subsequent detection (due to endogenous activities, milk should not be used with alkaline phosphatase or biotin–avidin detection methods).

(d) While the material to be probed may be crude cell lysates, the probe itself should be highly purified, unless a method of detection for specific proteins is available. Recombinant proteins make good probes because specific tag sequences can be incorporated for purification and detection, for example an epitope for detection with antibody or a consensus phosphorylation site for incorporation of ^{32}P. However, recombinant protein probes may be unstable, incorrectly folded, or missing post-translational modifications that are important for binding. In these cases one should try using native or *in vitro* translated protein. If specific antibodies are used for detection, it is possible to use a crude preparation of probe (e.g. an extract of cultured cells expressing the protein) as long as proper controls are performed (e.g. similar overlays using extracts that do not contain the probe protein).

(e) The probe can be labelled in a variety of ways. Perhaps the most direct method is to radiolabel the protein with ^{125}I. Phosphorylation of proteins using specific protein kinases and [^{32}P]ATP has also been widely successful (e.g. *Figure 1* and *Protocol 4*). Other methods utilize protein probes conjugated to enzymes, biotin, fluorescein, or digoxigenin with appropriate secondary detection reagents (13–15). Keep in mind, however, that any of these modifications can affect the binding activity of the probe.

(f) As described in *Protocol 5*, the probe does not have to be labelled at all provided that specific antibodies are available for detection. This may be antibody directed against either the protein itself or an affinity tag in the case of recombinant probes. When using antibodies for detection, however, it is critical to perform control overlays in the absence of protein probe to determine background (e.g. endogenous probe protein and cross-reacting proteins). For small binding regions, biotinylated synthetic peptides may be used as probes. It is worth mentioning that the probes created for protein overlays are useful in a variety of other experiments including screening cDNA expression libraries for clones encoding binding proteins and *in situ* overlay of fixed cells for subcellular localization studies (15, 16).

(g) The protein overlay is often performed in blocking buffer (*Protocols 4* and *5*) but other factors (e.g. divalent cations or detergents) may also be required to facilitate complex formation. Excess unlabelled probe or volume excluding polymers (e.g. PEG) may help to drive the reaction forward.

(h) A fixation step (e.g. 1% paraformaldehyde) can be included after probe addition to stabilize the formed complexes, but such experiments must be carefully performed and interpreted.

3.2 Two-dimensional diagonal analysis

Chemical cross-linking can be used in combination with two-dimensional electrophoresis to determine which components are directly associated in a large multiprotein complex (*Protocol 6*). In this type of experiment, a purified complex is partially cross-linked *in vitro* using a thiol-cleavable reagent. As illustrated in *Figure 2*, the samples are separated in the first dimension using SDS–PAGE without reducing agents. The gel is then treated with reducing agent to cleave the cross-linker and the second dimension of SDS–PAGE is performed. Because proteins are separated in both dimensions by molecular weight, uncross-linked components will migrate in a diagonal line across the second gel, while those that are cross-linked will be retarded in the first dimension and therefore run off the diagonal when separated in the second gel. The cross-linked components, identified by their molecular weight, are then mapped to adjacent positions within the architecture of the complex. Several experiments may be performed using cross-linkers of different length and reactivity to characterize various protein interactions within the complex.

Vincent M. Coghlan

Protocol 6. Analysis of chemically cross-linked complexes by two-dimensional electrophoresis

Reagents

- Thiol-cleavable, amine-selective cross-linking reagent (e.g. Traut's reagent or Lomant's reagent; Pierce)
- 7.5% discontinuous SDS–polyacrylamide slab gels (see Chapter 1); two slab gels are required
- Purified protein complex (20–50 µg)

- 4 × non-reducing sample buffer (see *Protocol 3*)
- Electrode buffer (see *Protocol 3*)
- Electrode buffer containing 3% 2-mercaptoethanol
- Stacking buffer: 125 mM Tris–HCl pH 6.8, 0.1% SDS

Method

1. Treat 20–50 µg of purified protein complex with the thiol-cleavable cross-linking reagent following the procedure supplied by the manufacturer.

2. Add 0.25 vol. of 4 × non-reducing sample buffer (which contains Tris, a primary amine) to quench the reaction.

3. Separate the cross-linked sample on a 7.5% SDS–polyacrylamide slab gel, using electrode buffer.

4. After completion of electrophoresis, remove the gel, and soak in electrode buffer containing 3% 2-mercaptoethanol at 55°C for 30 min to cleave the cross-link.[a]

5. Wash the gel three times (10 min each) in stacking buffer.

6. Cut out the entire sample lane for embedding in the second gel (7.5% SDS–polyacrylamide slab gel) and perform the second dimension of electrophoresis (see Chapter 7, Section 5 for general procedures for using gel strips in a two-dimensional separation).

7. Silver stain the gel to detect the separated proteins (see Chapter 2, Section 1.3) or transfer the gel to a suitable membrane for Western blot analysis (Chapter 2, Section 2).

[a] Alternatively, the soaking step in thiol reagent can be omitted and 2-mercaptoethanol added instead to the stacking buffer in the second gel so that thiol cleavage occurs during electrophoresis in the second dimension.

---

4. Characterizing protein interactions

When purified binding partners are available, gel electrophoresis can be used to study their interaction in detail, under a variety of conditions. Complex formation is usually evident on non-denaturing polyacrylamide gels by a shift in molecular weight providing a versatile assay for examining protein:protein interactions.

306

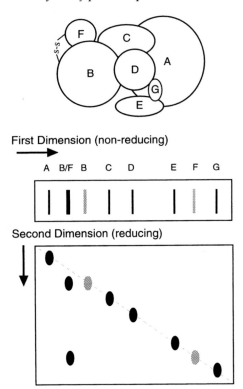

Figure 2. Analysis of cross-linked complex using two-dimensional electrophoresis. In this hypothetical multiprotein complex, proteins B and F are linked with a thiol-cleavable cross-linker. The two proteins migrate as a complex in the first gel, which is run under non-reducing conditions, and run off the diagonal when separated under reducing conditions in the second dimension (see text).

4.1 Mobility-shift assay

Mobility-shift or gel-retardation experiments are performed using low percentage polyacrylamide gels to separate free proteins from those bound in complexes. In a simple experiment, two purified proteins are incubated together in a suitable binding buffer and the mixture is loaded into one lane of a slab gel. Separate control lanes are loaded with each protein in buffer by themselves. As the gel runs, free proteins migrate as bands according to their individual charge and size while those bound in complexes are retarded due to their increased size. The caging effect of the gel is thought to keep local protein concentrations high, favouring association. The gel is fixed and stained to identify bands corresponding to free and bound protein. An example of this type of experiment is shown in *Figure 3*.

Experimental situations where the mobility-shift assay is of particular use include determination of the role of exogenously added cellular factors,

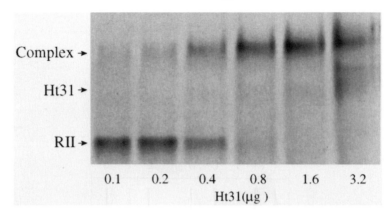

Complex →
Ht31 →
RII →

| 0.1 | 0.2 | 0.4 | 0.8 | 1.6 | 3.2 |

Ht31(μg)

Figure 3. Mobility-shift assay. In this experiment, 2 μg of the regulatory subunit of the cAMP-dependent protein kinase (RII) was incubated with the indicated amounts of the AKAP Ht31 and samples were separated on a 6% native polyacrylamide gel. A photograph of the Coomassie stained gel is shown (courtesy of Dr Zachary Hausken, Vollum Institute, Portland, OR).

investigation into the effects of protein phosphorylation, and comparison of wild-type and mutant proteins. The technique can also be used to measure relative binding affinities (see Section 4.2). Preparative gels (Chapter 3) can be used to purify shifted complexes for microsequencing or other analyses. While the general mobility-shift procedure (*Protocol 7*) gives satisfactory results with many complexes, the following points should be considered:

(a) Binding buffers should consist of reagents that encourage complex formation but do not dramatically affect electrophoretic mobility. For this reason, high concentrations of salts or reagents that form precipitates with the gel buffer should be avoided. BSA (0.1 mg/ml) is often included as a non-specific carrier protein.

(b) Some proteins may not form distinguishable bands on native gels for several reasons (sample heterogeneity, conformational dynamics, buffer effects, etc.). This problem can often be alleviated by varying the acrylamide concentration, the buffering system (pH), or the ionic strength of the gel. Inclusion of a discontinuous stacking gel may increase resolution in some cases while causing precipitates to form in others. For very large complexes or complexes containing lipoproteins, composite gels of agarose and low percentages of acrylamide may be useful (17).

(c) Tris–glycine and Tris–borate are good running buffer systems to try in initial experiments. For basic proteins, an acidic buffer system is recommended. To prevent denaturation of proteins, gels should be run in the cold (4 °C) at low current (20 mA/gel) and electrophoresis continued until the dye has run ~ 0.75 the way towards the bottom of the gel (or 5 mA overnight).

(d) If Western blotting is not to be carried out, the gel should be fixed and stained. Coomassie blue staining is useful when sufficient amounts of protein are loaded. The sensitivity can be increased by radiolabelling one of the binding partners and using autoradiographic detection. This is especially useful when assaying complex mixtures of proteins. For probes of high specific activity, excess unlabelled protein is added to help drive the reaction towards complex formation.

Protocol 7. Mobility-shift assay

Reagents

- Purified protein of interest
- 4 × interaction buffer: 100 mM Tris–HCl pH 6.8, 100 mM KCl, 40% glycerol, 0.05% bromophenol blue
- 4 × stacking gel buffer: 0.5 M Tris–HCl pH 6.8

- Electrophoresis buffer: 25 mM Tris, 192 mM glycine
- 4 × resolving gel buffer: 1.5 M Tris–HCl pH 8.8

Method

1. Prepare a 6% non-denaturing polyacrylamide gel with 3% stacking gel using the stacking and resolving gel buffers.[a] General procedures for gel preparation are described in Chapter 1.

2. In a 0.5 ml microcentrifuge tube, mix the binding proteins in a minimum volume (< 50 μl) of 1 × interaction buffer.[b] In separate tubes, add each protein by itself to interaction buffer as controls. Leave all tubes on ice for 30 min.

3. Load the samples onto the polyacrylamide gel and run in electrophoresis buffer at low current (< 20 mA) at 4°C.

4. Stain the gel using Coomassie blue or silver stain (see Chapter 2).

[a] Some complexes precipitate when using a discontinuous buffer system. In these cases, omit the stacking gel and/or use a Tris–borate buffering system (e.g. 0.5 × TBE: 50 mM Tris, 50 mM borate, 1.25 mM EDTA pH 8.3).
[b] Additional interaction buffer components may be required to ensure protein stability and optimal interaction.

4.2 Quantitative methods

An important part of the characterization of interacting proteins is to measure their affinity for one another. The simple interaction of two proteins represented by: A + B $\rightleftharpoons$ AB is usually defined in terms of a dissociation constant (K_d) defined by:

$$K_d = \frac{[A]\,[B]}{[AB]}$$

where [A] and [B] are the free concentrations of the two proteins and [AB] is the concentration of the complex at equilibrium. Thus, the dissociation constant can be calculated if bound and free proteins can be separately measured. For practical measurements, experiments are often performed using a concentration of one binding partner, A for example, that greatly exceeds (> 100-fold) the concentration of B (represented as [A] >> [B]). Thus, complex formation does not significantly reduce the amount of free A. Under these conditions, the dissociation constant becomes approximately equal to the concentration at which 50% of B is found in the complex. Another way to conduct binding experiments is in a competitive fashion where protein B is radiolabelled and the equilibrium between free and complexed proteins is perturbed using varying amounts of unlabelled B as a competitor.

Gel-based methods can be used to investigate binding equilibria. Thus, as discussed in Section 4.1, non-denaturing gel electrophoresis is a convenient means to separate bound and free proteins. Slab gels are typically used so that multiple lanes can be loaded with samples containing varying amounts of one binding partner (e.g. *Figure 3*). Densitometric analysis of the gels after staining is then used to quantitate the ratios of free and bound protein. Alternatively, experiments can be performed using radiolabelled protein and bands can either be excised from gels and counted or scanned from autoradiographs and quantified. If available, a digital PhosphoImager can be helpful when signals do not all lie within the linear response range of X-ray film. Competitive gel-shifts can be performed using different ratios of labelled and unlabelled binding protein in a manner similar to that used to study protein:DNA interactions (18).

Competitive protein overlays have also been used to examine equilibrium parameters. In these experiments, a series of similar protein overlays are performed using radiolabelled protein in the presence of varying amounts of unlabelled competitor to determine the concentration of competitor that produces half-maximal binding.

5. Limitations

Complete characterization of a protein interaction requires the application of multiple approaches. While the gel-based methods presented in this chapter are valuable tools to achieve this goal, they provide little insight into the functional relevance of a given interaction. Indeed, unrelated proteins have been found to associate with high affinity *in vitro*, forming complexes with no apparent physiological role (e.g. actin and DNase I). The kinetics of an interaction must also be considered as dissociation rates can be too fast to detect the complex regardless of affinity. Finally, failure to detect an interaction *in vitro* may result from the absence of critical cellular factors or regulatory components. Successful characterization most often results from a combination of gel-based methods, biophysical techniques, and cellular/genetic approaches.

References

1. Mathews, C.K. (1993). *J. Bacteriol.*, **175**, 6377.
2. Phizicky, E.M. and Fields, S. (1995). *Microbiol. Rev.*, **59**, 94.
3. Srere, P.A. and Matthews, C.K. (1990). In *Methods in enzymology* Vol. 182, p. 539. Academic Press, London.
4. Jarvis, T.C., Ring, D.M., Daube, S.S., and von Hippel, P.H. (1990). *J. Biol. Chem.*, **265**, 15160.
5. Lu, T., Van Dyke, M., and Sawadogo, M. (1993). *Anal. Biochem.*, **213**, 318.
6. Hathaway, D.R., Adelstein, R.S., and Klee, C.B. (1981). *J. Biol. Chem.*, **256**, 8183.
7. Coghlan, V.M., Perrino, B.A., Howard, M., Langeberg, L.K., Hicks, J.B., Gallatin, W.M., *et al.* (1995). *Science*, **267**, 108.
8. Gamby, C., Waage, M.C., Allen, R.G., and Baizer, L. (1996). *J. Biol. Chem.*, **271**, 26698.
9. Coghlan, V.M., Bergeson, S.E., Langeberg, L.K., Nilaver, G., and Scott, J.D. (1993). *Mol. Cell. Biochem.*, **128**, 309.
10. Hausken, Z.E., Coghlan, V.M., Schafer-Hastings, C.A., Reimann, E.M., and Scott, J.D. (1994). *J. Biol. Chem.*, **269**, 24245.
11. Hausken, Z.E., Dell'Acqua, M.L., Coghlan, V.M., and Scott, J.D. (1996). *J. Biol. Chem.*, **271**, 29016.
12. Vinson, C.R., LaMarco, K.L., Johnson, P.F., Landschultz, W.H., and McKnight, S.L. (1988). *Genes Dev.*, **2**, 801.
13. Soutar, A.K. and Wade, D.P. (1990). In *Protein function: a practical approach* (ed. T.E. Creighton), pp. 55–76. IRL Press, Oxford.
14. Gray, P.C., Tibbs, V.C., Catterall, W.A., and Murphy, B.J. (1997). *J. Biol. Chem.*, **272**, 6297.
15. Coghlan, V.M., Langeberg, L.K., Fernandez, A., Lamb, N.J.C., and Scott, J.D. (1994). *J. Biol. Chem.*, **269**, 7658.
16. Lohmann, S.M., DeCamilli, P., and Walter, U. (1988). In *Methods in enzymology* Vol. 159, p. 184. Academic Press, London.
17. Varelas, J.B., Zenarosa, N.R., and Froelich, C.J. (1991). *Anal. Biochem.*, **197**, 396.
18. Garner, M.M. and Revzin, A. (1981). *Nucleic Acids Res.*, **9**, 3047.
19. Smith, R.J., Capaldi, R.A., Muchmore, D., and Dahlquist, F. (1978). *Biochemistry*, **17**, 3719.
20. Rebois, R.V., Omedeo-Sale, F., and Fishman, P.H. (1981). *Proc. Natl. Acad. Sci. USA*, **78**, 2086.
21. Waugh, S.M., DiBella, E.E., and Pilch, P.F. (1989). *Biochemistry*, **28**, 3448.
22. Hermanson, G.T. (1996). In *Bioconjugate techniques* p. 214. Academic Press, San Diego.
23. Yi, F., Denker, B.M., and Neer, E.J. (1991). *J. Biol. Chem.*, **266**, 3900.
24. Carlsson, J., Drevin, H., and Axen, R. (1978). *Biochem. J.*, **173**, 723.
25. Bieniarz, C., Husain, M., Barnes, G., King, C.A., and Welch, C.J. (1996). *Bioconjugate Chem.*, **7**, 88.
26. Iwai, K., Fukuoka, S.I., Fushiki, T., Kido, K., Sengoku, Y., and Semba, T. (1988). *Anal. Biochem.*, **171**, 277.

A1

List of suppliers

Aldrich Chemical Co., The Old Brickyard, New Road, Gillingham, Dorset SP8 4JL, UK.

Amersham
Amersham International plc., Lincoln Place, Green End, Aylesbury, Buckinghamshire HP20 2TP, UK.
Amersham Corporation, 2636 South Clearbrook Drive, Arlington Heights, IL 60005, USA.

Anderman and Co. Ltd., 145 London Road, Kingston-Upon-Thames, Surrey KT17 7NH, UK.

Aristo Grid Lamp Products Inc., 35 Lumber Road, Roslyn, NY 11576, USA.

AstroCam Ltd., Innovation Centre, Cambridge Science Park, Milton Road, Cambridge CB4 4GS, UK.

Beckman Instruments
Beckman Instruments UK Ltd., Progress Road, Sands Industrial Estate, High Wycombe, Buckinghamshire HP12 4JL, UK.
Beckman Instruments Inc., PO Box 3100, 2500 Harbor Boulevard, Fullerton, CA 92634, USA.

Becton Dickinson
Becton Dickinson and Co., Between Towns Road, Cowley, Oxford OX4 3LY, UK.
Becton Dickinson and Co., 2 Bridgewater Lane, Lincoln Park, NJ 07035, USA.
Becton Dickinson Primary Care Diagnostics, 7 Loveton Circle, Sparks, MD 21152, USA.

Bio
Bio 101 Inc., c/o Statech Scientific Ltd., 61–63 Dudley Street, Luton, Bedfordshire LU2 0HP, UK.
Bio 101 Inc., PO Box 2284, La Jolla, CA 92038-2284, USA.

Bio-Rad Laboratories
Bio-Rad Laboratories Ltd., Bio-Rad House, Maylands Avenue, Hemel Hempstead HP2 7TD, UK.
Bio-Rad Laboratories, Division Headquarters, 3300 Regatta Boulevard, Richmond, CA 94804, USA.
Bio-Rad Laboratories, 2000 Alfred Nobel Drive, Hercules, CA 94547, USA.

Boehringer Mannheim

Boehringer Mannheim UK (Diagnostics and Biochemicals) Ltd., Bell Lane, Lewes, East Sussex BN17 1LG, UK.

Boehringer Mannheim Corporation, Biochemical Products, 9115 Hague Road, PO Box 504 Indianopolis, IN 46250-0414, USA.

Boehringer Mannheim Biochemica, GmbH, Sandhofer Str. 116, Postfach 310120 D-6800 Ma 31, Germany.

British Drug Houses (BDH) Ltd., Poole, Dorset, UK.

Calbiochem-Novabiochem

Calbiochem-Novabiochem (UK) Ltd., Boulevard Industrial Park, Padge Road, Beeston, Nottingham NG9 2JR, UK.

Calbiochem-Novabiochem Corporation., 10394 Pacific Center Court, San Diego, CA 92121, USA.

Dextra Laboratories Ltd., The Innovation Centre, The University, Reading, Berks RG6 6BX, UK.

Diagen, Hilden, Germany.

Difco Laboratories

Difco Laboratories Ltd., PO Box 14B, Central Avenue, West Molesey, Surrey KT8 2SE, UK.

Difco Laboratories, PO Box 331058, Detroit, MI 48232-7058, USA.

Digital Pixel Ltd., PO Box 625, Brighton, Sussex BN1 SJT, UK.

Du Pont

Dupont (UK) Ltd. (Industrial Products Division), Wedgwood Way, Stevenage, Hertfordshire SG1 4Q, UK.

Du Pont Co. (Biotechnology Systems Division), PO Box 80024, Wilmington, DE 19880-002, USA.

Eppendorf, Netheler-Hinz-GmbH, Barkhausenweg 1, Hamburg 2000, Germany.

European Collection of Animal Cell Culture, Division of Biologics, PHLS Centre for Applied Microbiology and Research, Porton Down, Salisbury, Wiltshire SP4 0JG, UK.

Falcon (Falcon is a registered trademark of Becton Dickinson and Co.)

Fisher Scientific Co., 711 Forbest Avenue, Pittsburgh, PA 15219-4785, USA.

Flow Laboratories, Woodcock Hill, Harefield Road, Rickmansworth, Hertfordshire WD3 1PQ, UK.

Flowgen Instruments Ltd., Lynn Lane, Shenstone, Lichfield, Staffs WS14 0EE, UK.

Fluka

Fluka-Chemie AG, CH-9470, Buchs, Switzerland.

Fluka Chemicals Ltd., The Old Brickyard, New Road, Gillingham, Dorset SP8 4JL, UK.

Gibco BRL

Gibco BRL (Life Technologies Ltd.), Trident House, Renfrew Road, Paisley PA3 4EF, UK.

Gibco BRL (Life Technologies Inc.), 3175 Staler Road, Grand Island, NY 14072-0068, USA.

Gibco-BRL, 8400 Helgerman Court, Gaithersburg, MD 20877, USA. ??

Glen Spectra Ltd., 2–4 Wigton Gardens, Stanmore, Middlesex HA7 1BG, UK.

Hoefer Scientific Instruments, 654 Minnesota Street, PO Box 77387, San Francisco, CA 94107, USA.

Arnold R. Horwell, 73 Maygrove Road, West Hampstead, London NW6 2BP, UK.

Hybaid

Hybaid Ltd., 111–113 Waldegrave Road, Teddington, Middlesex TW11 8LL, UK.

Hybaid, National Labnet Corporation, PO Box 841, Woodbridge, NJ 07095, USA.

HyClone Laboratories, 1725 South HyClone Road, Logan, UT 84321, USA.

International Biotechnologies Inc., 25 Science Park, New Haven, Connecticut 06535, USA.

Invitrogen Corporation

Invitrogen Corporation, 3985 B Sorrenton Valley Building, San Diego, CA 92121, USA.

Invitrogen Corporation, c/o British Biotechnology Products Ltd., 4–10 The Quadrant, Barton Lane, Abingdon, Oxon OX14 3YS, UK.

Janssen Life Science Products, Janssen Pharmaceutica Inc., 1125 Trenton-Harbourton Road, Titusville, New Jersey 08560, USA.

Jouan Ltd., 130 Western Road, Tring, Herts HP23 4BU, UK.

Kodak: Eastman Fine Chemicals, 343 State Street, Rochester, NY, USA.

Lambda Fluoreszenzetechnologie Gmbh, Grottenhof Strasse 3, A8053 Graz, Austria.

Life Technologies Inc., 8451 Helgerman Court, Gaithersburg, MN 20877, USA.

Mallinckrodt Chemical Inc., Performance and Laboratory Chemicals Division, 16305 Swingley Ridge Drive, Chesterfield, MO 63017, USA.

Merck

Merck Industries Inc., 5 Skyline Drive, Nawthorne, NY 10532, USA.

Merck, Frankfurter Strasse, 250, Postfach 4119, D-64293, Germany.

Merck Ltd., Hunter Boulevard, Magna Park, Lutterworth, Leics. LE17 4XN, UK.

Millipore

Millipore (UK) Ltd., The Boulevard, Blackmoor Lane, Watford, Hertfordshire WD1 8YW, UK.

Millipore Corp./Biosearch, PO Box 255, 80 Ashby Road, Bedford, MA 01730, USA.

New England Biolabs (NBL)

New England Biolabs (NBL), 32 Tozer Road, Beverley, MA 01915-5510, USA.

New England BioLabs (UK) Ltd., 67 Knowl Place, Wilbury Way, Hitchin, Hertfordshire SG4 OTY, UK.

Nikon Corporation, Fuji Building, 2–3 Marunouchi 3-chome, Chiyoda-ku, Tokyo, Japan.

Novagen Inc., 597 Science Drive, Madison, WI 53711, USA.

NOVEX, 11040 Roselle Street, San Diego, CA, USA.

Omega Optical Inc., PO Box 573, Brattleburg, Vermont VT 05302, USA.

Oxford Glycosystems Ltd., Hitching Court, Blacklands Way, Abingdon, Oxford OX14 1RG, UK.

Perkin-Elmer

Perkin Elmer Ltd., Post Office Lane, Beaconsfield, Buckinghamshire HP9 1QA, UK.

Perkin Elmer-Cetus (The Perkin-Elmer Corporation), 761 Main Avenue, Norwalk, CT 0689, USA.

Pharmacia-Amersham, PO Box 1327, 800 Centennial Avenue, Piscataway, NJ 08855-1327, USA.

Pharmacia

Pharmacia Biotech Europe, Procordia EuroCentre, Rue de la Fuse-e 62, B-1130 Brussels, Belgium.

Pharmacia Biotech Ltd., 23 Grosvenor Road, St Albans, Hertfordshire AL1 3AW, UK.

Pharmacia Biosystems Ltd. (Biotechnology Division), Davy Avenue, Knowl-hill, Milton Keynes MK5 8PH, UK.

Pierce

Pierce Europe BV, PO Box 1512, 3260 BA Oud Beijerland, The Netherlands.

Pierce, 3747 N. Meridian Road, PO Box 117, Rockford, IL 61105, USA.

Promega

Promega Ltd., Delta House, Enterprise Road, Chilworth Research Centre, Southampton, UK.

Promega Corporation, 2800 Woods Hollow Road, Madison, WI 53711-5399, USA.

Qiagen

Qiagen Inc., c/o Hybaid, 111–113 Waldegrave Road, Teddington, Middlesex TW11 8LL, UK.

Qiagen Inc., 9259 Eton Avenue, Chatsworth, CA 91311, USA.

Schleicher and Schuell

Schleicher and Schuell Inc., Keene, NH 03431A, USA.

Schleicher and Schuell Inc., D-3354 Dassel, Germany.

Schleicher and Schuell Inc., c/o Andermann and Co. Ltd., UK.

Seikagaku Corporation, 1-800-237-4512 Tokyo Yakugyo Building, 1–5 Nihon-bashi-honcho 2-chome, Chuo-ku, Tokyo 103, Japan.

Shandon Scientific Ltd., Chadwick Road, Astmoor, Runcorn, Cheshire WA7 1PR, UK.

Sigma Chemical Company
Sigma Chemical Company (UK), Fancy Road, Poole, Dorset BH17 7NH, UK.
Sigma Chemical Company, 3050 Spruce Street, PO Box 14508, St. Louis, MO
  63178-9916, USA.
Soham Scientific, Unit 6, Mereside, Soham, Cambridgeshire CB7 SEE, UK.
Sorvall DuPont Company, Biotechnology Division, PO Box 80022, Wilming-
  ton, DE 19880-0022, USA.
Stratagene
Stratagene Ltd., Unit 140, Cambridge Innovation Centre, Milton Road, Cam-
  bridge CB4 4FG,UK.
Stratagene Inc., 11011 North Torrey Pines Road, La Jolla, CA 92037, USA.
Unichem Ltd., Mayflower Close, Chandler's Ford Industrial Estate, East-
  leigh, Hampshire SO5 3AR, UK.
United States Biochemical, PO Box 22400, Cleveland, OH 44122, USA.
UVP
UVP Inc., 2066 W. 11th Street, Upland, CA 91786, USA.
UVP International Ltd., Unit 18 Cambridge Science Park, Milton Road, Cam-
  bridge CB4 4GS, UK.
VWR Scientific Products, Philadelphia Regional Distribution Center, 200
  Center Square Road, Bridgeport, NJ 08014, USA.
VYDAC/The Separations Group, 17434 Mojave Street, Hesperio, CA 92345,
  USA.
Wellcome Reagents, Langley Court, Beckenham, Kent BR3 3BS, UK.
Whatman International Ltd., Whatman House, St. Leonard's Road, Maid-
  stone, Kent ME16 0LS, UK.
Wright Instruments Ltd., Unit 10, 26 Queensway, Enfield, Middlesex EN4,
  UK.

A2

Bibliography of polypeptide detection methods

CARL R. MERRIL and KAREN M. WASHART

1. Detection of proteins in gels

1.1 General polypeptide detection methods

1.1.1 Organic dyes

Reviews

1. Wilson, C.M. (1983). In *Methods in enzymology* (ed. C.H.W. Hirs and S.N. Timasheff), Vol. 91, p. 236. Academic Press, New York.
2. Merril, C.R., Harasewych, M.G., and Harrington, M.G. (1986). In *Gel electrophoresis of proteins* (ed. M.J. Dunn). Wright, Bristol.
3. Wilson, C.M. (1979). *Anal. Biochem.*, **96**, 263.
4. Wirth, P.J. and Romano, A. (1995). *J. Chromatogr. A*, **698**, 123.
5. Beeley, J.A., Newman, F., Wilson, P.H., and Shimmin, I.C. (1996). *Electrophoresis*, **17**, 505.

Coomassie blue R-250 in methanol–acetic acid or methanol–TCA

1. Righetti, P.G. (1983). In *Laboratory techniques in biochemistry and molecular biology* (ed. T.S. Work and E. Work), Vol. 11, p. 148. North Holland, Amsterdam.
2. Wilson, C.M. (1979). *Anal. Biochem.*, **96**, 263.
3. Weber, K. and Osborn, M. (1969). *J. Biol. Chem.*, **244**, 4406.
4. Dibas, A.I. and Yorio, T. (1996). *Biochem. Biophys. Res. Commun.*, **220**, 929.

Coomassie blue R-250 in isopropanol–acetic acid

1. Reisner, A.H. (1984). In *Methods in enzymology* (ed. W.B. Jakoby), Vol. 104, p. 439. Academic Press, New York.
2. Andrews, A.T. (1986). In *Electrophoresis: theory, techniques and biochemical and clinical applications*, 2nd edn. Oxford University Press, Oxford.
3. Wilson, C.M. (1983). In *Methods in enzymology* (ed. C.H.W. Hirs and S.N. Timasheff), Vol. 91, p. 236. Academic Press, New York.
4. Blakesley, R. W. and Boezi, J. A. (1977). *Anal. Biochem.*, **82**, 580.
5. Chrambach, A., Reisfeld, R.A., Wyckoff, M., and Zaccari, J. (1967). *Anal. Biochem.*, **20**, 150.
6. Diezel, W., Kopperschläger, G., and Hoffman, E. (1972). *Anal. Biochem.*, **48**, 617.

7. Malik, N. and Berrie, A. (1972). *Anal. Biochem.*, **49**, 173.
8. Reisner, A.H., Nemes, P., and Bucholtz, C. (1975). *Anal. Biochem.*, **64**, 509.

Coomassie blue G-250
1. Neumann, U., Khalaf, H., and Rimpler, M. (1994). *Electrophoresis*, **15**, 916.

Staining proteins using Fast green
1. Allen, R.E., Masak, K.C., and McAllister, P.K. (1980). *Anal. Biochem.*, **104**, 494.
2. Wilson, C.M. (1983). In *Methods in enzymology* (ed. C.H.W. Hirs and S.N. Timasheff), Vol. 91, p. 236. Academic Press, New York.

Staining proteins with Amido black
1. Wilson, C.M. (1979). *Anal. Biochem.*, **96**, 263.
2. Wilson, C.M. (1983). In *Methods in enzymology* (ed. C.H.W. Hirs and S.N. Timasheff), Vol. 91, p. 236. Academic Press, New York.
3. McMaster-Kaye, R. and Kaye, J.S. (1974). *Anal. Biochem.*, **61**, 120.

Staining proteins with CPTS (copper phthalocyanine 3,4′,4″,4‴-tetrasulfonic acid tetrasodium salt)
1. Bickar, D. and Reid, P.D. (1992). *Anal. Biochem.*, **109**, 115.

Staining proteins prior to electrophoresis
1. Bosshard, H.F. and Datyner, A. (1977). *Anal. Biochem.*, **82**, 327.
2. Sun, S.M. and Hall, T. C. (1974). *Anal. Biochem.*, **61**, 237.
3. Schägger, H., Aquila, H., and Van Jagow, G. (1988). *Anal. Biochem.*, **173**, 201.

1.1.2 Silver stains

Reviews
1. Dunn, M.J. and Burghes, H.M. (1983). *Electrophoresis*, **4**, 173.
2. Merril, C.R., Goldman, D., and Van Keuren, M.L. (1984). In *Methods in enzymology* (ed. W.B. Jakoby), Vol. 104, p. 441. Academic Press, New York.

Methods
1. Merril, C.R., Switzer, R.C., and Van Keuren, M.L. (1979). *Proc. Natl. Acad. Sci. USA*, **76**, 4335.
2. Oakley, B.R., Kirsch, D.R., and Morris, N.R. (1980). *Anal. Biochem.*, **105**, 361.
3. Switzer, R.C., Merril, C.R., and Shifrin, S. (1979). *Anal. Biochem.*, **98**, 231.
4. Sammons, D.W., Adams, L.D., and Nishizawa, E.E. (1981). *Electrophoresis*, **2**, 135.
5. Morrissey, J. H. (1981). *Anal. Biochem.*, **117**, 307.
6. Wray, W., Bonlikas, T., Wray, V.P., and Hancock, R. (1981). *Anal. Biochem.*, **118**, 197.
7. Ohsawa, K. and Ebata, N. (1983). *Anal. Biochem.*, **98**, 231.
8. Hochstrasser, D.F. and Merril, C.R. (1988). *Appl. Theor. Electrophoresis*, **1**, 35.
9. Dunn, M.J. (1996). *Methods Mol. Biol.*, **59**, 363.
10. Swain, M. and Ross, N.W. (1995). *Electrophoresis*, **16**, 948.

Troubleshooting
1. Marshall, T. and Williams, K.M. (1983). *Anal. Biochem.*, **139**, 502.
2. Ochs, D. (1983). *Anal. Biochem.*, **135**, 470.
3. Merril, C.R., Switzer, R.C., and Van Keuren, M.C. (1979). *Proc. Natl. Acad. Sci. USA*, **76**, 4335.
4. Hallinan, F.U. (1983). *Electrophoresis*, **4**, 265.

1.1.3 Negative staining techniques

1. Lee, C., Levin, A., and Branton, D. (1987). *Anal. Biochem.*, **166**, 308.
2. Dzandu, J.K., Johnson, J.F., and Wise, G.F. (1988). *Anal. Biochem.*, **174**, 157.
3. Nelles, L.P. and Bamburg, J.R. (1976). *Anal. Biochem.*, **73**, 522.
4. Higgins, R.C. and Dahmus, M.E. (1979). *Anal. Biochem.*, **93**, 257.
5. Candiano, G., Porotto, M., Lanciotti, M., and Ghiggeri, G.M. (1996). *Anal. Biochem.*, **243**, 245.
6. Liu, R.H., Jacob, J., and Tennant, B. (1997). *BioTechniques*, **22**, 594.

1.1.4 Labelling with fluorophore prior to electrophoresis

Dansyl chloride

1. Schetter, H. and McLeod, B. (1979). *Anal. Biochem.*, **98**, 329.
2. Stephens, R.E. (1975). *Anal. Biochem.*, **65**, 369.
3. Tjissen, P. and Kurstak, E. (1979). *Anal. Biochem.*, **99**, 97.

Fluorescamine

1. Douglas, S.A., La Marca, M.E., and Mets, L.J. (1978). In *Electrophoresis '78* (ed. N. Catsimpoolas), Vol. 2, p. 155. Elsevier/North Holland, Amsterdam.
2. Eng, P.R. and Parker, C.O. (1974). *Anal. Biochem.*, **59**, 323.
3. Ragland, W.L., Benton, T.L., Pace, J.L., Beach, F.G., and Wade, A.E. (1978). In *Electrophoresis '78* (ed. N. Catsimpoolas), Vol. 2, p. 217. Elsevier/North Holland, Amsterdam.
4. Ragland, W.L., Pace, J.L., and Kemper, D.L. (1974). *Anal. Biochem.*, **59**, 24.

MDPF

1. Barger, B.O., White, F.C., Pace, J.L., Kemper, D.L., and Ragland, W.L. (1976). *Anal. Biochem.*, **70**, 327.
2. Douglas, S.A., La Marca, M.E., and Mets, L.J. (1978). In *Electrophoresis '78* (ed. N. Catsimpoolas), Vol. 2, p. 155. Elsevier/North Holland, Amsterdam.
3. Ragland, W.L., Benton, T.L., Pace, J.L., Beach, F.G., and Wade, A.E. (1978). In *Electrophoresis '78* (ed. N. Catsimpoolas), Vol. 2, p. 217. Elsevier/North Holland, Amsterdam.

DACM

1. Yamamoto, K., Okamoto, Y., and Sekine, T. (1978). *Anal. Biochem.*, **84**, 313.

o-Phthaldialdehyde

1. Weidekamm, E., Wallach, D.F.H., and Flückiger, R. (1973). *Anal. Biochem.*, **54**, 102.

1.1.5 Labelling with fluorophore after electrophoresis
Anilinonaphthalene sulfone (ANS)
1. Hartman, B.K. and Udenfriend, S. (1969). *Anal. Biochem.*, **30**, 391.
2. Kane, C.D. and Bernlohr, D.A. (1996). *Anal. Biochem.*, **233**, 197.
3. Cioni, P. and Strambini, G.B. (1996). *J. Mol. Biol.*, **263**, 789.

Bis-ANS
1. Horowitz, P.M. and Bowman, S. (1987). *Anal. Biochem.*, **165**, 430.

Fluorescamine
1. Jackowski, G. and Liew, C.C. (1980). *Anal. Biochem.*, **102**, 34.

p-Hydrazinoacridine
1. Carson, D.D. (1977). *Anal. Biochem.*, **78**, 428.

o-Phthaldialdehyde
1. Liebowitz, M.J. and Wang, R.W. (1984). *Anal. Biochem.*, **137**, 161.

SYPRO red and orange fluorescent stains
1. Steinberg, T.H., Jones, L.J., Haugland, R.P., and Singer, V.L. (1996). *Anal. Biochem.*, **239**, 223.
2. Steinberg, T.H., Haughland, R.P., and Singer, V.L. (1996). *Anal. Biochem.*, **239**, 238.

1.1.6 Direct detection methods
Via protein phosphorescence
1. Mardian, J.K.W. and Isenberg, I. (1976). *Anal. Biochem.*, **91**, 1.

Detection of SDS:polypeptides by chilling
1. Wallace, R.W., Yu, P.H., Dieckart, J.P., and Dieckart, J.W. (1974). *Anal. Biochem.*, **61**, 86.

Detection of SDS:protein complexes using pinacryptol yellow
1. Stoklosa, J.T. and Latz, H.W. (1974). *Biochem. Biophys. Res. Commun.*, **20**, 393.

Detection of SDS:polypeptides by precipitation with K⁺ ions
1. Nelles, L.P. and Dahmus, M.E. (1979). *Anal. Biochem.*, **73**, 522.

Detection of SDS:polypeptides using sodium acetate
1. Higgins, R.C. and Dahmus, M.E. (1979). *Anal. Biochem.*, **93**, 257.

Detection of SDS:polypeptides by reaction with cationic surfactant
1. Tagi, T., Kubo, K., and Isemura, T. (1977). *Anal. Biochem.*, **79**, 104.

1.2 Detection of radioactive proteins

Reviews

1. Dunbar, B.S. (1987). In *Two-dimensional electrophoresis and immunological techniques*, p. 103. Plenum Press, New York.
2. Latter, G.I., Burbank, S., Fleming, S., and Leavitt, J. (1984). *Clin. Chem.*, **30**, 1925.
3. Morrison, T.B. and Parkinson, S. (1994). *BioTechniques*, **17**, 922.
4. Quemeneur, E. and Simonnet, F. (1995). *BioTechniques*, **18**, 100.

Radiolabelling proteins *in vitro* prior to electrophoresis

1. Dunbar, B.S. (1987). In *Two-dimensional electrophoresis and immunological techniques*, p. 103. Plenum Press, New York.
2. Bailey, G.S. (1994). *Methods Mol. Biol.*, **32**, 441.
3. Pollard, J.W. (1995). *Methods Mol. Biol.*, **32**, 67.

Radiolabelling proteins after gel electrophoresis

1. Christopher, A.R., Nagpal, M.L., Carrol, A.R., and Brown, J.C. (1978). *Anal. Biochem.*, **85**, 404.
2. Elder, J.H., Pickelt, R.A., Hampton, J., and Lerner, R.A. (1977). *J. Biol. Chem.*, **252**, 6510.
3. Zapolski, E.J., Gersten, D.M., and Ledley, R.S. (1982). *Anal. Biochem.*, **123**, 325.
4. Wegmann, R., Balmain, N., Ricard-Blum, S., and Guha, S. (1995). *Cell Mol. Biol. (Noisy-le-grand)*, **41**, 1.

Indirect autoradiography using an X-ray intensifying screen

1. Laskey, R.A. (1980). In *Methods in enzymology* (ed. L. Grossman and K. Moldave), Vol. 65, p. 363. Academic Press, New York.
2. Laskey, R.A. and Mills, A.D. (1977). *FEBS Lett.*, **82**, 314.
3. Bonner, W.M. (1983). In *Methods in enzymology* (ed. S. Fleischer and B. Fleischer), Vol. 96, p. 215. Academic Press, New York.

Fluorography using PPO in DMSO

1. Bonner, W.M. and Laskey, R.A. (1974). *Eur. J. Biochem.*, **46**, 83.
2. Laskey, R.A. (1980). In *Methods in enzymology* (ed. L. Grossman and K. Moldave), Vol. 65, p. 363. Academic Press, New York.
3. Laskey, R.A. and Mills, A.D. (1977). *FEBS Lett.*, **82**, 314.
4. Bonner, W.M. (1984). In *Methods in enzymology* (ed. W.B. Jakoby), Vol. 104, p. 460. Academic Press, New York.

Fluorography using PPO in glacial acetic acid

1. Skinner, K. and Griswold, M.D. (1983). *Biochem. J.*, **209**, 281.

Fluorography using sodium salicylate in polyacrylamide gels

1. Chamberlain, J.P. (1979). *Anal. Biochem.*, **98**, 132.
2. Bonner, W.M. (1984). In *Methods in enzymology* (ed. W.B. Jakoby), Vol. 104, p. 460. Academic Press, New York.

Fluorography using sodium salicylate in agarose gels

1. Heegard, N.H.H., Hebsgaard, K.P., and Bjerrum , O.J. (1984). *Electrophoresis*, **5**, 230.

Fluorography using commercial reagents

1. Roberts, P.L. (1985). *Anal. Biochem.*, **147**, 521.
2. McConkey, E.H. and Anderson, C. (1984). *Electrophoresis*, **5**, 230.

Quenching of radiolabelled proteins by gel conditions

1. Harding, C.R. and Scott, I.R. (1983). *Anal. Biochem.*, **129**, 371.
2. Van Keuran, M.L., Goldman, D., and Merril, C.R. (1981). *Anal. Biochem.*, **116**, 248.

Image intensification

1. Rigby, P.J.W. (1981). *Amersham Research News No. 13.*
2. Laskey, R.A. (1981). *Amersham Research News No. 23.*

Double-labelled detection using X-ray film

1. Gruenstein, E.I. and Pollard, A.L. (1976). *Anal. Biochem.*, **76**, 452.
2. Kroenenberg, L.H. (1979). *Anal. Biochem.*, **93**, 189.
3. McConkey, E.H. (1979). *Anal. Biochem.*, **96**, 39.
4. Walton, K.E., Styer, D., and Gruenstein, E. (1979). *J. Biol. Chem.*, **254**, 795.
5. Cooper, P.C. and Burgess, A.W. (1982). *Anal. Biochem.*, **126**, 301.

Electronic data capture

1. Davidson, J.B. and Case, A. (1982). *Science*, **215**, 1398.
2. Burbeck, S. (1983). *Electrophoresis*, **4**, 127.

Storage phosphorimaging

1. Amemiya, Y. and Miyahura, J. (1988). *Nature*, **336**, 89.
2. Johnston, R.F., Pickett, S.C., and Barker, D.L. (1990). *Electrophoresis*, **11**, 355.
3. Ito, T., Suzuki, T., Lim, D.K., Wellman, S.E., and Ho, I.K. (1995). *J. Neurosci Methods*, **59**, 265.
4. Bolt, M.W. and Mahoney, P.A. (1997). *Anal. Biochem.*, **247**, 185.

1.3 Immunological methods

Incubation of the gel with radiolabelled antibody

1. Burridge, K. (1978). In *Methods in enzymology* (ed. V. Ginsburg), Vol. 50, p. 54. Academic Press, New York.
2. Kasamatsu, H. and Flory, P.J. (1978). *Virology*, **86**, 344.

Incubation of the gel with unlabelled antibody then with [^{125}I]protein A

1. Burridge, K. (1978). In *Methods in enzymology* (ed. V. Ginsburg), Vol. 50, p. 54. Academic Press, New York.
2. Adair, W.S., Jurivich, D., and Goodenough, U.W. (1978). *J. Cell. Biol.*, **79**, 281.
3. Bigelis, R. and Burridge, K. (1978). *Biochem. Biophys. Res. Commun.*, **82**, 322.

4. Saltzgaber-Müller, J. and Schatz, G. (1978). *J. Biol. Chem.*, **253**, 305.
5. Breslav, M., McKinney, A., Becker, J.M., and Naider, F. (1996). *Anal. Biochem.*, **239**, 213.

Incubation of the gel with fluorescein-labelled antibody

1. Groschel-Stewart, U., Schreiber, J., Mahlmeister, C., and Weber, K. (1976). *Histochemistry*, **46**, 229.
2. Stumph, W.E., Elgin, S.C.R., and Hood, L. (1974). *J. Immunol.*, **113**, 1752.

1.4 Glycoproteins
Periodic acid–Schiff (PAS) method
Dansyl hydrazine
1. Eckhardt, A.E., Hayes, C.E., and Goldstein, I.E. (1976). *Anal. Biochem.*, **73**, 192.
2. Gander, J.E. (1984). In *Methods in enzymology* (ed. W.B. Jakoby), Vol. 104, p. 447. Academic Press, New York.
3. Furlan, M., Perret, B.A., and Beck, E.A. (1979). *Anal. Biochem.*, **96**, 208.
4. Thornton, D.J., Carlstedt, I., and Sheehan, J.K. (1994). *Methods Mol. Biol.*, **32**, 119.
5. Thornton, D.J., Carlstedt, I., and Sheehan, J.K. (1996). *Mol. Biotechnol.*, **5**, 171.
Fuchsin
1. Fairbanks, G., Steck, T.L., and Wallach, D.L.H. (1971). *Biochemistry*, **10**, 2026.
2. Zaccharias, R.J., Zell, T.E., Morrison, J.H., and Woodlock, J.J. (1969). *Anal. Biochem.*, **31**, 148.
Alcian blue
1. Wardi, A.H. and Michos, G.A. (1972). *Anal. Biochem.*, **49**, 607.

Thymol–sulfuric acid method

1. Rauchsen, D. (1979). *Anal. Biochem.*, **99**, 474.
2. Gander, J.E. (1984). In *Methods in enzymology* (ed. W.B. Jakoby), Vol. 104, p. 447. Academic Press, New York.

Periodic acid–silver stain

1. Dubray, G. and Bezard, G. (1982). *Anal. Biochem.*, **119**, 325.

Stains-all

1. Green, M.R. and Pastewka, J.V. (1975). *Anal. Biochem.*, **65**, 66.
2. King, L.E. and Morrison, M. (1976). *Anal. Biochem.*, **71**, 223.
3. Goldberg, H.A. and Warner, K.J. (1997). *Anal. Biochem.*, **251**, 227.

Fluorescent lectins

1. Furlan, M., Perret, B.A., and Beck, E.A. (1979). *Anal. Biochem.*, **96**, 208.
2. Cotrufo, R., Monsurro, M.R., Delfino, G., and Geraci, G. (1983). *Anal. Biochem.*, **134**, 313.
3. Gander, J.E. (1984). In *Methods in enzymology* (ed. W.B. Jakoby), Vol. 104, p. 447. Academic Press, New York.

Lectins with covalently bound enzymes

1. Avigad, G. (1978). *Anal. Biochem.*, **86**, 443.
2. Wood, J.G. and Sarinana, F.O. (1971). *Anal. Biochem.*, **69**, 320.
3. Moroi, M. and Jung, S.M. (1984). *Biochim. Biophys. Acta*, **798**, 295.
4. Op De Beeck, L., Verlooy, J.E., Van Buul-Offers, S.C., and Du Caju, M.V. (1997). *J. Endocrinol.*, **154**, 1.

Radiolabelled lectins

1. Burridge, K. (1978). In *Methods in enzymology* (ed. V. Ginsburg), Vol. 50, p. 54. Academic Press, New York.
2. Rostas, J.A.P., Kelley, P.T., and Cotman, C.W. (1977). *Anal. Biochem.*, **80**, 366.
3. Koch, G.L.E. and Smith, M.J. (1982). *Eur. J. Biochem.*, **128**, 107.
4. Gershoni, J.M. and Palade, G. (1982). *Anal. Biochem.*, **127**, 396.
5. Dupuis, G. and Doucet, J.P. (1981). *Biochim. Biophys. Acta*, **669**, 171.
6. Leong, D.K., Oliva, L., and Butterworth, R.F. (1996). *Alcohol Clin. Exp. Res.*, **20**, 601.

Double-staining techniques

1. Dzandu, J.K., Deh, M.E., Barratt, D.L., and Wise, G.E. (1984). *Proc. Natl. Acad. Sci. USA*, **81**, 1733.
2. Dzandu, J.K. (1989). *Appl. Theor. Electrophoresis*, **1**, 137.

1.5 Phosphoproteins

Entrapment of liberated phosphate (ELP) using methyl green

1. Cutting, J.A. and Roth, T.F. (1973). *Anal. Biochem.*, **54**, 386.
2. Cutting, J.A. (1984). In *Methods in enzymology* (ed. W.B. Jakoby), Vol. 104, p. 451. Academic Press, New York.

ELP method using rhodamine B

1. Debruyne, I. (1983). *Anal. Biochem.*, **133**, 110.

Stains-all

1. Green, M.R., Pastewka, J.V., and Peacock, A.C. (1973). *Anal. Biochem.*, **56**, 43.

Silver stain

1. Satoh, K. and Busch, H. (1981). *Cell Biol. Int. Rep.*, **5**, 857.

Using electroblotting

1. Cantor, L., Lamy, F., and Lecocq, R.E. (1987). *Anal. Biochem.*, **160**, 414.

1.6 Lipoproteins

Staining prior to electrophoresis

1. Ressler, N., Springgate, R., and Kaufman, J. (1961). *J. Chromatogr.*, **6**, 409.

Staining after electrophoresis

1. Prat, J.P., Lamy, J.N., and Weill, J.D. (1969). *Bull. Soc. Chim. Biol.*, **51**, 1367.

Silver staining

1. Tsai, C.M. and Frasch, C.E. (1982). *Anal. Biochem.*, **119**, 115.
2. Goldman, D., Merril, C.R., and Ebert, M.H. (1980). *Clin. Chem.*, **26**, 1317.

1.7 Proteins with available thiol groups

1. Yamamoto, K., Okamoto, Y., and Sekine, T. (1978). *Anal. Biochem.*, **84**, 313.
2. Zelazowski, A.J. (1980). *Anal. Biochem.*, **103**, 307.

1.8 Cadmium-containing proteins using dipyridyl-ferrous iodide

1. Zelazowski, A.J. (1980). *Anal. Biochem.*, **103**, 307.

1.9 Specific enzymes

See Section 4 of this Appendix.

2. Detection of proteins on electroblots

2.1 Reviews

1. Langone, J.J. (1982). *J. Immunol. Methods*, **55**, 277.
2. Gershoni, J.M. and Palade, G. (1982). *Anal. Biochem.*, **127**, 396.
3. Towbin, H. and Gordon, J. (1984). *J. Immunol. Methods*, **72**, 313.
4. Symington, J. (1984). In *Two-dimensional electrophoresis of proteins: methods and applications* (ed. J.E. Celis and R. Bravo), p. 127. Academic Press, New York.
5. Bers, G. and Garfin, D. (1985). *BioTechniques*, **3**, 276.
6. Merril, C.R., Harasewych, M.G., and Harrington, M.G. (1986). In *Gel electrophoresis of proteins* (ed. M.J. Dunn), p. 313. Wright, Bristol.
7. Soutar, A.K. and Wade, D.P. (1989). In *Protein function: a practical approach* (ed. T.E. Creighton), p. 55. Oxford University Press, Oxford.
8. Dunn, M.J. (1997). *Methods Mol. Biol.*, **64**, 37.
9. Liang, F.T., Granstrom, D.E., Timoney, J.F., and Shi, Y.F. (1997). *Anal. Biochem.*, **250**, 61.
10. Suck, R.W. and Krupinska, K. (1996). *BioTechniques*, **21**, 418.

2.2 General detection methods

Amido black

1. Towbin, H., Staehelin, T., and Gordon, J. (1979). *Proc. Natl. Acad. Sci. USA*, **76**, 4350.
2. Gershoni, J.M. and Palade, G. (1982). *Anal. Biochem.*, **127**, 396.
3. Soutar, A.K. and Wade, D.P. (1989). In *Protein function: a practical approach* (ed. T.E. Creighton), p. 55. Oxford University Press, Oxford.
4. Gentile, F., Bali, E., and Pignalosa, G. (1997). *Anal. Biochem.*, **245**, 260.
5. Mahoney, C.W., Nakanishi, N., and Ohashi, M. (1997). *Anal. Biochem.*, **248**, 182.

Coomassie blue

1. Burnette, W.N. (1981). *Anal. Biochem.*, **112**, 195.
2. Houen, G., Bruun, L., and Barkholt, V. (1997). *Electrophoresis*, **18**, 701.

Fast green

1. Reinhart, M.P. and Malmud, D. (1982). *Anal. Biochem.*, **123**, 229.
2. Parchment, R.E., Ewing, C.M., and Shaper, J.H. (1986). *Anal. Biochem.*, **154**, 460.

Staining proteins with CPTS (copper phthalocyanine 3,4′,4″,4‴-tetrasulfonic acid tetrasodium salt

1. Bickar, D. and Reid, P.D. (1992). *Anal. Biochem.*, **203**, 109.

India ink

1. Hancock, K. and Tsang, V.C.W. (1983). *Anal. Biochem.*, **133**, 157.
2. Tsang, V.C.W., Hancock, K., Maddison, S.E., Beatty, A.L., and Moss, D.M. (1984). *J. Immunol.*, **132**, 2607.
3. Hailat, N.Q. and Hanash, S.M. (1995). *Indian J. Biochem. Biophys.*, **32**, 240.

Silver staining

1. Merril, C.R. and Pratt, M.E. (1986). *Anal. Biochem.*, **156**, 96.
2. Merril, C.R., Harrington, M., and Alley, V. (1984). *Electrophoresis*, **5**, 289.
3. Yuen, K.C.C., Johnson, T.K., Dennell, R.E., and Consigli, R.A. (1982). *Anal. Biochem.*, **126**, 398.
4. Brada, D. and Roth, J. (1984). *Anal. Biochem.*, **142**, 79.
5. Hailat, N.Q. and Hanash, S.M. (1995). *Indian J. Biochem. Biophys.*, **32**, 240.

Using colloidal gold

1. Moeremans, M., Daneels, G., and DeMey, J. (1986). *Anal. Biochem.*, **153**, 18.
2. Rohringer, R. and Holden, D.W. (1985). *Anal. Biochem.*, **144**, 118.

Using colloidal iron

1. Moeremans, M., Daneels, G., and DeMey, J. (1986). *Anal. Biochem.*, **153**, 18.
2. Patton, W.F., Lam, L., Su, Q., Lui, M., Erdjument-Bromage, H., and Tempst, P. (1994). *Anal. Biochem.*, **220**, 324.

Using BPSA (bathophenanthroline disulfonate)

1. Lim, M.J., Patton, W.F., Lopez, M.F., Spofford, K.H., Shojaee, N., and Shepro, D. (1997). *Anal. Biochem.*, **245**, 184.

Protein tagging with dinitrophenol (DNP)

1. Wojtkowiak, Z., Briggs, R.C., and Hnilica, L.S. (1983). *Anal. Biochem.*, **129**, 486.

Protein tagging with pyridoxil phosphate

1. Kittler, J.M., Meisler, N.T., Viceps-Madore, D., Cidlowski, J.A., and Thanassi, J.W. (1984). *Anal. Biochem.*, **137**, 210.

Protein tagging with biotin

1. Bio Radiations. (1985). From *Bio-Rad Laboratories, Ltd., No. 56 EG.*

2.3 Radioactive proteins

1. Towbin, H., Staehelin, T., and Gordon, J. (1979). *Proc. Natl. Acad. Sci. USA*, **76**, 4350.
2. Roberts, P.L. (1985). *Anal. Biochem.*, **147**, 521.

2.4 Glycoproteins

General procedures

1. Gershoni, J.M., Bayer, E.A., and Wilchek, M. (1985). *Anal. Biochem.*, **146**, 59.
2. Keren, Z., Berke, G., and Gershoni, J.M. (1986). *Anal. Biochem.*, **155**, 182.
3. Thornton, D.J., Carlstedt, I., and Sheehan, J.K. (1996). *Mol. Biotechnol.*, **5**, 171.
4. Thornton, D.J., Carlstedt, I., and Sheehan, J.K. (1994). *Methods Mol. Biol.*, **32**, 119.

Radiolabelled lectins

1. Gershoni, J.M. and Palade, G. (1982). *Anal. Biochem.*, **127**, 396.
2. Dunbar, B.S. (1987). In *Two-dimensional electrophoresis and immunological techniques*, p. 345. Plenum Press, New York.

Peroxidase labelled lectins

1. Moroi, M. and Jung, S.M. (1984). *Biochim. Biophys. Acta*, **798**, 295.
2. Carpenter, G.H., Proctor, G.B., Pankhurst, C.L., Linden, R.W., Shori, D.K., and Zhang, D.S. (1996). *Electrophoresis*, **17**, 91.

Using anti-lectin antibody

1. Glass, W.F., Briggs, R.C., and Hnilica, L.S. (1981). *Anal. Biochem.*, **115**, 219.

2.5 Immunological detection

Using radiolabelled antibody

1. Towbin, H., Staehelin, T., and Gordon, J. (1979). *Proc. Natl. Acad. Sci. USA*, **76**, 4350.
2. Dunbar, B.S. (1987). In *Two-dimensional electrophoresis and immunological techniques*, p. 341. Plenum Press, New York.
3. Hussain, A.A., Jona, J.A., Yamada, A., and Dittert, L.W. (1995). *Anal. Biochem.*, **224**, 221.
4. Bailey, G.S. (1994). *Methods Mol. Biol.*, **32**, 449.

Using fluorescent labelled antibody

Fluorescein

1. Towbin, H., Staehelin, T., and Gordon, J. (1979). *Proc. Natl. Acad. Sci. USA*, **76**, 4350.
2. The, T.H. and Feltkamp, T.E.W. (1970). *Immunology*, **18**, 865.

MDPF

1. Weigele, M., De Bernado, S., Leimgruber, W., Cleeland, R., and Grunber, E. (1973). *Biochem. Biophys. Res. Commun.*, **54**, 899.

Rhodamine
1. Towbin, H., Staehelin, T., and Gordon, J. (1979). *Proc. Natl. Acad. Sci. USA*, **76**, 4350.

Using enzyme-conjugated antibody

1. Blake, M.S., Johnson, K.H., Russel-Jones, G.J., and Gotschlich, E.C. (1984). *Anal. Biochem.*, **136**, 175.
2. Knecht, D.A. and Dimond, R.L. (1984). *Anal. Biochem.*, **136**, 180.
3. Towbin, H. and Gordon, J. (1984). *J. Immunol. Methods*, **72**, 313.

Using biotin-conjugated antibody

1. Dunbar, B.S. (1987). In *Two-dimensional electrophoresis and immunological techniques*, p. 343. Plenum Press, New York.
2. Hsu, S.M., Raine, L., and Fanger, H. (1981). *Am. J. Clin. Pathol.*, **75**, 816.

Using radiolabelled *S. aureus* protein A

1. Dunbar, B.S. (1987). In *Two-dimensional electrophoresis and immunological techniques*, p. 341. Plenum Press, New York.
2. Renart, J. and Sandoval, I.V. (1984). In *Methods in enzymology* (ed. W.B. Jakoby), Vol. 104, p. 455. Academic Press, New York.
3. Burnette, W.N. (1981). *Anal. Biochem.*, **112**, 195.
4. Brada, D. and Roth, J. (1984). *Anal. Biochem.*, **142**, 79.
5. Renart, J., Reiser, J., and Stark, G.R. (1979). *Proc. Natl. Acad. Sci. USA*, **76**, 3116.

Using protein G

1. Hu, G.R., Harrop, P., Warlow, R.S., Gacis, M.L., and Walls, R.S. (1997). *J. Immunol. Methods*, **202**, 113.

Immunogold staining

1. Hsu, Y. (1984). *Anal. Biochem.*, **142**, 221.
2. Surek, B. and Latzko, E. (1984). *Biochem. Biophys. Res. Commun.*, **121**, 284.

Immunodetection of pre-stained proteins

Coomassie blue
1. Jackson, P. and Thompson, R.J. (1984). *Electrophoresis*, **5**, 35.
Silver stain
1. Yuen, K.C.C., Johnson, T.K., Dennell, R.E., and Consigli, R.A. (1982). *Anal. Biochem.*, **126**, 398.

3. Quantitation of proteins

1. Goldring, J.P. and Ravaioli, L. (1996). *Anal. Biochem.*, **242**, 197.
2. Klein, D., Kern, R.M., and Sokol, R.Z. (1995). *Biochem. Mol. Biol. Int.*, **36**, 59.
3. Gillespie, P.G. and Gillespie, S.K. (1997). *Anal. Biochem.*, **246**, 239.

4. Enzyme localization

Enzyme	Reference
Acid phosphatase	50, 58, 81, 98, 188, 228, 233
Aconitase	98, 228, 233, 235
Adenine phosphoribosyl transferase (AMP pyrophosphorylase)	98, 175, 252
Adenosine deaminase	92, 182, 233, 242
Adenosine kinase	98, 139
Adenylate kinase	98, 228, 233, 270
ADP-glycogen transferase	80, 81
Alanine amino transferase	37, 98, 164, 228
Alcohol dehydrogenase	81, 98, 228, 233, 236, 238
Aldolase	41, 98, 148, 228, 233
Alkaline phosphatase	58, 81, 98, 188, 228, 246
Amine oxidase	81, 158
Amino acid oxidase	11, 81, 98, 102, 182
Aminoacyl tRNA synthetase	33
5-Aminolaevulinate synthase	53, 54
Aminopeptidase	244
AMP deaminase	8, 98, 182
Amylase	27, 29, 81, 98, 104, 112, 143, 178
Anthranilate phosphoribosyl transferase	105
Anthranilate synthetase	95
Arginase	71, 98, 182
Arginosuccinase	98, 182
Aromatic amino acid decarboxylase	145
Aromatic amino acid transaminase	228
Arylamidases	81
Arylsulfatases	34, 98, 114, 228
Aspartate amino transferase	51, 81, 98, 220, 228, 233, 271
Aspartate carbamoyl transferase	14, 92, 98
Aspartate oxidase	182
ATP pyrophosphatase	212, 222
Carbonic anydrase	61, 98, 114, 228
Carboxypeptidase	159, 197
Catalase	81, 93, 98, 228, 233, 159
Catechol oxidase	97
Cathepsin b	156, 174
Cellobiose phosphorylase	81
Cellulases	16, 23, 40, 66, 159
Cholinesterase	81, 99, 159
Chorismate synthase	159, 188
Chymotrypsin	7, 88, 159, 194, 240
Citrate synthase	46, 98
Creatine kinase	81, 90, 95, 98, 228, 272, 279
3′5′ Cyclic AMP phosphodiesterase	81, 98, 171
3′5′ Cylic nucleotide phosphodiesterase	254

Bibliography of polypeptide detection methods

Enzyme	Reference
Cystathione-β-synthase	268
Cytidine deaminase	69, 98, 117, 126, 268
DAHP synthase	188
Deoxyribonuclease	25, 111, 130, 136, 204
Dextranase	159, 178
Dextran sucrose	169
Dihydrouracil dehydrogenase	96
Dipeptidase	245
DNase	25, 111, 130, 136, 204
DNA helicases	231
DNA nucleotidyl transferase	283
DNA polymerase	81, 123, 159
Elastase and pro-elastase	60, 159, 194, 267
Enolase	98, 159, 227, 233
3-Enol pyruvoylshikimate-5-phosphate synthase	188
Esterases	42, 50, 81, 98, 107, 159, 214, 228, 235, 278, 273
Ferrisidophore reductase	172
Folate reductase	81
β-D-Fructofuranosidase	9, 72, 159, 178, 233, 218
Fructose-1,6-biphosphate	45, 114, 159, 188, 228
Fructose-5-dehydrogenase	6
Fructosyl transferase	217
L-Fucose dehydrogenase	159, 221
α-Fucosidase	81, 98, 159, 255
Fumarase	64, 98, 228, 233
Fumarate hydratase	159, 211, 253
Galactokinase	81, 98, 159, 255
Galactose-6-phosphate dehydrogenase	228
Galactose-1-phosphate uridyl transferase	81, 98, 159, 184
α-Galactosidase	18, 20, 98, 159, 173, 233, 257
β-Galactosidase	4, 5, 67, 81, 98, 159, 189
Galactosyl transferases	202
1,4-α-D-Glucan branching enzyme	218, 277
β-1,3-glucanase	124
Glucose oxidase	67, 81, 159
Glucose-6-phosphate dehydrogenase	48, 51, 98, 159, 228, 233
Glucose phosphate isomerase	159, 233
Glucose-1-phosphate uridylyl transferase	81, 98, 161, 162, 188
α-Glucosidase	81, 98, 159, 247
β-Glucosidase	21, 81, 98, 159
Glucosyl transferase	178
β-Glucuronidase	1, 78, 29, 98, 159, 228, 233
Glutamate dehydrogenase	81, 98, 137, 159, 182, 228
Glutamate-oxaloacetate transaminase	159, 181
Glutamate-pyruvate transaminase	38, 159
Glutaminase	55, 159

Bibliography of polypeptide detection methods

Enzyme	Reference
Glutamine synthetase	159, 167, 170, 239
γ-Glutamyl transferase	121, 159
Glutathione peroxidase	19, 98, 159
Glutathione reductase	98, 126, 211, 228, 273
Glutathione S-transferase	26, 134, 159, 210
Glyceraldehyde-3-phosphate dehydrogenase	35, 81, 98, 159, 188, 228, 233
Glycerol kinase	81, 159, 252
Glycerol-3-phosphate dehydrogenase	35, 81, 98, 147, 159, 228, 233
Glycogen phosphorylase	82
Glycogen synthase	141
Glycolate oxidase	62, 98, 159
Glycosidases	49, 81, 82, 159, 176–178, 206
Glycosyl transferases	159, 176, 177
Glyoxalase	35, 98, 140, 159, 198
Guanine deaminase (guanase)	22, 81, 98, 159
Guanylate kinase	98, 122, 159
Hexokinase	81, 98, 131, 159, 213, 233, 252
Hexosaminidase	159
Homoserine dehydrogenase	81, 159, 193
Hydroxyacyl coenzyme A dehydrogenase	47, 98, 159
D(-)3-Hydroxybutyrate dehydrogenase	159, 228
Hydroxysteroid dehydrogenase	81, 159, 189, 250
Hypoxanthine-guanine phosphoribosyl transferase	9, 22, 81, 98, 159, 228, 252, 259
Inorganic pyrophosphatase	9, 76, 81, 98, 228, 233, 256
Inosine triphosphatase	98
Invertase	9
Isocitrate dehydrogenase	81, 98, 105, 159, 187, 228, 233, 256
Isocitrate lyase	159, 209
α-Ketoglutarate semialdehye dehydrogenase	81
α-Ketoisocaproate dehydrogenase	81
α-Keto-β-methyl valeriate dehydrogenase	81
Lactate dehydrogenase	2, 32, 81, 98, 159, 228, 233, 234, 242, 266
Lactose synthase	202
Leucine aminopeptidase	159, 228, 244, 248
Levan sucrose	169
Lipase	81, 112, 179
Lipoyl dehydrogenase	81, 168
Lipoxygenase	59, 159
Malate dehydrogenase	3, 81, 98, 159, 200, 205, 211, 228, 233, 281
Malate synthase	159, 261
Malic enzyme	44, 98, 205, 233
Mannitol dehydrogenase	159, 260
Mannose-phosphate isomerase	98, 159, 186, 233
α-Mannosidase	82, 98, 159, 203, 233
Monophenol monoxygenase (tyrosinase)	251

Bibliography of polypeptide detection methods

Enzyme	Reference
β-*N*-acetyl hexosaminidase	98, 195, 233
NAD(HP) cytochrome reductase	See cytochrome reductase
NAD(P) nucleosidase	77, 98, 208
Nitrate reductase	81, 119, 159, 257
Nitrite reductase	115, 159, 257
Nucleases (see DNases and RNases)	144, 215
Nucleoside phosphorylase	65, 98, 159, 241
Nucleotidase	63, 81, 159, 282
Nucleotide pyrophosphatase	10, 159
Oestradiol 17β-dehydrogenase	132
Ornithine carbamoyl transferase	13, 70, 92, 98, 159
Oxidases	73, 81
PEP carboxylase	224
Peptidases (see dipeptidase, tripeptidase)	91, 149, 159, 207, 228, 233
Peroxidase	50, 150, 159, 228, 230, 233, 257
Phosphatases	152, 155, 188
Phosphodiesterases	84, 101, 110, 111, 159
Phosphofructokinase	28, 88, 91, 159, 187, 262
Phosphoglucoisomerase	56, 98, 228, 233
Phosphoglucomutase	81, 98, 142, 159, 228, 233
Phosphogluconate dehydrogenase	74, 81, 98, 228, 233
Phosphoglycerate kinase	17, 98, 233
Phosphoglyceromutase	36, 98, 159, 196, 233
Phosphoglycollate phosphatase	12, 98, 159
Phospholipase	154, 159, 229, 237
Phosphotransferases	252
Plasminogen activators	108, 159, 194
Polynucleotide phosphorylase	88, 138
Proteases	183, 258
Proteinase (see peptidases, subtilisin, chymotrypsin, trypsin, elastase, plasminogen activators)	7, 30, 68, 108, 112, 133, 151, 156, 159, 191, 194, 258
Proteinase inhibitors	75
Protein kinase	83, 109, 120, 125, 146, 155, 159, 201
Protein tyrosine phosphatase	31, 86
Pullalanase	275
Pyridoxine kinase	39, 98, 159
Pyrophosphatase	See inorganic pyrophosphatase
Pyruvate carboxylase	159, 224
Pyruvate decarboxylase	159, 188, 280
Pyruvate kinase	98, 118, 159, 165, 233, 257
Retinol dehydrogenase	159, 228
Ribonuclease	24, 25, 81, 116, 129, 159, 180, 228
RNA polymerase	159, 188
RNase	4, 25, 81, 116, 129, 180, 228
Serine dehydratase	85, 159, 274
Sorbitol dehydrogenase	35, 98, 159, 228

Bibliography of polypeptide detection methods

Enzyme	Reference
Subtilisin	135, 157
Sucrose phosphorylase	81, 159
Sulfite oxidase	43, 159
Superoxide dismutase	14, 15, 57, 87, 98, 159
Tetrahydrofolate dehydrogenase	189
Thermolysin	60
Threonine deaminase	100
Thymidine kinase	123
Transamidases	153, 243
Trehalase	128, 135, 159, 245
Triose-phosphate isomerase	1, 98, 128, 223, 233, 238
Tripeptide aminopeptidase	245
Trypsin	7, 88, 156, 159, 194, 265
Tyrosinase	251
Tyrosine amino transferase	159, 232, 264
UDPG dehydrogenase	81, 159
UDPG pyrophosphorylase	81, 98, 161, 162, 188
UMP kinase	89, 98
Urease	81, 159, 163, 225
Urokinase	91, 108, 159, 194, 263
Xanthine dehydrogenase	228
β-Xylanase	23

References

1. Aebersold, R., Winans, G.A., Teel, D.J., Milner, G.B., and Utter, F.M. (1987). *Manual for starch gel electrophoresis: a method for the detection of genetic variation*. NOAA Technical Report NMFS 61, US Dept of Commerce, National Marine Fisheries Service, Seattle, WA.
2. Allen, J.M. (1961). *Ann. N. Y. Acad. Sci.*, **94**, 937.
3. Allen, S.L. (1968). *Ann. N. Y. Acad. Sci.*, **151**, 190.
4. Alpers, D.H. (1969). *J. Biol. Chem.*, **244**, 1238.
5. Alpers, D.H., Steers, E., Shifrin, S., and Tomkins, G. (1968). *Ann. N. Y. Acad. Sci.*, **151**, 545.
6. Ameyama, M., Shinigawa, E., Matsushita, K., and Adachi, O. (1981). *J. Bacteriol.*, **145**, 184.
7. Andary, T.J. and Dabich, D. (1974). *Anal. Biochem.*, **57**, 457.
8. Anderson, J.E., Teng, Y.S., and Liblett, E.R. (1975). In *Birth defects: original article series*, Vol. 11, p. 295. The National Foundation, March of Dimes, New York.
9. Babczinski, P. (1980). *Anal. Biochem.*, **105**, 328.
10. Balakrishnan, C.V., Ravindranath, D.D., and Appaji, R.N. (1974). *Arch. Biochem. Biophys.*, **164**, 156.
11. Barker, R.F. and Hopkinson, D.A. (1977). *Ann. Hum. Genet.*, **41**, 27.
12. Barker, R.F. and Hopkinson, D.A. (1978). *Ann. Hum. Genet.*, **42**, 1.
13. Baron, D.N. and Buttery, J.E. (1972). *J. Clin. Pathol.*, **25**, 415.
14. Beauchamp, C. and Fridovitch, I. (1971). *Anal. Biochem.*, **44**, 276.

15. Beckman, G., Lundgren, A., and Tarnvik, A. (1973). *Hum. Hered.*, **23**, 338.
16. Beguin, P. (1983). *Anal. Biochem.*, **131**, 333.
17. Beutler, E. (1969). *Biochem. Genet.*, **3**, 189.
18. Beutler, E. and Kuhl, W. (1972). *J. Biol. Chem.*, **247**, 7195.
19. Beutler, E. and West, C. (1974). *Am. J. Hum. Genet.*, **26**, 255.
20. Beutler, E., Guinto, E., and Kuhl, W. (1973). *Am. J. Hum. Genet.*, **25**, 42.
21. Beutler, E., Kuhl, W., Trinidad, F., Teplitz, R., and Nadler, H. (1971). *Am. J. Hum. Genet.*, **23**, 62.
22. Bieber, A.C. (1974). *Anal. Biochem.*, **60**, 206.
23. Biely, P., Markovic, O., and Mislovicová, D. (1985). *Anal. Biochem.*, **144**, 147.
24. Biswas, S. and Hollander, V.P. (1969). *J. Biol. Chem.*, **244**, 4185.
25. Blank, A., Suigiyama, R.H., and Dekker, C.A. (1982). *Anal. Biochem.*, **120**, 267.
26. Board, P.G. (1980). *Anal. Biochem.*, **105**, 147.
27. Boettcher, D. and De La Lande, F.A. (1969). *Anal. Biochem.*, **28**, 510.
28. Brock, D.J.H. (1969). *Biochem. J.*, **113**, 235.
29. Brown, T.L., Yet, M.G., and Wold, F. (1982). *Anal. Biochem.*, **122**, 164.
30. Burdett, P.E., Kipps, A.E., and Whitehead, P.H. (1976). *Anal. Biochem.*, **72**, 315.
31. Burridge, K. and Nelson, A. (1995). *Anal. Biochem.*, **232**, 56.
32. Carda-Abella, P., Perez-Cuadrada, S., Laru-Baruque, S., Gil-Grande, L., and Nunez-Puertas, A. (1982). *Cancer*, **49**, 80.
33. Chang, G.-G., Denq, R.-Y., and Pan, F. (1985). *Anal. Biochem.*, **149**, 474.
34. Chang, P.L., Ballantyne, S.R., and Davidson, R.G. (1979). *Anal. Biochem.*, **97**, 36.
35. Charlesworth, D. (1972). *Ann. Hum. Genet.*, **35**, 477.
36. Chen, S.H., Anderson, J., Giblett, E.R., and Lewis, M. (1974). *Am. J. Hum. Genet.*, **26**, 73.
37. Chen, S.H., Giblett, E.R., Anderson, J.E., and Fosuum, B.L.G. (1972). *Ann. Hum. Genet.*, **35**, 401.
38. Chen, S-H. and Giblett, E.R. (1971). *Science*, **173**, 148.
39. Chern, C.J. and Beutler, E. (1976). *Ann. Hum. Genet.*, **28**, 9.
40. Chernoglazov, V.M., Ermolova, O.V., Vozny, Y.V., and Klyosov, A.A. (1989). *Anal.Biochem.*, **182**, 250.
41. Christen, P. and Gasser, A. (1980). *Anal. Biochem.*, **109**, 270.
42. Coates, P.M., Mestriner, M.A., and Hopkinson, D.A. (1975). *Ann. Hum. Genet.*, **39**, 1.
43. Cohen, H.J. (1973). *Anal. Biochem.*, **53**, 208.
44. Cohen, P.T.W. and Omenn, G.S. (1972). *Biochem. Genet.*, **7**, 303.
45. Colombo, G. and Marcus, F. (1973). *FEBS Lett.*, **40**, 37.
46. Craig, I. (1973). *Biochem. Genet.*, **9**, 351.
47. Craig, I., Tolley, E., and Borrow, M. (1976). In *Birth defects: original article series*, Vol. 12, p. 114. The National Foundation, March of Dimes, New York.
48. Criss, W.E. and McKerns, K.W. (1968). *Biochemistry*, **7**, 125.
49. Cruickshank, R.H. and Wade, G.C. (1980). *Anal. Biochem.*, **107**, 177.
50. Cullis, C.A. and Kolodynska, K. (1975). *Biochem. Genet.*, **13**, 687.
51. Dao, M.L., Johnson, B.C., and Hartman, P.E. (1980). *Proc. Natl. Acad. Sci. USA*, **79**, 2860.
52. Davidson, R.G., Cortner, J.A., Rattazi, M.C., Ruddle, F.H., and Lubs, H.A. (1970). *Science*, **169**, 391.
53. Davies, R.C. and Neuberger, A. (1973). *Biochem. J.*, **133**, 471.

54. Davies, R.C. and Neuberger, A. (1979). *Biochem. J.*, **177**, 649.
55. Davis, J.N. and Prusiner, S. (1973). *Anal. Biochem.*, **54**, 272.
56. De Lorenzo, R.J. and Ruddle, F.H. (1969). *Biochem. Genet.*, **3**, 151.
57. De Rosa, G., Duncan, D.S., Keen, C.L., and Hurley, L.S. (1979). *Biochim. Biophys. Acta*, **566**, 32.
58. Debruyne, I. (1983). *Anal. Biochem.*, **133**, 110.
59. DeLumen, B.O. and Kazeniac, S.J. (1976). *Anal. Biochem.*, **72**, 428.
60. Dijkhof, J. and Poort, C. (1977). *Anal. Biochem.*, **83**, 315.
61. Drescher, D.G. (1978). *Anal. Biochem.*, **90**, 349.
62. Duley, J. and Holmes, R.S. (1974). *Genetics*, **76**, 93.
63. Dvorak, H.F. and Heppel, L.A. (1968). *J. Biol. Chem.*, **243**, 2647.
64. Edwards, Y.H. and Hopkinson, D.A. (1978). *Ann. Hum. Genet.*, **42**, 303.
65. Edwards, Y.H., Hopkinson, D.A., and Harris, H. (1971). *Ann. Hum. Genet.*, **34**, 395.
66. Erickson, K.E. and Petterson, B. (1973). *Anal. Biochem.*, **56**, 618.
67. Evangelista, R.A., Pollak, A., and Templeton, E.F.G. (1991). *Anal. Biochem.*, **197**, 213.
68. Every, D. (1981). *Anal. Biochem.*, **116**, 519.
69. Fan, L.L. and Masters, B.S.S. (1974). *Arch. Biochem. Biophys.*, **165**, 665.
70. Farkas, D.H., Skombra, C.J., Anderson, G.R., and Hughes, R.G., Jr. (1987). *Anal. Biochem.*, **160**, 421.
71. Farron, F. (1973). *Anal. Biochem.*, **53**, 264.
72. Faye, L. (1981). *Anal. Biochem.*, **112**, 90.
73. Feinstein, R.N. and Lindahl, R. (1973). *Anal. Biochem.*, **56**, 353.
74. Fildes, R.A. and Parr, C.W. (1963). *Nature*, **200**, 890.
75. Filho, J.X. and De Azevedo Moreira, R. (1978). *Anal. Biochem.*, **84**, 296.
76. Fischer, R.A., Turner, B.M., Dorkin, H.L., and Harris, H. (1974). *Ann. Hum. Genet.*, **37**, 341.
77. Flechner, L., Hirschhorn, S., and Bekjerkunst, A. (1968). *Life Sci.*, **7**, 1327.
78. Fondo, E.Y. and Bartalos, M. (1969). *Biochem. Genet.*, **3**, 591.
79. Franke, U. (1976). *Am. J. Hum. Genet.*, **28**, 357.
80. Frederick, J.F. (1968). *Ann. N. Y. Acad. Sci.*, **151**, 413.
81. Gabriel, O. (1971). In *Methods in enzymology* (ed. W.B. Jakoby), Vol. 22, p. 578. Academic Press, New York.
82. Gabriel, O. and Wang, S.F. (1969). *Anal. Biochem.*, **27**, 545.
83. Gagelman, M., Pyerin, W., Kübler, D., and Kinzel, V. (1979). *Anal. Biochem.*, **93**, 52.
84. Gangyi, H. (1990). *Anal. Biochem.*, **185**, 90.
85. Gannon, F. and Jones, K.M. (1977). *Anal. Biochem.*, **79**, 594.
86. Gates, R.E., Miller, J.L., and Kinge, L.E., Jr. (1996). *Anal. Biochem.*, **237**, 208.
87. Geller, B.L. and Winge, D.R. (1983). *Anal. Biochem.*, **128**, 86.
88. Gertler, A., Trencer, Y., and Tinman, G. (1973). *Anal. Biochem.*, **54**, 270.
89. Giblett, E.R., Anderson, J.A., Chen, S.H., Teng, Y.S., and Cohen, F. (1974). *Am. J. Hum. Genet.*, **26**, 627.
90. Graeber, G.M., Reardon, M.J., Fleming, A.W., Head, H.D., Zajtchuk, R., Brott, W.H., *et al.* (1981). *Ann. Thor. Surg.*, **32**, 230.
91. Granelli-Piperano, A. and Reich, E. (1978). *J. Exp. Med.*, **148**, 223.
92. Grayson, J.E., Yon, R.P., and Buterworth, P.J. (1979). *Biochem. J.*, **183**, 239.
93. Gregory, E.M. and Fridovich, E.M. (1974). *Anal. Biochem.*, **58**, 57.

94. Grove, T.H. and Levy, H.R. (1975). *Anal. Biochem.*, **65**, 458.
95. Hall, N. and DeLuca, M. (1976). *Anal. Biochem.*, **76**, 561.
96. Hallock, R.O. and Yamada, E.W. (1973). *Anal. Biochem.*, **56**, 84.
97. Hanold, G.R. and Stahmann, N. (1968). *Cereal Chem.*, **45**, 99.
98. Harris, H. and Hopkinson, D.A. (1976). *Handbook of enzyme electrophoresis in human genetics.* North Holland, Amsterdam, and supplements (1977) and (1978).
99. Harris, H., Hopkinson, D.A., and Robson, E.B. (1962). *Nature*, **196**, 1296.
100. Hatfield, G.W. and Umbarger, H.E. (1980). *J. Biol. Chem.*, **245**, 1736.
101. Hawley, D.M., Tsou, K.C., and Hodes, M.E. (1981). *Anal. Biochem.*, **117**, 18.
102. Hayes, M.B. and Wellner, D. (1969). *J. Biol. Chem.*, **244**, 6636.
103. Heeb, M.J. and Gabriel, O. (1984). In *Methods in enzymology* (ed. W.B. Jakoby), Vol. 104, p. 416. Academic Press, New York.
104. Heller, H. and Kulka, R.G. (1968). *Biochim. Biophys. Acta*, **165**, 393.
105. Henderson, C.J., Nagano, H., Zalkin, H., and Hurang, L.H. (1970). *J. Biol. Chem.*, **245**, 1416.
106. Henderson, N.S. (1968). *Ann. N. Y. Acad. Sci.*, **151**, 429.
107. Herd, J.K. and Tschida, J. (1975). *Anal. Biochem.*, **68**, 218.
108. Heussen, C. and Dowdle, E.B. (1980). *Anal. Biochem.*, **102**, 192.
109. Hirsch, A. and Rosen, M. (1974). *Anal. Biochem.*, **60**, 389.
110. Hodes, M.E. and Retz, J.E. (1981). *Anal. Biochem.*, **110**, 150.
111. Hodes, M.E., Crisp, M., and Gelb, E. (1977). *Anal. Biochem.*, **80**, 239.
112. Hofelmann, M., Kittsteiner-Eberle, R., and Schreier, P. (1983). *Anal. Biochem.*, **128**, 217.
113. Hopkinson, D.A., Coppock, J.S., Muhlemann, M.F., and Edwards, Y.H. (1974). *Ann. Hum. Genet.*, **38**, 155.
114. Hubert, E. and Marcus, F. (1974). *FEBS Lett.*, **40**, 37.
115. Hucklesby, D.P. and Hageman, R.H. (1973). *Anal. Biochem.*, **56**, 591.
116. Huet, J., Sentenac, A., and Fromageot, P. (1978). *FEBS Lett.*, **94**, 28.
117. Ichihara, K., Kusunose, E., and Kusunose, M. (1973). *Eur. J. Biochem.*, **38**, 463.
118. Imamura, K. and Tanaka, T. (1972). *J. Biochem.*, **71**, 1043.
119. Ingle, J. (1968). *Biochem. J.*, **108**, 715.
120. Isbell, J.C., Christian, S.T., Mashburn, N.A., and Bell, P.D. (1995). *Life Sci.*, **57**, 1701.
121. Izumi, M. and Taketa, K. (1981). In *Electrophoresis '81* (ed. R.C. Allen and P. Arnaud), p. 709. De Gruyter, Berlin.
122. Jamil, T., Fisher, R.A., and Harris, H. (1976). *Hum. Hered.*, **25**, 402.
123. Jovin, T.M., Englund, P.T., and Bertsch, L.L. (1969). *J. Biol. Chem.*, **244**, 2996.
124. Kalix, S. and Buchenauer, H. (1995). *Electrophoresis*, **16**, 1016.
125. Kameshita, I. and Fujisawa, H. (1989). *Anal. Biochem.*, **183**, 139.
126. Kaplan, J.C. and Beutler, E. (1967). *Biochem. Biophys. Res. Commun.*, **29**, 605.
127. Kaplan, J.C. and Beutler, E. (1968). *Nature*, **217**, 256.
128. Kaplan, J.C., Teeple, L., Shore, N., and Beutler, E. (1968). *Biochem. Biophys. Res. Commun.*, **31**, 768.
129. Karn, R.C., Crisp, M., Yount, E.A., and Hodes, M.E. (1979). *Anal. Biochem.*, **96**, 464.
130. Karpetsky, T., Brown, G.E., McFarland, E., Rahman, A., Rictro, K., Roth, W., *et al.* (1981). In *Electrophoresis '81* (ed. R.C. Allen and P. Arnaud), p. 674. De Gruyter, Berlin.

131. Katzen, H.M. and Schimke, R.T. (1965). *Proc. Natl. Acad. Sci. USA*, **54**, 1218.
132. Kavanolas, H.J., Baedecker, M.L., and Engel, L.L. (1970). *J. Biol. Chem.*, **245**, 4722.
133. Kelleher, P.J. and Juliano, R.C. (1984). *Anal. Biochem.*, **136**, 470.
134. Kenney, W.C. and Boyer, T.D. (1981). *Anal. Biochem.*, **116**, 344.
135. Killick, K.A. and Wang, L.W. (1980). *Anal. Biochem.*, **106**, 367.
136. Kim, H.S. and Liao, T.H. (1982). *Anal. Biochem.*, **119**, 62.
137. Kimura, K., Miyakawa, A., Imai, T., and Sasakawa, T. (1977). *J. Biochem.*, **81**, 467.
138. Klee, C.B. (1969). *J. Biol. Chem.*, **244**, 2558.
139. Klobucher, L.A., Nichols, E.A., Kucherlapati, R.S., and Ruddle, F.H. (1976). In *Birth defects: original article series*, Vol. 12, p. 171. The National Foundation, March of Dimes, New York.
140. Kompf, J., Bissbort, S., Gussman, S., and Ritter, H. (1975). *Humangenetik*, **27**, 141.
141. Krisman, C.R. and Blumenfeld, M.L. (1986). *Anal. Biochem.*, **154**, 409.
142. Kühn, P., Schmidtmann, U., and Spielmann, W. (1977). *Hum. Genet.*, **35**, 219.
143. Lacks, S.A. and Springhorn, S.S. (1980). *J. Biol. Chem.*, **255**, 7467.
144. Lacks, S.A., Springhorn, S.S., and Rosenthal, A.L. (1979). *Anal. Biochem.*, **100**, 357.
145. Landon, M. (1977). *Anal. Biochem.*, **77**, 293.
146. Lehel, C., Daniel-Issakani, S., Brasseur, M., and Strulovici, B. (1997). *Anal. Biochem.*, **244**, 340.
147. Leibenguth, F. (1975). *Biochem. Genet.*, **13**, 263.
148. Lewinski, N.D. and Dekker, E.E. (1978). *Anal. Biochem.*, **87**, 56.
149. Lewis, W.H.P. and Harris, H. (1967). *Nature*, **215**, 315.
150. Liu, E.H. and Gibson, D.M. (1977). *Anal. Biochem.*, **79**, 597.
151. Lo, K.W., Aoyagi, S., and Tsou, K.C. (1981). *Anal. Biochem.*, **117**, 24.
152. Lomholt, B. (1975). *Anal. Biochem.*, **65**, 569.
153. Lorand, L., Seifring, G.E., Tong, Y.S., Bruner-Lorand, J., and Gray, A.J. (1979). *Anal. Biochem.*, **93**, 453.
154. Lucas, M., Sanchez-Margalet, V., Pedrera, C., and Bellido, M.L. (1995). *Anal. Biochem.*, **231**, 277.
155. Lutz, M.P., Pinon, D.I., and Miller, L.J. (1994). *Anal. Biochem.*, **220**, 268.
156. Lynn, K.R. and Clevette-Radford, N.A. (1981). *Anal. Biochem.*, **117**, 280.
157. Lyublinskaya, L.A., Belyaev, S.V., Strongin, A.Ya., Matyash, L.F., and Stepanov, V.M. (1974). *Anal. Biochem.*, **62**, 371.
158. Ma Lin, A.W. and Castell, D.O. (1975). *Anal. Biochem.*, **69**, 637.
159. Manchenko, G.P. (1994). *Handbook of detection of enzymes on electrophoretic gels*. CRC Press, Boca Raton, FL.
160. Mannowitz, P., Goldstein, L., and Bellomo, F. (1978). *Anal. Biochem.*, **89**, 423.
161. Manrow, R.E. and Dottin, R.P. (1982). *Anal. Biochem.*, **120**, 181.
162. Manrow, R.F. and Dottin, R.P. (1980). *Proc. Natl. Acad. Sci. USA*, **77**, 730.
163. Martin de Llano, J.J. (1989). *Anal. Biochem.*, **177**, 37.
164. Matsuzawa, T., Kobayashi, T., Ogawa, H., and Kasahara, M. (1997). *Biochim. Biophys. Acta*, **1340**, 115.
165. Melendez-Heria, C., Corzo, J., and Perez, J. (1981). In *Electrophoresis '81* (ed. R.C. Allen and P. Arnaud), p. 693. De Gruyter, Berlin.
166. Merril, C.R., Harasewych, M.G., and Harrington, M.G. (1986). In *Gel electrophoresis of proteins* (ed. M.J. Dunn), p. 323. Wright, Bristol.

167. Meyer, J.M. and Stadtman, E.R. (1981). *J. Bacteriol.*, **146**, 705.
168. Millard, S.A., Kubose, A., and Gal, E.M. (1969). *J. Biol. Chem.*, **244**, 2511.
169. Miller, A.W. and Robyt, J.F. (1986). *Anal. Biochem.*, **156**, 357.
170. Miller, R.E., Shelton, E., and Stadtman, E.R. (1974). *Arch. Biochem. Biophys.*, **163**, 155.
171. Monn, E. and Christiansen, R.O. (1971). *Science*, **173**, 540.
172. Moody, M.D. and Dailey, H.A. (1983). *Anal. Biochem.*, **134**, 235.
173. Morizot, D.C. and Schmidt, M.E. (1990). In *Electrophoretic and isoelectric focusing techniques in fisheries management* (ed. D.H. Whitmore). CRC Press, Boca Raton, FL.
174. Mort, J.S. and Leduc, M. (1982). *Anal. Biochem.*, **119**, 148.
175. Mowbray, S., Watson, B., and Harris, H. (1972). *Ann. Hum. Genet.*, **36**, 153.
176. Mukasa, H., Tsumori, H., and Uezono, Y. (1994). *Electrophoresis*, **15**, 255.
177. Mukasa, H., Tsumori, H., and Takeda, H. (1994). *Electrophoresis*, **15**, 911.
178. Musaka, H., Shimura, A., and Tsumori, H. (1982). *Anal. Biochem.*, **123**, 276.
179. Nachlase, M.M., Morris, B., Rosenblatt, D., and Seligman, A.M. (1960). *J. Biophys. Biochem. Cytol.*, **7**, 261.
180. Nadano, D., Yasuda, T., Sawazaki, K., Takeshita, H., and Kishi, K. (1996). *Electrophoresis*, **17**, 104.
181. Nealon, D.A. and Rej, R. (1987). *Anal. Biochem.*, **161**, 64.
182. Nelson, R.L., Povey, S., Hopkinson, D.A., and Harris, H. (1977). *Biochem. Genet.*, **15**, 1023.
183. Nestler, H.P. and Doseff, A. (1997). *Anal. Biochem.*, **251**, 122.
184. Ng, W.G., Bergren, W.R., Fields, M., and Donnell, G.N. (1969). *Biochem. Biophys. Res. Commun.*, **37**, 354.
185. Nicholls, E.A., Elsevier, S.M., and Ruddle, F.H. (1974). *Cytogenet. Cell Genet.*, **13**, 275.
186. Nichols, E.A., Chapman, V.M., and Ruddle, F.H. (1973). *Biochem. Genet.*, **8**, 47.
187. Niessner, N. and Beutler, E. (1974). *Biochem. Med.*, **8**, 73.
188. Nimmo, H.G. and Nimmo, G.A. (1982). *Anal. Biochem.*, **121**, 17.
189. Nixon, P.F. and Blakely, R.C. (1968). *J. Biol.Chem.*, **243**, 4722.
190. Norden, A.G.W. and O'Brien, J.S. (1975). *Proc. Natl. Acad. Sci. USA*, **72**, 240.
191. North, M.J. and Harwood, J.M. (1979). *Biochim. Biophys. Acta*, **566**, 222.
192. O'Conner, J.L. (1977). *Anal. Biochem.*, **78**, 205.
193. Ogilvie, J.W., Sightler, J.H., and Clark, R.B. (1969). *Biochemistry*, **8**, 3557.
194. Ohlsson, B.G., Weström, B.R., and Karlsson, B.W. (1986). *Anal. Biochem.*, **152**, 239.
195. Okada, S., Veath, M.L., Lerov, J., and O'Brien, J.S. (1971). *Am. J. Hum. Genet.*, **23**, 55.
196. Omenn, G.S. and Cheung, S.C.Y. (1974). *Am. J. Hum. Genet.*, **26**, 393.
197. Pacaud, M. and Uriel, J. (1971). *Eur. J. Biochem.*, **23**, 435.
198. Parr, C.W., Bagster, I.A., and Welch, S.G. (1977). *Biochem. Genet.*, **15**, 109.
199. Paszkiewicz-Gadek, A., Gindzienski, A., and Porowska, H. (1995). *Anal. Biochem.*, **226**, 263.
200. Peterson, A.C. (1974). *Nature*, **248**, 561.
201. Phan-Dinh-Tuy, F., Weber, A., Henry, J., Cotterau, D., and Kahn, A. (1982). *Anal. Biochem.*, **127**, 73.
202. Pierce, M., Cummings, R.D., and Roth, S. (1980). *Anal. Biochem.*, **102**, 441.

203. Poenaru, L. and Dreyfus, J.C. (1973). *Biochim. Biophys. Acta*, **303**, 171.
204. Porter, A.L.G. (1971). *Anal. Biochem.*, **117**, 28.
205. Povey, S., Wilson, D.E., Harris, H., Gormley, I.P., Perry, P., and Buckton, K.E. (1975). *Ann. Hum. Genet.*, **39**, 203.
206. Price, R.G. and Dance, N. (1967). *Biochem. J.*, **105**, 877.
207. Rapley, S., Lewis, W.H.P., and Harris, H. (1971). *Ann. Hum. Genet.*, **34**, 307.
208. Ravazzolo, R., Bruzzone, G., Garrè, C., and Ajmar, F. (1976). *Biochem. Genet.*, **14**, 877.
209. Reeves, H.C. and Volk, M.J. (1972). *Anal. Biochem.*, **48**, 437.
210. Ricci, G., Bello, M.L., Caccuri, A.M., Galiazzo, F., and Federici, G. (1984). *Anal. Biochem.*, **143**, 226.
211. Richardson, B.J., Baverstock, P.R., and Adams, M. (1986). In *Allozyme electrophoresis: a handbook for animal systematics and population studies.* Academic Press, Sydney.
212. Risi, S., Höckel, M., Hulla, F.W., and Dose, K. (1977). *Eur. J. Biochem.*, **81**, 103.
213. Rogers, P.A., Fisher, R.A., and Harris, H. (1975). *Biochem. Genet.*, **13**, 857.
214. Rosenberg, V., Roegner, V., and Becker, F.F. (1975). *Anal. Biochem.*, **66**, 206.
215. Rosenthal, A.L. and Lacks, S.A. (1977). *Anal. Biochem.*, **80**, 76.
216. Rothe, G.M. and Maurer, W.D. (1986). In *Gel electrophoresis of proteins* (ed. M.J. Dunn), p. 37. Wright, Bristol.
217. Russell, R.R.B. (1979). *Anal. Biochem.*, **97**, 173.
218. Satoh, K. and Sato, K. (1980). *Anal. Biochem.*, **108**, 16.
219. Satoh, K., Imai, F., and Sato, K. (1978). *FEBS Lett.*, **95**, 239.
220. Scandalios, J.G., Sorenson, J.C., and Ott, L.A. (1975). *Biochem. Genet.*, **13**, 759.
221. Schachter, H., Sarney, J., McGuire, E.J., and Roseman, S. (1969). *J. Biol. Chem.*, **244**, 4785.
222. Schäfer, H.J., Scheurich, P., and Rathgeber, G. (1978). *Hoppe–Seyler's Z. Physiol. Chem.*, **359**, 1441.
223. Scopes, R.K. (1964). *Nature*, **201**, 924.
224. Scrutton, M.C. and Fatebene, F. (1975). *Anal. Biochem.*, **69**, 247.
225. Shaik-M, M.B., Guy, A.L., and Pancholy, S.K. (1980). *Anal. Biochem.*, **103**, 140.
226. Shapira, E., De Gregorio, R.R., Matalon, R., and Nadler, H.R. (1975). *Biochem. Biophys. Res. Commun.*, **62**, 448.
227. Sharma, H.K. and Rothstein, M. (1979). *Anal. Biochem.*, **98**, 226.
228. Shaw, C.R. and Prasad, R. (1980). *Biochem. Genet.*, **4**, 297.
229. Shier, W.T. and Troffer, J.T. (1978). *Anal. Biochem.*, **87**, 604.
230. Shimoni, M. (1994). *Anal. Biochem.*, **220**, 36.
231. Shukla, S.K. and McCarthy, D. (1994). *Nucleic Acids Res.*, **22**, 1626.
232. Shuler, J.K. and Tryfiates, G.P. (1977). *Enzyme*, **22**, 262.
233. Siciliano, M.J. and Shaw, C.R. (1976). In *Chromatographic and electrophoretic techniques* (ed. I. Smith), Vol. 2, p. 185. William Heinemann Medical Books, Ltd., London.
234. Simon, K., Chaplin, E.R., and Diamond, I. (1977). *Anal. Biochem.*, **79**, 571.
235. Slaughter, C.A., Hopkinson, D.A., and Harris, H. (1975). *Ann. Hum. Genet.*, **39**, 193.
236. Smith, H., Hopkinson, D.A., and Harris, H. (1971). *Ann. Hum. Genet.*, **34**, 251.
237. Smyth, C.J. and Wadstrom, T. (1975). *Anal. Biochem.*, **65**, 137.

238. Sofer, W. and Ursprung, H. (1968). *J. Biol. Chem.*, **243**, 3110.
239. Soliman, A. and Nordlund, S. (1989). *Biochim. Biophys. Acta*, **994**, 138.
240. Solomon, S.S., Palazzolo, M., and King, L.E. (1977). *Diabetes*, **26**, 967.
241. Spencer, N., Hopkinson, D.A., and Harris, H. (1968). *Ann. Hum. Genet.*, **32**, 9.
242. Springell, P.H. and Lynch, T.A. (1976). *Anal. Biochem.*, **74**, 251.
243. Stenberg, P. and Stenflo, J. (1979). *Anal. Biochem.*, **93**, 445.
244. Strogin, A.Y., Azarenkova, N.M., Vaganov, T.I., Levin, A.D., and Stepanov, V.M. (1976). *Anal. Biochem.*, **74**, 597.
245. Suguira, M., Ho, Y., Hirano, K., and Sawaki, S. (1977). *Anal. Biochem.*, **81**, 481.
246. Sussman, H.H., Small, P.A., and Cotlove, E. (1968). *J. Biol. Chem.*, **243**, 160.
247. Swallow, D.M., Corney, G., Harris, H., and Hirschhorn, R. (1975). *Ann. Hum. Genet.*, **38**, 391.
248. Takamiya, S., Ohsima, T., Tanizawa, K., and Soda, K. (1983). *Anal. Biochem.*, **130**, 266.
249. Teng, Y.S., Anderson, J.E., and Giblett, E.R. (1975). *Am. J. Hum. Genet.*, **27**, 492.
250. Thomas, J.L., Strickler, R.C., and Evans, B.W. (1997). *Biochemistry*, **36**, 9029.
251. Thomas, P., Delincee, H., and Diehl, J.F. (1978). *Anal. Biochem.*, **88**, 138.
252. Tischfield, J.A., Bernhard, H.P., and Ruddle, F.H. (1973). *Anal. Biochem.*, **53**, 545.
253. Tolley, E. and Craig, I. (1975). *Biochem. Genet.*, **13**, 867.
254. Tsou, K.C., Lo, K.W., and Yip, K.F. (1974). *FEBS Lett.*, **45**, 47.
255. Turner, B.M., Beratis, N.G., Turner, V.S., and Hirschhorn, K. (1974). *Clin. Chim. Acta*, **57**, 29.
256. Turner, B.M., Fisher, R.A., and Harris, H. (1974). *Ann. Hum. Genet.*, **37**, 455.
257. Vallejos, C.E. (1983). In *Isozymes in plant genetics and breeding, part A* (ed. S.D. Tanskley and T.J. Orton). Elsevier, Amsterdam.
258. Vasquez-Peyronel, D. and Cantera, A.M. (1995). *Electrophoresis*, **16**, 1894.
259. Vasquez, B. and Bieber, A.L. (1978). *Anal. Biochem.*, **84**, 504.
260. Veng, S.T.H., Hartanowicz, P., Lewandoski, C., Keller, J., Holick, M., and McGuiness, T. (1976). *Biochemistry*, **15**, 1743.
261. Volk, M.J., Trelease, R.N., and Reeves, H.C. (1974). *Anal. Biochem.*, **58**, 315.
262. Vora, S., Wims, L.A., Durham, S., and Morrison, S.I. (1981). *Blood*, **58**, 823.
263. Wagner, O.F., Bergmann, I., and Binder, B.R. (1985). *Anal. Biochem.*, **151**, 7.
264. Walker, D.G. and Khan, H.H. (1968). *Biochem. J.*, **108**, 169.
265. Ward, C.W. (1976). *Anal. Biochem.*, **74**, 242.
266. Werthamer, S., Freiberg, A., and Amaral, L. (1973). *Clin. Chim. Acta*, **45**, 5.
267. Westergaard, J.C. and Roberts, R.C. (1981). In *Electrophoresis '81* (ed. R.C. Allen and P. Arnaud), p. 674. De Gruyter, Berlin.
268. Willhardt, I. and Wiedernanders, B. (1975). *Anal. Biochem.*, **63**, 263.
269. Williams, L. and Hopkinson, D.A. (1975). *Hum. Hered.*, **25**, 567.
270. Wilson, D.E., Povey, S., and Harris, H. (1976). *Ann. Hum. Genet.*, **39**, 305.
271. Yagi, T., Kagamiyama, H., and Nozaki, M. (1971). *Anal. Biochem.*, **110**, 146.
272. Yamashita, T. and Utoh, H. (1978). *Anal. Biochem.*, **84**, 304.
273. Yamazaki, Y., Kageyama, Y., and Okuno, H. (1995). *Anal. Biochem.*, **231**, 295.
274. Yanagi, S., Tsutsumi, T., Saheki, S., Saheki, K., and Yamamoto, N. (1982). *Enzyme*, **28**, 400.
275. Yang, S.S. and Coleman, D. (1987). *Anal. Biochem.*, **160**, 480.

276. Ye, B., Gitler, C., and Gressel, J. (1997). *Anal. Biochem.*, **246**, 159.
277. Yonezawa, S. and Hori, S.H. (1975). *J. Histochem. Cytochem.*, **23**, 745.
278. Yourno, J. and Mastropaolo, W. (1981). *Blood*, **58**, 939.
279. Yue, R.H., Jacobs, H.K., Okabe, K., Keutel, H.J., and Kuby, S.A. (1968). *Biochemistry*, **7**, 4291.
280. Zehender, H., Trescher, D., and Ullrich, J. (1983). *Anal. Biochem.*, **135**, 16.
281. Zink, M.W. and Katz, J.S. (1973). *Can. J. Microbiol.*, **19**, 1187.
282. Zlotnick, G.W. and Gottlieb, M. (1986). *Anal. Biochem.*, **153**, 121.
283. Zöllner, E.J., Müller, W.E.G., and Zahn, R.K. (1973). *Z. Naturforsch*, **28C**, 376.

Index

Index

Index